Advances in Automated Driving Systems

Advances in Automated Driving Systems

Editors

Arno Eichberger
Zsolt Szalay
Martin Fellendorf
Henry Liu

MDPI • Basel • Beijing • Wuhan • Barcelona • Belgrade • Manchester • Tokyo • Cluj • Tianjin

Editors

Arno Eichberger
Institute of Automotive
Engineering, Graz University
of Technology,
Inffeldgasse 11/2,
8010 GRAZ, Austria

Zsolt Szalay
Department of Automotive
Technologies, Faculty of
Transportation Engineering
and Vehicle Engineering,
Budapest University of
Technology and Economics,
Building J, 6 Stoczek Street,
1111 Budapest, Hungary

Martin Fellendorf
Transport Planning and
Traffic Engineering, Graz
University of Technology,
Inffeldgasse 11/2,
8010 GRAZ, Austria

Henry Liu
Civil and Environmental
Engineering,
University of Michigan,
Ann Arbor, MI 48109, USA

Editorial Office
MDPI
St. Alban-Anlage 66
4052 Basel, Switzerland

This is a reprint of articles from the Special Issue published online in the open access journal *Energies* (ISSN 1996-1073) (available at: https://www.mdpi.com/journal/energies/special_issues/ Automated_Driving_System).

For citation purposes, cite each article independently as indicated on the article page online and as indicated below:

LastName, A.A.; LastName, B.B.; LastName, C.C. Article Title. *Journal Name* **Year**, *Volume Number*, Page Range.

ISBN 978-3-0365-4503-5 (Hbk)
ISBN 978-3-0365-4504-2 (PDF)

Cover image courtesy of Arno Eichberger.

Contents

About the Editors

Arno Eichberger

Arno Eichberger (Assoc.-Prof. Dr.), born in 1969, studied Mechanical Engineering at the University of Technology Graz, where he graduated in 1995. He received a doctorate degree in technical sciences in 1998 with a distinction. The doctoral thesis dealt with active head restraints for protection of the cervical spine in rear-end collisions. From 1998 to 2007, Arno Eichberger was employed at MAGNA STEYR Fahrzeugtechnik AG&Co, where he started as a simulation and development engineer in the Vehicle Safety department. He then moved to a position as a group leader for advanced development in the same department and later was assigned head of the innovation area of Intelligent Safety, being responsible for advanced development in the area of active and passive vehicle safety. Since 2007, Arno Eichberger has been employed at the Institute of Automotive Engineering (University of Technology Graz) as vice-director of the Institute and is head of the research area of Vehicle Dynamics, leading a group of about 15 scientists. His research includes the development and testing of automated driving, human–machine interaction, vehicle dynamics control and suspension development. Since 2012, he has been Associate Professor holding a venia docendi in Automotive Engineering. His research has been published in about 200 publications, including several books, journals, conference proceedings and patents. He is an international reviewer of scientific contributions in his area.

Zsolt Szalay

Zsolt SZALAY (Dr.), is an automotive enthusiast. He is an Electrical Engineer with a Ph.D. in Mechanical Engineering, with a strong business mind-set, solid industrial background and an M.Sc. degree in Business Administration. Founder and leader of the BME Automated Drive Laboratory. His main research interest is the testing and validation methodologies of connected and automated mobility systems. Author or co-author of several scientific papers and books, many of them directly related to the unique Hungarian proving ground for CAV testing and validation: ZalaZONE.

Martin Fellendorf

Martin Fellendorf (Prof. Dr.), Since 2005, Martin Fellendorf has been a full professor at the Graz University of Technology, Austria, heading the institute of transport planning and highway engineering within the faculty of civil engineering. His research interests focus on Transport Modelling and Intelligent Transport Systems issues, mainly traffic control in urban areas and motorway control. In his research, he looks at the interaction between all road-users and vehicle types. Currently, he and his team of Ph.D. candidates and researchers are involved in several national and European projects on activity-based demand modelling, vehicle emission modelling and the simulation of automated driving schemes. Before joining the university, he worked for 17 years at PTV in Karlsruhe, developing traffic engineering and traffic management software (PTV Visum and PTV Vissim). He was responsible for international product placement and conducted training sessions at various universities worldwide. Martin Fellendorf led multiple simulation studies optimizing traffic control in the USA , China and the United Arabic Emirates. Martin holds a Master's in Industrial Engineering and received his PhD (Dr.-Ing.) 1991 in Civil Engineering from the University of Karlsruhe, Germany.

Henry Liu

Henry Liu (Prof. Dr.) is a tenured professor in the Department of Civil and Environmental Engineering and the Director of Mcity at the University of Michigan, Ann Arbor. He is also a Research Professor at the University of Michigan Transportation Research Institute and the Director for the Center for Connected and Automated Transportation (USDOT Region 5 University Transportation Center). From August 2017 to August 2019, Prof. Liu served as the Chief Scientist on Smart Transportation for DiDi Global, Inc., one of the leading mobility service providers in the world. While he was with DiDi, he established and led the Urban Transportation Business Unit. Prior to joining the University of Michigan, Prof. Liu was an Associate Professor of Civil Engineering at the University of Minnesota, Twin Cities. Prof. Liu received his Ph.D. degree in Civil and Environmental Engineering from the University of Wisconsin at Madison in 2000 and his bachelor's degree in Automotive Engineering from Tsinghua University in China in 1993. Prof. Liu conducts interdisciplinary research at the interface of transportation engineering, automotive engineering, and artificial intelligence. Specifically, his scholarly interests concern traffic flow monitoring, modeling, and control, as well as the testing and evaluation of connected and automated vehicles. He has published more than 120 refereed journal papers. Professor Liu has nurtured a new generation of scholars, and some of his PhD students and postdocs have joined first-class universities, such as Columbia University, Purdue University, RPI, etc. Prof. Liu is the managing editor of the *Journal of Intelligent Transportation Systems*.

Editorial

Advances in Automated Driving Systems

Arno Eichberger [1,*], **Zsolt Szalay** [2], **Martin Fellendorf** [3] **and Henry Liu** [4]

1   Institute of Automotive Engineering, Graz University of Technology, 8010 Graz, Austria
2   Department of Automotive Technologies, Faculty of Transportation Engineering and Vehicle Engineering, Budapest University of Technology and Economic, 1111 Budapest, Hungary; szalay.zsolt@kjk.bme.hu
3   Institute of Transport Planning and Traffic Engineering, Graz University of Technology, 8010 Graz, Austria; martin.fellendorf@tugraz.at
4   Civil and Environmental Engineering, University of Michigan, Ann Arbor, MI 48109, USA; henryliu@umich.edu
*   Correspondence: arno.eichberger@tugraz.at; Tel.: +43-316-873-35210

1. Introduction

Electrification, automation of vehicle control, digitalization and new mobility are the mega trends in automotive engineering and they are strongly connected to each other. Whereas many demonstrations for highly automated vehicles have been made world-wide, many challenges remain to bring automated vehicles on the market for private and commercial use.

The main challenges related to automated vehicle control are:

1.   Reliable machine perception; accepted standards for vehicle approval and homologation;
2.   verification and validation of the functional safety especially at SAE level 3+ systems;
3.   legal and ethical implications;
4.   acceptance of vehicle automation by occupants and society;
5.   interaction between automated- and human-controlled vehicles in mixed traffic;
6.   human–machine interaction and usability;
7.   manipulation, misuse and cyber-security;
8.   but also the system costs for hard- and software and development effort.

These challenges mainly relate to the complex interaction between the human occupants, the automated vehicle and the environment the vehicle is operated in (see Figure 1). The main system components and the related challenges are elaborated in the following:

Citation: Eichberger, A.; Szalay, Z.; Fellendorf, M.; Liu, H. Advances in Automated Driving Systems. *Energies* **2022**, *15*, 3476. https://doi.org/10.3390/en15103476

Received: 20 April 2022
Accepted: 6 May 2022
Published: 10 May 2022

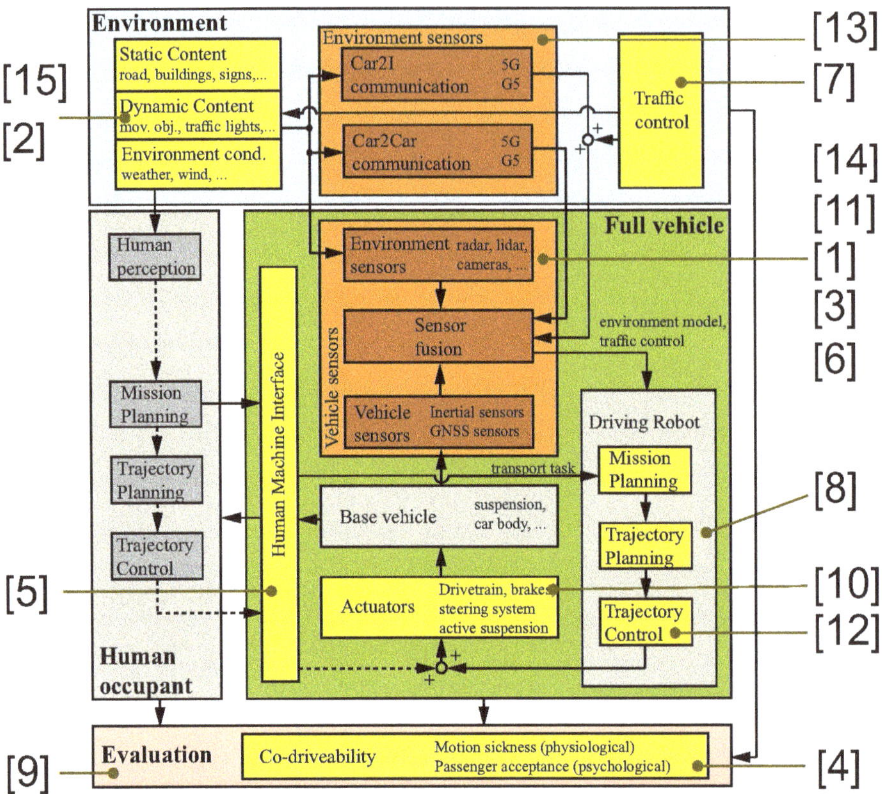

Figure 1. Traffic system in automated driving. The figure shows the complex interaction of the main system components in automated driving systems and how the articles of the Special Issue are thematically classified.

1.1. Environment

The environment, comprised of the static and dynamic content, as well as the ambient conditions, roadside infrastructure and traffic control systems are essential aspects in driving automation. It provides information to the driver with lane markings and traffic signs that were initially developed for human perception. However, requirements of the future road infrastructure need to consider opportunities and limitations of perception sensors.

1.2. Perception

Machine perception, highlighted in orange in Figure 1, is traditionally vehicle-based. Here, many challenges arise because of the increasingly complex algorithms to process raw sensor data into a reliable digital environment that allows planning of the vehicle trajectory. In addition, environmental conditions such as rain, fog or lighting conditions can deteriorate the machine perception, leading to enhance machine perception with data fused from different sensors.

Due to the high cost for the components and their integration in the vehicle, increasingly sensors located in the roadside environment communicate with the automated vehicle. They shall provide additional information about the static contents such as the road network, roadside infrastructure and buildings, as well as the dynamic content such as moving objects. For the vehicle-to-X communication (V2X) we see different technologies such as dedicated short range communication and mobile communication to allow for data exchange of a huge amount of data at minimum delay, while maintaining data security and privacy.

1.3. Vehicle Guidance

Visualized in grey, the driving robot (right side of Figure 1) increasingly performs tasks of the human driver (left side), consisting of mission and trajectory planning and control. The mission planning is something that is an intrinsic human task but widely supported by machine navigation systems. Here, external traffic control systems located in the road infrastructure or cloud-based services can additionally support to optimize transportation tasks by re-routing. However, the most difficult task is the trajectory planning using the horizon offered by the field of view of the human or machine perception. Instead of the traditional approach, namely to deterministically program driving tasks such as for adaptive cruise control, modern methods of artificial intelligence (AI) offer a data driven approach to handle complex and maybe even situations not being experienced before. Nevertheless, the safety validation of AI base trajectory planning is a not solved issue. Vehicle control, usually handled by traditional methods of automation and control, aims to minimize the error in planed and driven trajectories. Here, they need to cooperate with vehicle dynamics control (VDC) systems. Implementing intelligence in the road infrastructure allows for advanced traffic control that maybe even perform trajectory planning as the most delicate step in vehicle control.

1.4. Base Vehicle

The vehicle, depicted in green in Figure 1, is based on a traditional vehicle but enhanced with actuators, which will evolve from classical steering, power train and braking systems to advanced X-by-wire systems offering new levels of vehicle control.

1.5. Human Machine Interface

The human–machine interface (HMI) is a delicate component that needs to be designed carefully in order to improve the already high level of reliability in human vehicle control. Literature reports that billions of kilometers need to be driven with an automated vehicle in order to prove statistical significance of a superior behavior of a driving robot. As long as we have the human driver as an operator that needs to perform tasks in vehicle guidance, such as observation of the environment and fallback in case of system failure, the HMI is essential to avoid distraction or inappropriate behavior of the human driver.

1.6. Evaluation

Formerly the evaluation of driving behavior focused on the driver, and included criteria related to controllability, disturbance behavior, observability and parameter insensitivity. These criteria could be evaluated in a manageable amount of testing on proving grounds, often with open-loop maneuvers to exclude human vehicle guidance. In addition, human impression of the driving behavior and comfort was rated with different subjective and objective methods, leading into an evaluation of drivability of the vehicle. In driving automation, the driver increasingly transforms into a passenger, so we need to take into account the human as a co-driver or even a passenger, so rating becomes more of a co-drivability feature. Additional focus has to be put on other aspects such as perceived trust, safety and acceptance of the human occupant, which happens on a psychological level. However, the physiological aspect also has to be taken into account; for example, motion sickness as experienced often from passengers.

However, the sheer infinite amount of possible driving scenarios call for innovative methods for evaluating not only the safe behavior of an automated vehicle, but also a high rating of co-drivability.

2. Articles of the Special Issue

This Special Issue deals with recent advances related to the technological aspects of the aforementioned challenges:

- Machine perception for SAE L3+ driving automation;
- trajectory planning and decision making in complex traffic situations;
- X-by-wire system components;
- verification and validation of SAE L3+ systems;
- misuse, manipulation and cybersecurity;
- human–machine interaction, driver monitoring and driver intention recognition;
- road infrastructure measures for introduction of SAE L3+ systems;
- solutions for interactions of vehicles human and machine controlled in mixed traffic.

The collection includes 15 articles that deal with the aforementioned challenges. In Figure 1, the thematic classification of the different articles related to the different system components is illustrated. Not surprisingly, it illustrates that many studies are focused on reliable human perception.

Article [1] deals with a methodology to quantify the performance of sensor models in virtual validation and verification (V & V) of automated driving functions, an important step towards reduction of on-road testing. The effect of automation on traffic flow during the transition phase in mixed traffic was investigated by [2]. Article [3] deals with the quality of ground truth annotation data to improve the transfer of on-road testing results into simulation. The evaluation of perceived trust was examined in [4], demonstrated in a driving simulator study. The topic of drowsiness classification in the context of driving automation was investigated in [5]. In simulation of Automated Driving (AD) functions, modelling of camera sensors is often carried out with physical modelling; however, research in [6] presented an alternative with phenomenological modeling. Article [7] introduced a conflicted management framework, especially focusing on aiming at managing urban and peri-urban traffic. The potential of implementing AI into vehicle guidance, demonstrated on the safety of ACC, was investigated in [8]. Article [9] deals with insufficiencies during the decomposition of testing of ADAS functions from the system to lower levels, and defining rules for testing of modules to dispense with system tests. Improvement of vehicle control for wheel loaders was investigated in [10] using a deep learning-based prediction model of the throttle valve. The difficulties in reliable detection of pedestrians is addressed in [11], based on convolutional neural network algorithms applied on images manipulated with inverse gamma correction. Vehicle control at handling limits was investigated in [12], introducing a model-predictive controller that is able to initiate and maintain steady-state drifting. Article [13] deals with a functional prototype of a cooperative perception system aiming at future cloud-based services of automated driving functions, focusing on motorway use. A field study dealing with the capability of a market-introduced traffic sign recognition system was conducted in [14], revealing deficits by misreading of signs. A method to introduce automated driving functions in traffic flow simulation for virtual V & V was introduced by [15], based on a co-simulation framework between multi-body and traffic flow simulation.

As the Special Issue is dedicated to this topic, future research will continue in the development of the individual system components and their complex interaction, constantly rising the level of autonomy while providing an acceptable behavior for the individual and the society, superior compared to human vehicle guidance.

Author Contributions: Conceptualization, A.E., Z.S., M.F. and H.L.; writing—original draft preparation, A.E.; writing—review and editing, A.E., Z.S., M.F. and H.L.; visualization, A.E.; project administration, A.E. All authors have read and agreed to the published version of the manuscript.

Funding: This research received no external funding.

Institutional Review Board Statement: Not applicable.

Informed Consent Statement: Not applicable.

Data Availability Statement: Not applicable.

Acknowledgments: The editors express their thanks to the excellent and elaborative work of the international reviewers in evaluating the articles of this Special Issue.

References

1. Magosi, Z.F.; Wellershaus, C.; Tihanyi, V.R.; Luley, P.; Eichberger, A. Evaluation Methodology for Physical Radar Perception Sensor Models Based on On-Road Measurements for the Testing and Validation of Automated Driving. *Energies* **2022**, *15*, 2545. [CrossRef]
2. Fang, X.; Li, H.; Tettamanti, T.; Eichberger, A.; Fellendorf, M. Effects of Automated Vehicle Models at the Mixed Traffic Situation on a Motorway Scenario. *Energies* **2022**, *15*, 2008. [CrossRef]
3. Holder, M.; Elster, L.; Winner, H. Digitalize the Twin: A Method for Calibration of Reference Data for Transfer Real-World Test Drives into Simulation. *Energies* **2022**, *15*, 989. [CrossRef]
4. Clement, P.; Veledar, O.; Könczöl, C.; Danzinger, H.; Posch, M.; Eichberger, A.; Macher, G. Enhancing Acceptance and Trust in Automated Driving trough Virtual Experience on a Driving Simulator. *Energies* **2022**, *15*, 781. [CrossRef]
5. Arefnezhad, S.; Eichberger, A.; Frühwirth, M.; Kaufmann, C.; Moser, M.; Koglbauer, I.V. Driver Monitoring of Automated Vehicles by Classification of Driver Drowsiness Using a Deep Convolutional Neural Network Trained by Scalograms of ECG Signals. *Energies* **2022**, *15*, 480. [CrossRef]
6. Li, H.; Tarik, K.; Arefnezhad, S.; Magosi, Z.F.; Wellershaus, C.; Babic, D.; Babic, D.; Tihanyi, V.; Eichberger, A.; Baunach, M.C. Phenomenological Modelling of Camera Performance for Road Marking Detection. *Energies* **2022**, *15*, 194. [CrossRef]
7. Sziroczák, D.; Rohács, D. Automated Conflict Management Framework Development for Autonomous Aerial and Ground Vehicles. *Energies* **2021**, *14*, 8344. [CrossRef]
8. Jurj, S.L.; Grundt, D.; Werner, T.; Borchers, P.; Rothemann, K.; Möhlmann, E. Increasing the Safety of Adaptive Cruise Control Using Physics-Guided Reinforcement Learning. *Energies* **2021**, *14*, 7572. [CrossRef]
9. Klamann, B.; Winner, H. Comparing Different Levels of Technical Systems for a Modular Safety Approval—Why the State of the Art Does Not Dispense with System Tests Yet. *Energies* **2021**, *14*, 7516. [CrossRef]
10. Huang, J.; Cheng, X.; Shen, Y.; Kong, D.; Wang, J. Deep Learning-Based Prediction of Throttle Value and State for Wheel Loaders. *Energies* **2021**, *14*, 7202. [CrossRef]
11. Junaid, M.; Szalay, Z.; Török, Á. Evaluation of Non-Classical Decision-Making Methods in Self Driving Cars: Pedestrian Detection Testing on Cluster of Images with Different Luminance Conditions. *Energies* **2021**, *14*, 7172. [CrossRef]
12. Czibere, S.; Domina, Á.; Bárdos, Á.; Szalay, Z. Model Predictive Controller Design for Vehicle Motion Control at Handling Limits in Multiple Equilibria on Varying Road Surfaces. *Energies* **2021**, *14*, 6667. [CrossRef]
13. Tihanyi, V.; Rövid, A.; Remeli, V.; Vincze, Z.; Csonthó, M.; Pethő, Z.; Szalai, M.; Varga, B.; Khalil, A.; Szalay, Z. Towards Cooperative Perception Services for ITS: Digital Twin in the Automotive Edge Cloud. *Energies* **2021**, *14*, 5930. [CrossRef]
14. Babić, D.; Babić, D.; Fiolić, M.; Šarić, Ž. Analysis of Market-Ready Traffic Sign Recognition Systems in Cars: A Test Field Study. *Energies* **2021**, *14*, 3697. [CrossRef]
15. Nalic, D.; Pandurevic, A.; Eichberger, A.; Fellendorf, M.; Rogic, B. Software Framework for Testing of Automated Driving Systems in the Traffic Environment of Vissim. *Energies* **2021**, *14*, 3135. [CrossRef]

Article

Software Framework for Testing of Automated Driving Systems in the Traffic Environment of Vissim

Demin Nalic [1,*], Aleksa Pandurevic [1], Arno Eichberger [1], Martin Fellendorf [2] and Branko Rogic [3]

[1] Institute of Automotive Engineering, TU Graz, 8010 Graz, Austria; pandurevic@tugraz.at (A.P.); arno.eichberger@tugraz.at (A.E.)
[2] Institute of Highway Engineering and Transport Planning, TU Graz, 8010 Graz, Austria; martin.fellendorf@tugraz.at
[3] MAGNA Steyr Fahrzeugtechnik AG Co. & KG, 8045 Graz, Austria; branko.rogic@magna.com
* Correspondence: demin.nalic@tugraz.at

Abstract: As the complexity of automated driving systemss (ADSs) with automation levels above level 3 is rising, virtual testing for such systems is inevitable and necessary. The complexity of testing these levels lies in the modeling and calculation demands for the virtual environment, which consists of roads, traffic, static and dynamic objects, as well as the modeling of the car itself. An essential part of the safety and performance analysis of ADSs is the modeling and consideration of dynamic road traffic participants. There are multiple forms of traffic flow simulation software (TFSS), which are used to reproduce realistic traffic behavior and are integrated directly or over interfaces with vehicle simulation software environments. In this paper we focus on the TFSS from PTV Vissim in a co-simulation framework which combines Vissim and CarMaker. As it is a commonly used software in industry and research, it also provides complex driver models and interfaces to manipulate and develop customized traffic participants. Using the driver model DLL interface (DMDI) from Vissim it is possible to manipulate traffic participants or adjust driver models in a defined manner. Based on the DMDI, we extended the code and developed a framework for the manipulation and testing of ADSs in the traffic environment of Vissim. The efficiency and performance of the developed software framework are evaluated using the co-simulation framework for the testing of ADSs, which is based on Vissim and CarMaker.

Keywords: automated driving; scenario-based testing; software framework

Citation: Nalic, D.; Pandurevic, A.; Eichberger, A.; Fellendorf, M.; Rogic, B. Software Framework for Testing of Automated Driving Systems in the Traffic Environment of Vissim. *Energies* **2021**, *14*, 3135. https://doi.org/10.3390/en14113135

Academic Editor: Wiseman Yair

Received: 2 April 2021
Accepted: 21 May 2021
Published: 27 May 2021

Publisher's Note: MDPI stays neutral with regard to jurisdictional claims in published maps and institutional affiliations.

1. Introduction

The use of TFSS in automotive engineering has significantly improved the scope of the virtual testing of ADS. It is mostly used in co-simulation with other software tools for vehicle testing and simulation. There are various co-simulation platformss (CSPs) for the testing of ADSs in complex traffic environments. Hallerbach presented in [1] a simulation-based tool-chain to identify critical scenarios using a SUMO and a vehicle dynamic software. A framework coupling SUMO with vehicle dynamic software VTD for the development of ADSs is presented in [2]. In [3], a human-driven car from SILAB interacted over an interface with SUMO traffic participants in order to evaluate human interactions and the effect of ADSs in traffic. Implementing automated driving functions in MATLAB and coupling this with Vissim, an impact analysis of ADSs is performed in [4]. The CSP used in this work is based on the co-simulation between Vissim and CarMaker and is explained in greater detail below; see [5]. Common to all these interfaces is the fact that the vehicle under test has been developed separately from a certain vehicle simulation software. The traffic is created externally and imported by means of TFSS. In this study, the vehicle being tested is referred to as an EGO vehicle.

As a form of TFSS, Vissim provides comprehensive traffic flow modeling options and the possibility to manipulate traffic participants, making it suitable for the testing of ADSs and the generation of safety-relevant scenarios. The term "scenario" is defined in [6]. Focusing on safety-critical scenarioss (SCSs), a variety of research works have offered possible definitions and approaches to determining the criticality and safety relevance of a given scenario, for example [7–9]. The aim of generating and finding SCSs in the realistic and stochastic traffic environment of Vissim is based on the idea that the EGO vehicle drives through the traffic simulated by the TFSS and possibly encounters situations that cannot be resolved by the implemented ADS. The approach based on this idea is beneficial for the virtual testing of an EGO vehicle, since it reflects normal driving in traffic and takes in account all the effects and factors which could potentially occur during a ride. The difficulty of using Vissim or any other TFSS is that it is not guaranteed that a significant amount of SCSs for ADS testing will be generated. To provide a reference and an approximate estimation of the amount and relevance of SCSs, that will occur, distance-based approaches can be used. In [10,11], distance-based testing approaches for ADS are introduced. Based on accident data, the average distance between two accidents was statistically analyzed in order to determine how many kilometers an ADS should be tested in order to achieve the same safety level as a human driver. In both works, the number of kilometers, depending on the accident considered, lies in millions of real-world testing kilometers, which are needed to prove the safety of ADS compared to a human driver. In [12], the distance-based testing is reduced to scenario-based testing. In [12] a statistical method is introduced in order to calculate the number of scenarios required for the same evidence as the approaches presented in [10,11]. These accident rates and scenario amounts can hardly be reached via Vissim because the Vissim driver model relies on tactical driving behavior. This is due to the fact that traffic participants plan their actions with a temporal and spatial horizon; see [13]. In such trajectory planning, the neighboring vehicle is taken into account, as well as vehicles that are far in front of the EGO vehicle. This means that the vehicles in the Vissim traffic simulation have enough of a planning horizon to avoid conflict areas and conflict situations, which is comparable to human driving behavior. Due to this face, the cars collide with each other extremely rarely, and do not make unpredictable movements and maneuvers, which corresponds to the real-life situation. Nonetheless, human driving behavior is in rare cases incorrect, resulting in conflict situations and accidents. Accident statistics suggest that up to 90% of police-reported accidents are mainly caused by human drivers [14]. On the one hand, TFSS-based driving behavior correspondsto the realistic planning of real drivers. However, on the other hand, if we use, e.g., Vissim as a TFSS for the testing and validation of ADSs, as described in [5,15], this would not yield a satisfactory number of SCSs. Another recent study in [16] introduced a method based on deep reinforcement learning to train traffic participants with naturalistic driving data. The main goal of this approach is to train traffic participants in such a way that they produce SCSs and reach accident rates corresponding to those occurring in the real world. With a similar approach and objective, but using a more deterministic approach, we present a software framework for researchers using Vissim for the generation of SCSs.The main objective of this software framework is to offer an appropriate and adjustable environment for testing purposes of ADSs and, more specifically, for the generation of SCSs. The so-called driver model framework (DMF) is based on the DMDI software code provided by Vissim. This code allows Vissim users to manipulate traffic participants in a defined manner. Since the provided Vissim interface and the code lacks explanations, it is complex to use. Therefore, the DMF provides a structured C++ code with a class architecture and useful methods for accessing and setting different vehicle parameters for traffic participants. This provides an optimized environment for the testing and development of ADSs in the complex traffic environment of Vissim. As the main case study to describe the functionality of the DMF, the co-simulation between CarMaker and Vissim described in [4] was used in this study. This framework was adjusted by means of the DMF to utilize the co-simulation for the generation of SCSs. This utilization was carried out by implementing the stress testing method (STM) presented in [17]. The

DMF code itself can also be used independently of Vissim and is available to users for testing purposes and other related applications.

2. Software Description

For the DMF, we used the DMDI provided by PTV Vissim (see [18]). This C++ code contains three essential functions from Vissim: DriverModelSetValue, DriverModelGetValue and DriverModelExecuteCommand. These functions are used by Vissim to provide and retrieve the data from the DMF. The Vissim participants, which are controlled by the DMDI, are called DLL driver modelss (DDMs) hereinafter. The Vissim DriverModelExecuteCommand value is passed upon each simulation step, denoting the action to be taken, such as initialization, driver creation, driver deletion and the driver movement of each traffic participant controlled by the DMDI in the TFSS. In between the commands, multiple calls of DriverModelSetValue and DriverModelGetValue are made for each vehicle controlled by the DDM. The function DriverModelSetValue provides the DDM with the current vehicle values, which can be stored, processed and modified. In order to provide Vissim with new values, which will be used for the vehicle's movement, multiple calls of DriverModelGetValue for each traffic participant are necessary. Building on top of the provided DMDI, DriverModelSetValue and DriverModelGetValue are encapsulated into setInjectorData and getInjectorData, member functions of InjectorAbstract which is further described in the software architecture section.

In these functions, all the necessary logic for extracting, storing and updating of selected vehicles is contained. As the DMF has been created to manipulate traffic participants in the surrounding area of the EGO vehicle, one vehicle in the Vissim traffic is set to be the EGO vehicle by means of the vehicle ID, which is provided by the DMDI. This vehicle has an individual ID, which can be freely defined and set. Using the DMDI in the testing framework presented in [5], the EGO vehicle ID equals 1 and is fixed within the co-simulation between Vissim and CarMaker. For the DMF, which is described in this paper, the area of interest is the surrounding area of the EGO vehicle, which consists of other traffic participants, referred to as nearby vehicles in this work. The DMF concept is depicted in Figure 1. First, all the surrounding area has to be defined, stored and updated continuously through each Vissim simulation step. This is done with the DMF method capture, in which certain nearby vehicles are selected as target vehicles by means of user-defined rules. To manipulate the traffic participants around the EGO vehicle, an action method is defined, which activates a user-defined critical maneuver for the Vissim vehicle. This critical maneuver is performed by the traffic participant from Vissim, which is referred to as the target vehicle and is activated close to the EGO vehicle. The software implementation of the process logic of the DMF is shown in Figure 2. Each Vissim call to the setInjectorData or getInjectorData method passes multiple arguments, one of which always denotes the value which is going to be sent or achieved. After setInjectorData is executed for each value of the EGO vehicle, multiple setInjectorData calls follow, with the values of the nearby vehicles. Thus, information about EGO vehicles and detected nearby vehicles is obtained and can be processed and prepared if needed. The whole setInjectorData and getInjectorData routine is performed for each vehicle controlled by the DMF in each simulation step. The DMF provides the user with the basic methods necessary to read and manipulate the traffic participants relevant to the EGO vehicle being tested. The user has to implement two DMF methods, capture and action, in order to create a usable DLL. A description of the methods capture and action is given in the following points:

- In the capture method, it is guaranteed that nearby vehicles in relation to the EGO vehicle with current values can be queried. The user then defines the rules by which certain or all nearby vehicles are selected as target vehicles and which action is applied during defined time intervals. Duration and pauses between those intervals are also user-configurable.
- In the action method, the user defines an action which will be applied to the target vehicles. Examples of such actions can be a braking maneuver carried out by a traffic

participant which is directly in front of the EGO vehicle, or a cut-in maneuver by the traffic participant in front of the EGO vehicle.

The obvious advantage of the DMF is that the user does not care about monitoring all nearby vehicles, taking care of vehicle IDs, or observing if the nearby vehicles have current values. The tasks of the user are merely to select vehicles of interest and define the action that will be applied.

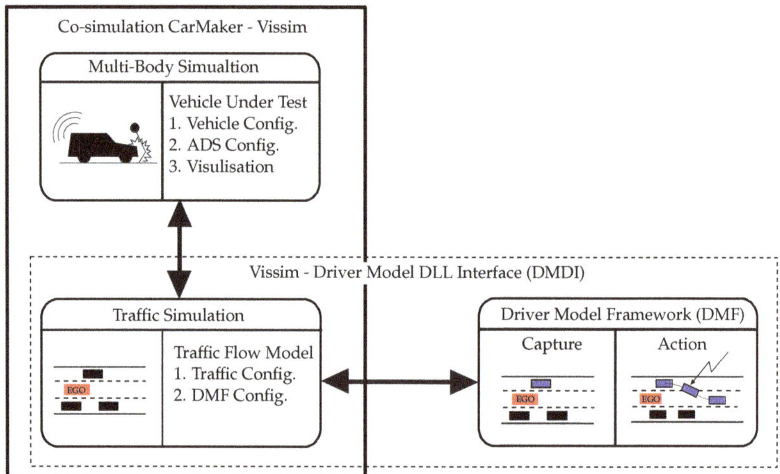

Figure 1. The co-simulation between Vissim and CarMaker of the framework concept in [5] is adapted with the DMF.

2.1. Software Architecture

The DMF is written in C++, encapsulates the provided Vissim code, and provides a simple interface for the user. The class hierarchy can be seen in Figure 3. The InjectorAbstract class comprises the majority of the DMF logic, including setInjectorData and getInjectorData methods, which are called from Vissim. They are required to be a part of InjectorAbstract, since the whole DMF logic depends on them, reading out the data, storing it, scheduling the execution of capture and action methods and passing new data to Vissim. The Injector class is a child class of the InjectorClass, which implements the pure virtual methods capture and action. This inheritance serves to keep users away from the basic Vissim DLL code of the framework and provides them with simple vehicle manipulation. The user-code is separated from the DMF logic, and it is simple for the user to write and organize. Aside from that, the user is free to extend the functionality of NearbyVehicle and EgoVehicle classes, which contain the data of the respective vehicle categories.

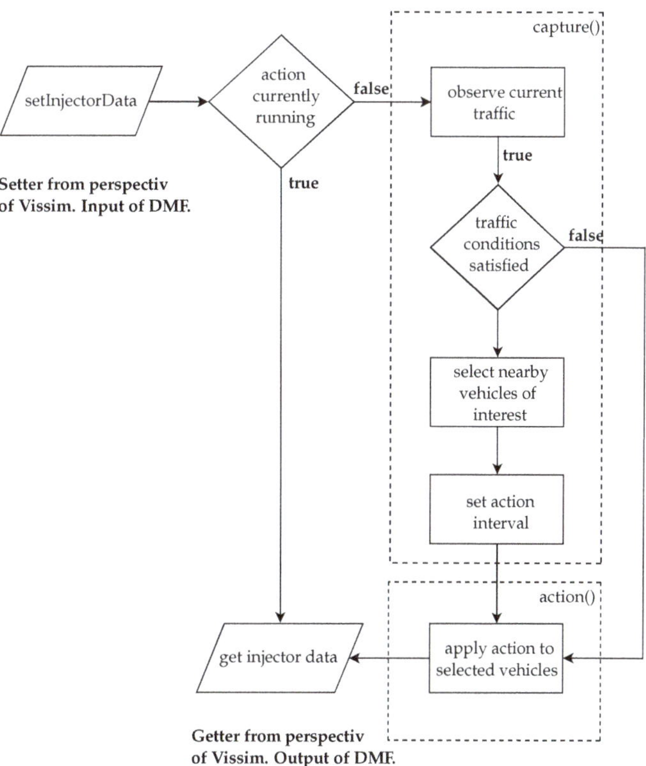

Figure 2. The process logic and the main components of the DMF.

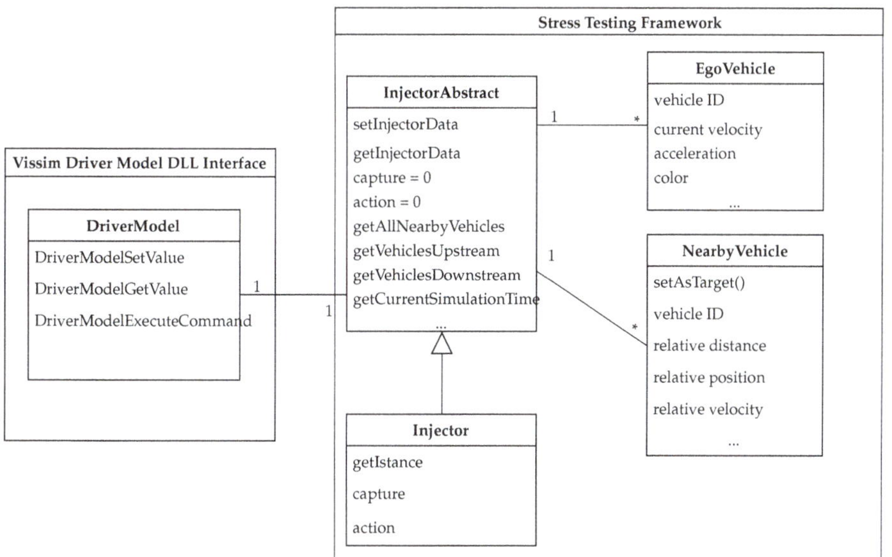

Figure 3. A simplified view of the framework class hierarchy.

2.2. Software Functionalities

As mentioned in the software description, the user has to fill in the code for the capture and action methods. The methods provided by the DMF upon which the user depends are:

- getAllNearbyVehicles
- getVehiclesDownstream
- getVehiclesUpstream
- getCurrentSimTime

The first three methods return a vector containing nearby vehicles either all of them or only vehicles in front of vehicles behind the EGO vehicle. Each vehicle is represented by the NearbyVehicle object, which contains all the relevant information about the vehicle, such as relative distance and speed to the EGO vehicle, acceleration, relative position, and other vehicle states or parameters. These methods are meant to be used in the capture method, in which the user is observing the traffic and labeling relevant vehicles as target vehicles. When the user-defined criteria are fulfilled, the startAction method is to be implemented, with a specified action duration time and an optionally selected pause time after the action's end. During the time of the respective action, the takes control of the target vehicles. After the action time expires, the TFSS internal model takes back the control of the target vehicles. Through these methods, the DMF provides the user with the following main functionalities:

- Providing the user with the updated list of nearby vehicles;
- Offering the user the ability to select target vehicles and manipulate them;
- Keeping the list of target vehicles updated during the action;
- Scheduling the *capture* and *action* methods in time;
- Taking and releasing the control of the vehicles from and to the user;

Another essential feature of the DMF is that it has two separate modes of operation. In both of these modes, previously defined functionalities are present, but they require different approaches and implementations for different modes of operation. The DMF can operate both in Vissim only and in co-simulation between CarMaker and Vissim.

3. Use Case Application of the DMF

As introduced in Section 2, the use case for the DMF refers to the co-simulation between CarMaker and Vissim software, adjusting Vissim with the use of the DMF. The concept of the co-simulation framework is depicted in Figure 4 and the adjustment of Vissim using the DMF has already been introduced in Figure 1.

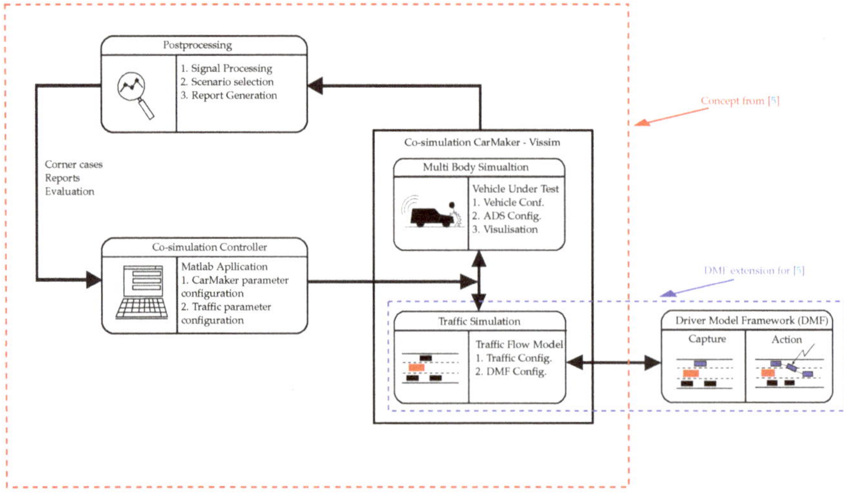

Figure 4. The concept of the co-simulation framework from [5] adjusted with the DMF from Figure 1.

This adjustment using the DMF was realized by implementing the STM which was introduced in [17]. Defined maneuvers were provoked in the vicinity of an EGO vehicle in order to force the tested vehicle into a challenging situation for the ADS. Using statistical data from accidents in Austria, the maneuvers which were provoked were classified into lateral and longitudinal maneuvers. These longitudinal and lateral maneuvers were implemented using the DMF, in which they could be parameterized and used for the generation of SCSs in the vicinity of the EGO vehicle. In [17] the comparison between a testing procedure with and without the STM is shown. In this testing procedure, an EGO vehicle equipped with adaptive cruise control and an automated lance change algorithm was tested on 10,000 simulation kilometers. The SCSs considered for evaluation purposes were collisions, and the criticality assessment criteria were those defined in [8]. It was observed that collisions could be generated, and very critical and critical scenarios from [8] increased by 1859 and 2320 over the course of 10,000 simulation kilometers, respectively. The second use case of the framework involves using the DMF only with Vissim. As a result of this, a vehicle with automated driving functions could be developed and implemented in the traffic environment, using the same DMDI from Vissim. In [19], a longitudinal control unit was developed using the Vissim DMDI in order to test the performance of automated vehicles on a single-lane road. A similar approach is presented in [20], where the impact of an emergency control function on mobility and safety was evaluated. The research work presented in [21,22] emphasizes the usage of the DLL interface for the analysis and evaluation of connected and autonomous vehicles in traffic. Using it for the purpose of testing an ADS, the advantage of this approach lies in the avoidance of couplings with other vehicle simulation software tools, such as CarMaker, VTD and others. In this case, the simulation times, implementation efforts and the need for additional tools could be decreased. A possible disadvantage of this approach is the simple point mass models of vehicles provided in Vissim. This issue can also be solved by developing and integrating single-track or more complex vehicle models using the same DMDI interface. Table 1 provides the software specifications of the the DMF in the default version. For the DMF with the implementation of the STM, the code information is provided in Table 2. The first use case requires IPG CarMaker, and for the second use case it is possible to implement the provided DMDI directly in Vissim on any test road. The simple braking example with the STM and DMF are provided for research and development purposes.

Table 1. DMF software information and code link for the default DMF.

Nr.	Code Metadata Description	Description
C1	Current code version	v1
C2	Permanent link to code/repository used for this code version	https://github.com/ftgTUGraz/DriverModel_Framework (accessed on 14 February 2021)
C3	Code Ocean compute capsule	none
C4	Legal Code License	GPL-3.0 License
C5	Code versioning system used	git
C6	Software code languages, tools, and services used	C++, PTV Vissim 11.00-14
C7	Compilation requirements, operating environments & dependencies	Visual Studio 2019

Table 2. DMF software information and code link for the implementation of the STM.

Nr.	Code Metadata Description	Description
C1	Current code version	v1
C2	Permanent link to code/repository used for this code version	https://github.com/ftgTUGraz/DriverModel_STM (accessed on 14 February 2021)
C3	Code Ocean compute capsule	none
C4	Legal Code License	GPL-3.0 License
C5	Code versioning system used	git
C6	Software code languages, tools, and services used	C++, IPG CarMaker 8.1.1 (optional), PTV Vissim 11.00-14
C7	Compilation requirements, operating environments, & dependencies	Visual Studio 2019

4. Conclusions

With the presented driver model framework, the user can implement and adjust driver models for traffic participants using the traffic flow simulation software Vissim. By that, traffic participants for testing purposes of a particular automated driving system of the vehicle under test can be tested on a virtual basis. Using the co-simulation between Vissim and CarMaker, a use case of the presented software framework was shown. An upgrade of the co-simulation with the presented framework for testing automated driving systems increases the benefit of the Vissim and CarMaker co-simulation environment. For future research, the software framework will be adjusted for implementing more realistic vehicle dynamics to the neighboring traffic vehicles, including vehicle based environment sensors as well for improved model validity and Vehicle-to-X communication for impact analysis of automated driving systems on traffic.

Author Contributions: Conceptualization, D.N.; methodology, D.N. and A.P.; software, D.N. and A.P.; validation, A.E., B.R., and D.N.; formal analysis, D.N. and B.R.; investigation, D.N.; resources, D.N., M.F., and B.R.; data curation, D.N.; writing—original draft preparation, D.N., M.F., and A.P.; writing—review and editing, A.E., M.F., and B.R.; visualization, D.N.; supervision, A.E. and B.R. All authors have read and agreed to the published version of the manuscript.

Funding: This work is founded by the Austrian Federal Ministry of Transport, Innovation and Technology as part of the FFG Program "EFREtop".

Data Availability Statement: Not applicable.

Acknowledgments: Open Access Funding by the Graz University of Technology

Conflicts of Interest: No conflict of interest exists: We wish to confirm that there are no known conflicts of interest associated with this publication and there has been no significant financial support for this work that could have influenced its outcome.

References

1. Hallerbach, S.; Xia, Y.; Eberle, U.; Koester, F. Simulation-Based Identification of Critical Scenarios for Cooperative and Automated Vehicles. *SAE Int. J. Connect. Autom. Veh.* **2018**, *1*, 93–106. [CrossRef]
2. Semrau, M.; Erdmann, J. Simulation Framework for Testing ADAS in Chinese Traffic Situations. In Proceedings of the 2016 SUMO User Conference, Berlin, Germany, 23–25 May 2016; pp. 103–115.
3. Barthauer, M.; Hafner, A. Coupling traffic and driving simulation: Taking advantage of SUMO and SILAB together. *EPiC Ser. Eng.* **2018**, *2*, 56–66. [CrossRef]
4. Haberl, M.; Fellendorf, M.; Rudigier, M.; Kerschbaumer, A.; Eichberger, A.; Rogic, B.; Neuhold, R. Simulation assisted impact analyses of automated driving on motorways for different levels of automation and penetration. In Proceedings of the Mobil.TUM 2017, Munich, Germany, 4–5 July 2017.

5. Nalic, D.; Eichberger, A.; Hanzl, G.; Fellendorf, M.; Rogic, B. Development of a Co-Simulation Framework for Systematic Generation of Scenarios for Testing and Validation of Automated Driving Systems. In Proceedings of the IEEE Intelligent Transportation Systems Conference (ITSC), Auckland, New Zealand, 27–30 October 2019; pp. 1895–1901. [CrossRef]
6. Ulbrich, S.; Menzel, T.; Reschka, A.; Schuldt, F.; Maurer, M. Defining and Substantiating the Terms Scene, Situation, and Scenario for Automated Driving. In Proceedings of the 2015 IEEE 18th International Conference on Intelligent Transportation Systems, Las Palmas, Gran Canaria, Spain, 15–18 September 2015; pp. 982–988. [CrossRef]
7. Mugur, T. Enhancing ADAS Test and Validation with Automated Search for Critical Situations. In Proceedings of the DSC 2015, Berlin, Germany, 16–18 September 2015.
8. Junietz, P.; Bonakdar, F.; Klamann, B.; Winner, H. Criticality Metric for the Safety Validation of Automated Driving using Model Predictive Trajectory Optimization. In Proceedings of the IEEE Conference on Intelligent Transportation Systems, ITSC, Maui, HI, USA, 4–7 November 2018; pp. 60–65.
9. Feng, S.; Feng, Y.; Sun, H.; Bao, S.; Zhang, Y.; Liu, H.X. Testing scenario library generation for connected and automated vehicles, part II: Case studies. *IEEE Trans. Intell. Transp. Syst.* **2020**, 1–13. [CrossRef]
10. Kalra, N.; Paddock, S.M. *Driving to Safety: How Many Miles of Driving Would It Take to Demonstrate Autonomous Vehicle Reliability?* RAND Corporation: Santa Monica, CA, USA, 2016. Available online: https://www.rand.org/pubs/research_reports/RR1478.html (accessed on 14 February 2021).
11. Wachenfeld, W.; Winner, H. The Release of Autonomous Vehicles. In *Autonomous Driving: Technical Legal and Social Aspects*; Springer: Berlin/Heidelberg, Germany, 2016; pp. 425–449.
12. Amersbach, C.; Winner, H. Defining Required and Feasible Test Coverage for Scenario-Based Validation of Highly Automated Vehicles. In Proceedings of the 2019 IEEE Intelligent Transportation Systems Conference (ITSC), Auckland, New Zealand, 27–30 October 2019; pp. 425–430. [CrossRef]
13. Fellendorf, M.; Vortisch, P. Microscopic Traffc Flow Simulator VISSIM. In *Fundamentals of Traffic Simulation*, 1st ed.; International Series in Operations Research & Management Science; Springer Science + Business Media: Berlin, Germany, 2010; Volume 145, pp. 63–94.
14. Statistisches Bundesamt. Verkehr: Verkehrsunfälle. Technical Report Reihe 7, Statistisches Bundesamt. 2018. Available online: https://www.destatis.de/DE/Themen/Gesellschaft-Umwelt/Verkehrsunfaelle/Publikationen/Downloads-Verkehrsunfaelle/verkehrsunfaelle-jahr-2080700177004.pdf (accessed on 16 February 2021).
15. Nalic, D.; Pandurevic, A.; Eichberger, A.; Rogic, B. Design and Implementation of a Co-Simulation Framework for Testing of Automated Driving Systems. *Sustainability* **2020**, *12*, 10476. [CrossRef]
16. Feng, S.; Yan, X.; Sun, H.; Feng, Y.; Liu, H.X. Intelligent driving intelligence test for autonomous vehicles with naturalistic and adversarial environment. *Nat. Commun.* **2021**, *12*, 748. [CrossRef] [PubMed]
17. Nalic, D.; Li, H.; Eichberger, A.; Wellershaus, C.; Pandurevic, A.; Rogic, B. Stress Testing Method for Scenario-Based Testing of Automated Driving Systems. *IEEE Access* **2020**. [CrossRef]
18. Barceló, J. (Ed.) *Fundamentals of Traffic Simulation*; Springer: New York, NY, USA, 2010; Volume 145.
19. Wang, Y.; Wang, L. Autonomous vehicles' performance on single lane road: A simulation under VISSIM environment. In Proceedings of the 2017 10th International Congress on Image and Signal Processing, BioMedical Engineering and Informatics (CISP-BMEI), Shanghai, China, 14–16 October 2017; pp. 1–5. [CrossRef]
20. Thus, J.J.; Kang, J.; Park, S.; Park, I.; Lee, J. Automated emergency vehicle control strategy based on automated driving controls. *J. Adv. Transp.* **2020**, *2020*, 3867921.
21. Ard, T.; Dollar, R.A.; Vahidi, A.; Zhang, Y.; Karbowski, D. Microsimulation of energy and flow effects from optimal automated driving in mixed traffic. *Transp. Res. Part C Emerg. Technol.* **2020**, *120*, 102806. [CrossRef]
22. Papadoulis, A.; Quddus, M.; Imprialou, M. Evaluating the safety impact of connected and autonomous vehicles on motorways. *Accid. Anal. Prev.* **2019**, *124*, 12–22. [CrossRef]

Article

Analysis of Market-Ready Traffic Sign Recognition Systems in Cars: A Test Field Study

Darko Babić, Dario Babić *, Mario Fiolić and Željko Šarić

Faculty of Transport and Traffic Sciences, University of Zagreb, 10000 Zagreb, Croatia;
darko.babic@fpz.unizg.hr (D.B.); mario.fiolic@fpz.unizg.hr (M.F.); zeljko.saric@fpz.unizg.hr (Ž.Š.)
* Correspondence: dario.babic@fpz.unizg.hr

Abstract: Advanced Driver Assistance System (ADAS) represents a collection of vehicle-based intelligent safety systems. One in particular, Traffic Sign Recognition System (TSRS), is designed to detect and interpret roadside information in the form of signage. Even though TSRS has been on the market for more than a decade now, the available ones differ in hardware and software solutions they use, as well as in quantity and typology of signs they recognize. The aim of this study is to determine whether differences between detection and readability accuracy of market-ready TSRS exist and to what extent, as well as how different levels of "graphical changes" on the signs affect their accuracy. For this purpose, signs ("speed limit" and "prohibition of overtaking") were placed on a test field and 17 vehicles from 14 different car brands underwent testing. Overall, the results showed that sign detection and readability by TSRS differ between car brands and that even small changes in the design of signs can drastically affect TSRS accuracy. Even in a controlled environment where no sign has been altered, there has been a 5% margin of misread signs.

Keywords: traffic signs; ADAS; traffic sign recognition system; automated driving

Citation: Babić, D.; Babić, D.; Fiolić, M.; Šarić, Ž. Analysis of Market-Ready Traffic Sign Recognition Systems in Cars: A Test Field Study. *Energies* **2021**, *14*, 3697. https://doi.org/10.3390/en14123697

Academic Editors: Arno Eichberger, Zsolt Szalay, Martin Fellendorf and Henry Liu

Received: 12 May 2021
Accepted: 17 June 2021
Published: 21 June 2021

Publisher's Note: MDPI stays neutral with regard to jurisdictional claims in published maps and institutional affiliations.

1. Introduction

Road traffic accidents are a significant social problem and it is estimated that, depending on the country, their costs amount to 1% up to 3% of gross domestic product [1]. Although road safety is improving in most European countries, the progress remains slow and misaligned with the established targets. This slow progress is partially due to the dynamic and complex nature of road traffic, and safety performance depends on a number of interconnected factors related to roadway environment, vehicles and humans.

In the past decade, with new technological breakthroughs, a significant effort has been devoted to improving vehicle safety systems. These safety systems can be divided into two categories: passive safety systems and active safety systems. A passive safety system reduces injuries sustained by passengers when an accident occurs, while active ones try to keep a vehicle under control and avoid accidents [2,3].

In general, Advanced Driver Assistance System (ADAS) is a collection of numerous intelligent units integrated into the vehicle itself that perform different tasks and assist the human driver in driving. Common ADAS functions include adaptive speed control, lane departure warning, forward collision warning, automatic high beam assist, traffic sign recognition, pedestrian and object detection, automatic emergency braking, etc. All these functions base their operations on different cameras, RADARs, LIDARs and other sensors which "scan" the environment around the vehicle in order to gather the needed information. Since the efficiency of such systems majorly depends on the data collected from the surrounding environment, it is clear that different road infrastructure elements, such as traffic signs or road markings, provide necessary cues not only to human drivers but also to built-in vehicle technologies.

Traffic Sign Recognition System (TSRS) is designed to detect and interpret roadside information in the form of signage. Its basic infrastructure can be generalized into three specific components: visual sensor, image processor and vehicle display [4]. Acquiring information from traffic signs involves traffic sign detection (TSD), which consists of finding the location, orientation and size of traffic signs in natural scene images, and traffic sign recognition (TSR), or classifying the detected traffic signs into types and categories in order to extract the information they are providing to drivers [5]. In order to detect and recognize traffic signs, as mentioned before, vehicles are equipped with different technologies. Cameras are the most common sensors and can be used for TSR, TSD or both at the same time. LIDAR sensors have been used for TSD. Their 3D perceptive capabilities are useful to determine the position of a sign and its shape, and can also use the intensity of reflected light to improve detection accuracy based on the high reflectivity of traffic signs [5]. Potentially, the best solution presents the combination of LIDAR and cameras. Such fusion enables the collection of information from different sources, their comparison and analysis, and thus better detection and recognition of signs. Besides detecting and recognizing road signs, TSRS also use digital maps with already implemented signs (mainly related to the speed limit signs).

After sensors and front-facing cameras collect data, algorithms are used to segment and analyze the stimuli. This process includes shape, color and symbol detection as well as classification of signs based on the aforementioned characteristics [4]. A vast body of literature has analyzed the efficiency of different algorithms for segmentation and classification of signs [6–9]. Several review papers on this subject have also been published in the past few years [10–13]. Besides the overall review of the working procedures, studies identified main issues and challenges regarding the accuracy of TSRS. They are generally related to fading and blurring of traffic signs, visibility levels of signs in comparison to the environment, differences between existing traffic sign systems, multiple appearances of signs, damaged or partially obscured signs, correct location of signs, unavailability of public databases, electronic signs, etc. An on-road study conducted in Australia and New Zealand confirmed some of the aforementioned issues [4]. For the purpose of this study, authors used five cars with TSRS and drove a number of trials designed and conducted in order to identify key issues existing in the current TSRS and to investigate potential causes of the found issues. The study highlighted several applicable changes that could improve traffic sign readability, including electronic signs, installation and maintenance, sign positioning and location, sign face design, vehicle-mounted signs and other advisory and information signs.

From the literature review, one can conclude that general problems regarding TSRS are well known. However, different car manufacturers use different hardware and software solutions that may differ in traffic sign detection and readability accuracy. In addition, there are some differences between the signs being detected and presented to the driver. Namely, some cars can only read "speed limit" signs, while others in addition to "speed limit" signs can read other signs such as "prohibition of overtaking", "end of all restrictions", "start and end of highway", etc.

For this reason, the main aim of this study is to determine whether differences between detection and readability accuracy of market-ready TSRS exist, and to what extent. Moreover, since the problem of partially obscured signs was identified in literature as one of the main challenges of TSRS, the objective of the study was to test how simulated "graphical changes" affect their detection accuracy.

2. Materials and Methods

In this section, the research methodology is presented. The section consists of four subsections each describing a part of the methodology, from the testing track, vehicles, scenarios and test procedures to data analysis.

2.1. Testing Track

The experiment was conducted on a road inside the campus of the University of Zagreb. The road is a typical two-way road with almost no traffic since it is only used for connecting buildings on the campus. The total length of the road is 1.2 km and it consists of straight sections, four curves and five intersections with the right of way in the direction in which tests were conducted. Nine traffic signs were placed on straight sections of the testing track, as shown in Figure 1. Seven of them were used, i.e., their graphical image was changed according to scenario designs, while the remaining two (8 and 9) were placed for control purposes only and their "readings" were not recorded and analyzed. Only speed limit and prohibition of overtaking signs were used since the majority of current TSRS read only these signs. In total, four "speed limit" signs (50 km/h, 60 km/h, 90 km/h and 100 km/h) and three "prohibition of overtaking" signs were used. The speed limits (50 km/h, 60 km/h, 90 km/h) were chosen based on the fact that they are the most common and that their meaning could easily be altered. For example, with a very small modification, 60 km/h could easily be altered to 80 km/h. The 100 km/h speed limit was chosen since it contains three digits. All signs were newly made and fulfilled all technical properties (visibility, chromaticity, etc.) defined by the Croatian standard [14]. The design of the signs was also according to the Croatian standard, which is based on the Vienna convention. All of them are 60 cm in diameter and were made using class II sheeting with prismatic retroreflection. Furthermore, all signs were placed according to the Croatian standard [14], which implies a 30–75 cm distance from the edge of the road and a height between 1.2 to 1.5 m. The distance between each sign was set to a minimum of 100 m and the locations of the signs did not have any environmental "disturbances" which could affect the TSRS. Although vegetation next to the road is present, all signs were clearly visible, i.e., they were not covered with vegetation. The layout of the signs at each location was as follows: Location 1—50 km/h speed limit; Location 2—prohibition of overtaking; Location 3—60 km/h speed limit; Location 4—prohibition of overtaking; Location 5—90 km/h speed limit; Location 6—prohibition of overtaking; Location 7—100 km/h speed limit; Location 8—prohibition of overtaking (control sign); Location 9—60 km/h speed limit (control sign).

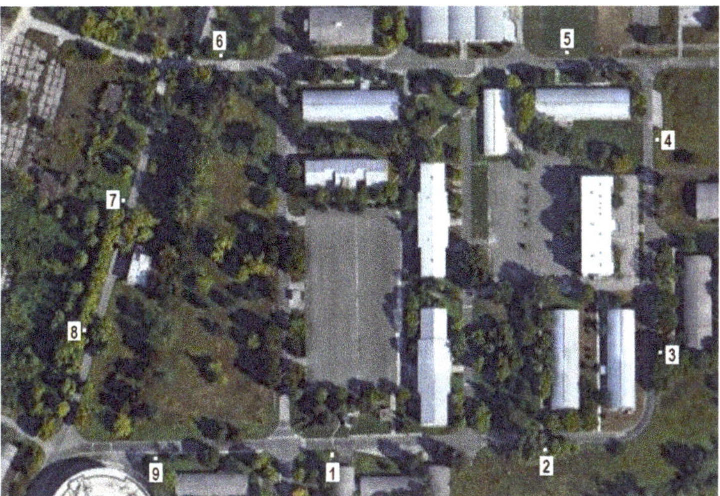

Figure 1. Layout of the test route and placement of the signs.

2.2. Vehicles

Seventeen (17) cars in total were included in the study. Fourteen of them were from different manufacturers while the remaining three were from the same manufacturer. Most of those cars are new vehicles (12 of them were produced in 2020), while the other five ranged from 2014 to 2019. Each vehicle has TSRS which is based on a camera located at the front windshield inside the vehicle. The vehicles were, for the sole purpose of the research, lent to us by their official distributors for their respective manufacturing brand in Croatia. It is worth noting that the technical aspects of the system were not provided to us. Since the aim of the study is to detect the differences between existing TSRS and not to determine which car brand has the "best" TSRS, brand names are in code.

2.3. Scenarios and Test Procedure

In order to test the accuracy of TSRS, nine scenarios were created. The first one was a control scenario in which signs did not have any graphical changes. In the second scenario, the red outline of the signs was covered with black paper. In the third and fourth scenarios, minor changes on symbols were made. For example, a piece of black paper was placed so that 60 km/h speed limit looks similar to 80 km/h or black lines were placed on the prohibition of overtaking signs. In the fifth, sixth and seventh scenarios, a half of the signs' face was covered with black paper. The difference between the scenarios was in the orientation of the coverage. In the eighth scenario, the last digit (0) on the speed limit sign was covered with white paper. On the two prohibition of overtaking signs, the symbols of cars were covered with white paper while on the third prohibition of overtaking sign, the symbols were mirrored. In the last scenario, symbols were covered with graffiti. The visual presentation of signs in each scenario is shown in Figure 2.

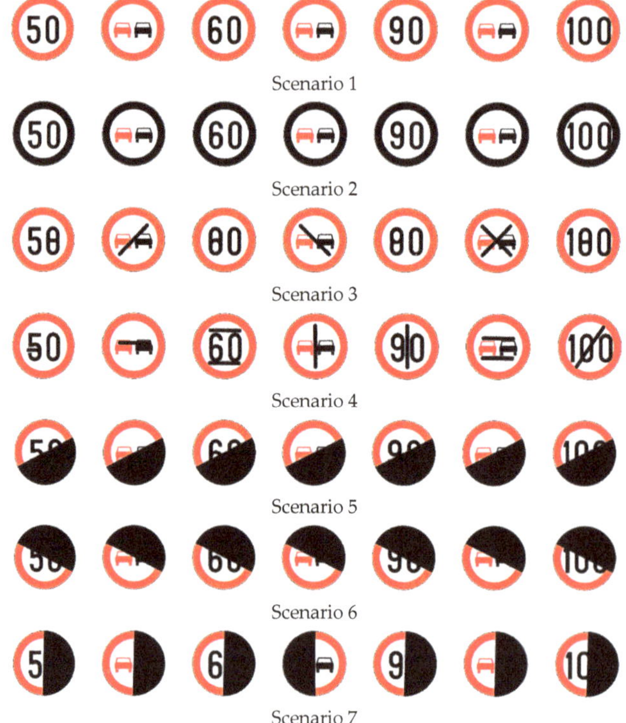

Figure 2. *Cont.*

Figure 2. Visual presentation of signs used in each scenario.

Tests were conducted during daytime between 9:00 a.m. and 3:00 p.m. in the period of three days. During all three days, weather conditions were almost the same, i.e., normal (sunny) weather. Before the start, we held a "practice" run with each vehicle in order to check if the TSRS were functioning properly. During the test, each vehicle passed through the test track once per each scenario. Alongside the driver, the researcher was present in the vehicle during tests, controlling and recording if the signs were or were not displayed to the driver and, if yes, what was displayed. Only one vehicle was permitted on the testing track during each run and other traffic was not present. Because of the road configuration (campus site), the driving speed was between 40 and 60 km/h.

2.4. Data Analysis

For the first part of the analysis, we grouped the data into three categories: (1) correctly recognized signs, (2) unrecognized signs and (3) wrongly recognized signs. For each category, a percentage distribution of the results was calculated and compared. For the second part, the results from groups (2) and (3) were joined into one group, unrecognized or wrongly recognized signs, since they all in all presented the error of TSRS. On top of that, scenarios were grouped based on the level of modification they were submitted to: control (scenario 1), minor changes (scenarios 2 + 3 + 4), medium changes (scenarios 8 + 9) and major changes (scenarios 5 + 6 + 7) in order to analyze the differences between the control scenario (1) and each level of graphical change. Mean values for each vehicle and group were computed and repeated measures ANOVA with Bonferroni adjustment (alpha level 0.05) were used to test the differences between the control condition and each group.

In the second part of the analysis, we tested the differences between vehicles per scenario. Since the values in the data set for this analysis were dichotomous, Cochran's Q test was used (alpha level 0.05). As one of the assumptions of Cochran's Q test is that a sample has two mutually exclusive categories, the analysis was conducted on the correctly recognized signs categories and not on unrecognized or wrongly recognized signs.

3. Results

In total, 1071 recognitions were possible (7 signs per scenario $\times$ 9 scenarios $\times$ 17 vehicles). In 21.20% of the cases, signs were recognized correctly. In 72.46%, signs were not recognized at all, while in 6.35%, signs were wrongly recognized by vehicles. Of course, these percentages vary between each scenario. As expected, in the control scenario (scenario 1), the highest level of correctly recognized signs was recorded (73.11%). On the other hand, in the same scenario, 21.58% of the signs were not recognized at all (mainly prohibition of overtaking, but also in some cases speed limit signs). This is due to the fact that the type of signs which TSRS recognize and display to the driver varies to some extent between brands. Moreover, some of the signs (5.04%) were wrongly recognized even in the control scenario (for example, 90 km/h was recognized as 30 km/h). In the second scenario (black outline of the signs), the number of unrecognized signs increased to 36.13%, while the percentage of wrongly recognized signs stayed the same as in scenario 1 (5.04%).

A major decrease in sign recognition occurred in scenarios 3 and 4 in which minor changes were made to the signs. Namely, the percentage of correctly recognized signs in scenarios 3 and 4 fell to 20.17% and 4.20%, respectively. In addition to the increase in unrecognized signs, there was also an increase in wrongly recognized signs—26.89% in scenario 3 and 14.29% in scenario 4.

When signs were half covered by black paper (scenarios 4, 5 and 6) as well as by graffiti (scenario 9), the percentage of unrecognized signs was between 98% and 100%. In scenario 8, 33.61% of the signs were correctly recognized (63.03% of signs unrecognized), while 3.36% were wrongly recognized (mainly speed limit signs).

A graphical presentation of the aforementioned results is shown in Figure 3 while the overall results for each vehicle per each scenario are presented in Appendix A.

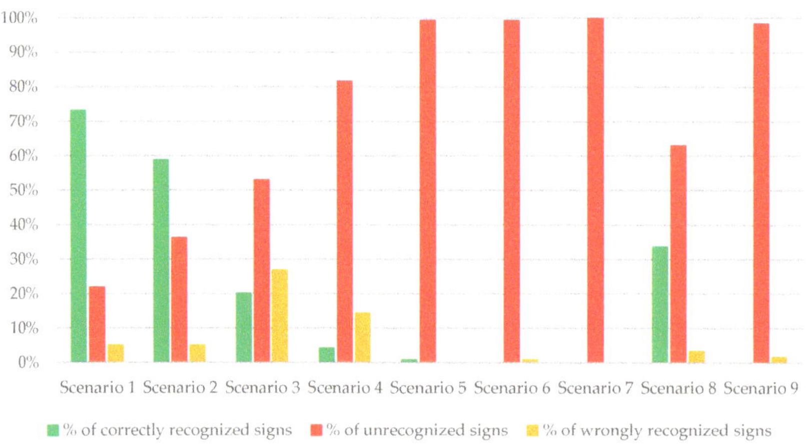

Figure 3. Percentages of "correctly recognized", "unrecognized" and "wrongly recognized" signs per scenario.

As mentioned in the Data Analysis section, the scenarios were grouped into four categories: control (scenario 1), minor changes (scenarios 2, 3 and 4), medium changes (scenarios 8 and 9) and major changes (scenarios 5, 6 and 7). This was done in order to analyze how different levels of changes on signs affect recognition accuracy. For each vehicle and group, mean value was computed. Overall, the decrease in correctly recognized signs between controlled conditions and minor, medium and major changes is 62%, 77% and 99%, respectively (Figure 4).

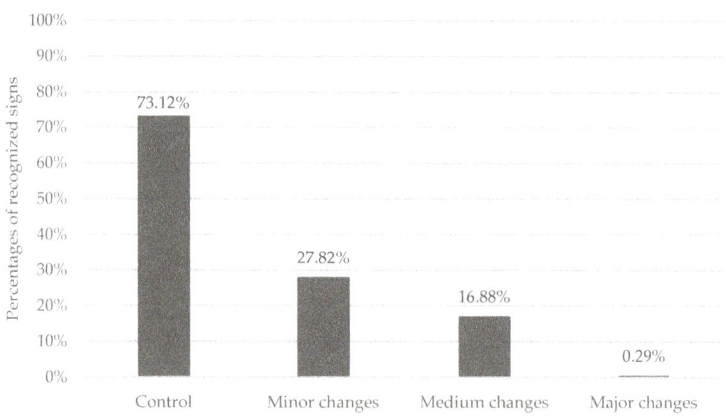

Figure 4. Percentages of recognized signs per each category.

Although differences between each category are evident, a repeated measure ANOVA with Bonferroni adjustment was used to test the differences between the control condition and each group. The results of ANOVA analysis shown in Table 1 confirms that a statistical difference ($p < 0.05$) between the control condition and each group representing different levels of graphical changes of signs exists.

Table 1. Results of ANOVA.

(I)	(J)	Mean Dif. (I–J)	Std. Error	p	95% Confidence Interval for Difference	
					Lower Bound	Upper Bound
Control	Minor changes	0.454	0.051	0.000	0.301	0.607
	Medium changes	0.563	0.065	0.000	0.368	0.758
	Major changes	0.728	0.059	0.000	0.550	0.906

Furthermore, we tested the difference between traffic sign recognition of each vehicle in each scenario with Cochran's Q test. Since scenarios 6, 7 and 9 after grouping did not have variations in values (no values for group "correctly recognized signs"), Cochran's Q test was not performed on them.

Overall, a statistical difference ($p < 0.05$) in traffic sign recognition between tested vehicles was recorded for scenarios 1, 2, 3 and 8, while for scenarios 4 and 5, no statistical difference was found ($p > 0.05$), as shown in Table 2.

Table 2. Results of Cochran's Q test—significant differences in traffic sign recognition between tested vehicles per analyzed scenarios.

Scenario	Asymptotic Sig.
1	0.002
2	0.000
3	0.041
4	0.699
5	0.453
8	0.000

4. Discussion

Due to their great potential in reducing road accidents, ADAS technologies have become one of the fastest-growing safety application areas. In the European Union, ADAS systems are mandatory for all new and certified vehicles starting from 2022 and all new registered vehicles by 2024 [15]. One of those systems is traffic sign recognition, which provides drivers with information about traffic signs.

However, the analysis of market-ready TSRS shows that the functionalities of systems differ to some extent between car brands, mainly in the type and the number of signs that can be recognized and displayed to the driver. Some of the cars display only speed limit signs, some besides speed limit display prohibition of overtaking, while some are, in addition to the above, displaying main warning signs such as "dangerous curve", "pedestrian crossing", etc. Furthermore, while the majority of TSRS signs are displayed in color, there are a few brands that display signs in black and white.

The results of this study show that TSRS accuracy differs between car brands and graphical clarity of the sign. In other words, each scenario with graphical changes on signs had significantly lower recognition level compared to the control condition (ranging from 62% to 99%). It is important to note that even the control condition did not have a 100% recognition and, contrary to expectations, had 5% of wrongly recognized signs. This is precisely due to the differences in TSRS between car brands. Moreover, between each category of graphical change and the control condition, there has been a statistically significant difference in the number of correctly recognized signs. The results suggest that even small changes in the design of a sign, such as the change of the outline color or a minor change in the symbol, can drastically affect TSRS accuracy. This result is in accordance with previous studies [4,12].

Although graphical changes on the signs were created artificially in this study, similar situations exist in real road conditions [16,17] and may result in TSRS errors. Such errors may affect the driver in different ways. Studies have shown that drivers in general perceive a relatively low amount of traffic signs [18–20], and thus an additional display of signs in the vehicle (especially speed limits) should increase overall compliance with traffic regulations. On the other hand, if the sign is not recognized by the TSRS, the driver will not have additional information, which may result in inappropriate and/or improper driving behavior. Furthermore, if TSRS recognizes the signs but wrongly (for example 60 km/h as 80 km/h), such error may confuse drivers and thus distract them, which again may lead to inappropriate and/or improper driving behavior. This result confirms the importance of proper maintenance of signs and their surroundings as highlighted in several recommendations and publications [4,21,22].

Since the general functioning and accuracy of TSRS significantly differ between car brands, a "standardization" of such systems is needed. Standardization, in this sense, implies defining the minimal number and types of signs which every TSRS should recognize, the way signs are displayed to the drivers and minimal levels of recognition accuracy at least for properly placed and maintained signs during daytime and low visibility conditions (night-time, rain, fog, etc.). Although the last requirement is difficult to define since recognition accuracy depends on the number of factors [10–13], the standardization of TSRS could potentially accelerate their development and possibly eliminate some of the current problems. In addition, recent studies emphasize that, although ADAS systems provide a significant progress in safety, they also may distract drivers [23,24]. Thus, the education of drivers regarding the functionality and limitations of TSRS is needed as well in order to avoid or at least decrease potential confusion of drivers. The education altogether with the standardization could decrease the driver's distraction caused by TSRS. Finally, in order to increase the accuracy of TSRS, a database of traffic signs for each country should be established. The database in this sense should have all the variations of signs and their meaning for each country. In the primal stage, it could be developed at least for speed limits so the algorithms used for sign classification and recognition can be "taught" and thus the accuracy of the whole system may be increased.

Although this study provides valuable results, there are some limitations. The study was conducted in controlled conditions (practically a closed road section, dry daytime weather, lack of other traffic, all new traffic signs, relatively low driving speed, etc.) that are not generally present in the real world. However, such controlled conditions represent the best-case scenario, meaning that TSRS accuracy would decrease even more in real world conditions. Moreover, the study included only 17 cars from 14 brands. Even though the used sample is relatively small, most of the major car brands were included and all cars had the best equipment provided by the manufacturer. Finally, due to unavailability of the technical data, proper comparison of the TSRS was not possible.

5. Conclusions

The main aim of this field study was to test whether differences between detection and readability accuracy of market-ready TSRS exist and to what extent, as well as how different "graphical changes" affect their accuracy. A total of 17 cars from 14 brands were tested. Fourteen cars were from different manufacturers while three were from the same manufacturer. The results confirm that TSRS accuracy differs between car brands and even between the same brand, although, due to a limited sample, this needs to be further tested. Furthermore, graphical changes significantly affected the accuracy of the TSRS in all vehicles compared to the control condition (ranging from 62% to 99%).

Based on the results and limitations of this study, future research should investigate TSRS accuracy on real roads with other traffic present and in different conditions as well as different types and quality of traffic signs. In order to develop a deeper insight, the technical data about the installed TSRS should be analyzed in order to evaluate each system and its limitations. Furthermore, a detailed elaboration of standardization methodology and principles should also be developed.

Author Contributions: Conceptualization, D.B. (Darko Babić) and D.B. (Dario Babić); Methodology, D.B. (Darko Babić), D.B. (Dario Babić), M.F. and Ž.Š.; Formal analysis, M.F., D.B. (Dario Babić) and Ž.Š.; Investigation, D.B. (Dario Babić) and M.F.; Resources, D.B. (Dario Babić), M.F. and Ž.Š.; Writing— D.B. (Darko Babić) and D.B. (Dario Babić); Supervision, D.B. (Darko Babić); Project administration, D.B. (Darko Babić); Funding acquisition, D.B. (Darko Babić). All authors have read and agreed to the published version of the manuscript.

Funding: This research was a part of the project entitled "Establishing a Methodology for Testing and Evaluating the ADAS Systems" funded by the University of Zagreb (Potpore za temeljno financiranje znanstvene i umjetničke djelatnosti Sveučilišta u Zagrebu u ak. god. 2019./2020.).

Data Availability Statement: The data used to support the findings of this study are available from the corresponding author upon request.

Conflicts of Interest: The authors declare no conflict of interest.

Appendix A Results for Each Vehicle per Each Scenario

S	TS	V1	V2	V3	V4	V5	V6	V7	V8	V9	V10	V11	V12	V13	V14	V15	V16	V17
											Vehicles							
1	1	1	1	1	1	1	1	1	1	1	1	1	1	1	1	1	1	1
	2	1	1	0	1	0	0	0	0	1	1	1	0	0	1	1	1	1
	3	1	1	1	0	0	1	1	1	1	1	1	1	1	1	1	1	1
	4	1	1	0	1	0	0	0	0	1	1	1	0	0	1	1	1	1
	5	1	0	−1	1	1	1	1	1	1	1	1	1	−1	−1	−1	1	1
	6	1	1	0	1	0	0	0	0	1	1	1	0	0	1	0	1	1
	7	1	1	1	1	1	1	1	1	1	1	1	1	−1	0	−1	1	1
2	1	1	1	0	0	0	1	1	1	1	0	1	1	1	1	1	1	1
	2	1	1	0	0	0	0	0	0	1	0	1	0	0	1	1	1	1
	3	1	1	0	0	0	1	1	1	1	0	1	1	1	1	1	1	1
	4	1	1	0	0	0	0	0	0	1	0	1	0	0	1	1	1	1
	5	1	1	0	0	0	1	1	1	1	0	1	1	−1	−1	−1	1	1
	6	1	1	0	0	0	0	0	0	1	0	1	0	0	1	1	1	1
	7	1	1	0	0	0	1	1	1	1	0	1	1	−1	−1	−1	1	1
3	1	1	1	1	1	1	1	1	1	0	1	1	0	1	1	0	1	1
	2	1	0	0	0	0	0	0	0	0	0	0	0	0	1	0	0	0
	3	−1	−1	−1	0	0	−1	−1	−1	0	0	0	−1	0	0	0	−1	−1
	4	1	0	0	0	0	0	0	0	0	0	1	0	0	1	0	1	0
	5	−1	0	−1	−1	−1	−1	−1	−1	0	−1	−1	−1	−1	−1	0	1	−1
	6	0	0	0	0	0	0	0	0	0	0	0	0	0	0	0	0	0
	7	−1	0	1	−1	−1	−1	−1	−1	0	−1	−1	1	0	−1	0	1	−1
4	1	−1	0	−1	−1	1	1	−1	−1	−1	−1	−1	−1	0	−1	−1	−1	−1
	2	1	0	0	0	0	0	0	0	0	0	−1	0	0	−1	1	0	0
	3	0	0	0	0	0	0	0	0	0	0	0	0	0	0	0	0	0
	4	0	0	0	0	0	0	0	0	0	0	0	0	0	0	0	0	0
	5	0	0	0	0	0	0	0	0	0	0	0	0	0	0	0	1	−1
	6	0	0	0	0	0	0	0	0	0	0	0	0	0	0	0	0	0
	7	0	0	0	0	0	0	0	0	0	0	−1	0	0	0	0	0	0

	1	0	0	0	0	0	0	0	0	0	0	0	0	0	0	1	0
	2	0	0	0	0	0	0	0	0	0	0	0	0	0	0	0	0
	3	0	0	0	0	0	0	0	0	0	0	0	0	0	0	0	0
5	4	0	0	0	0	0	0	0	0	0	0	0	0	0	0	0	0
	5	0	0	0	0	0	0	0	0	0	0	0	0	0	0	0	0
	6	0	0	0	0	0	0	0	0	0	0	0	0	0	0	0	0
	7	0	0	0	0	0	0	0	0	0	0	0	0	0	0	0	0

S—Scenario; TS—Traffic signs; V—Vehicle

References

1. World Health Organization. *Road Traffic Injuries*; World Health Organization: Geneva, Switzerland, 2017.
2. Lu, M.; Wevers, K.; van der Heijden, R. Technical feasibility of advanced driver assistance systems (ADAS) for road traffic safety. *Transp. Plan. Technol.* **2005**, *28*, 167–187. [CrossRef]
3. Dörterler, M.; Bay Ömer, F. Neural network based vehicular location prediction model for cooperative active safety systems. *Promet Traffic Transp.* **2018**, *30*, 205–215. [CrossRef]
4. Austroads. *Implications of Traffic Sign Recognition (TSR) Systems for Road Operators*; Technical report AP-R580-18; Austroads: Sydney, Australia, 2018.
5. Martí, E.; de Miguel, M.Á.; García, F.; Pérez, J. A review of sensor technologies for perception in automated driving. *IEEE Intell. Transp. Syst. Mag.* **2019**, *11*, 94–108. [CrossRef]
6. Cao, J.; Song, C.; Peng, S.; Xiao, F.; Song, S. Improved traffic sign detection and recognition algorithm for intelligent vehicles. *Sensors* **2019**, *19*, 4021. [CrossRef] [PubMed]
7. Dewi, C.; Chen, R.C.; Yu, H. Weight analysis for various prohibitory sign detection and recognition using deep learning. *Multimed. Tools Appl.* **2020**, *79*, 32897–32915. [CrossRef]
8. Gil Jiménez, P.; Bascón, S.M.; Moreno, H.G.; Arroyo, S.L.; Ferreras, F.L. Traffic sign shape classification and localization based on the normalized FFT of the signature of blobs and 2D homographies. *Signal. Process.* **2008**, *88*, 2943–2955. [CrossRef]
9. Tian, Y.; Gelernter, J.; Wang, X.; Li, J.; Yu, Y. Traffic sign detection using a multi-scale recurrent attention network. *IEEE Trans. Intell. Transp. Syst.* **2019**, *20*, 4466–4475. [CrossRef]
10. Gudigar, A.; Chokkadi, S.; Raghavendra, U. A review on automatic detection and recognition of traffic sign. *Multimed. Tools Appl.* **2016**, *75*, 333–364. [CrossRef]
11. Liu, C.; Li, S.; Chang, F.; Wang, Y. Machine vision based traffic sign detection methods: Review, analyses and perspectives. *IEEE Access* **2019**, *7*, 86578–86596. [CrossRef]
12. Wali, S.B.; Abdullah, M.A.; Hannan, M.A.; Hussain, A.; Samad, S.A.; Ker, P.J.; Bin Mansor, M. Vision-based traffic sign detection and recognition systems: Current trends and challenges. *Sensors* **2019**, *19*, 2093. [CrossRef] [PubMed]
13. Zourlidou, S.; Sester, M. Traffic regulator detection and identification from crowdsourced data—A systematic literature review. *ISPRS Int. J. Geo-Inf.* **2019**, *8*, 491. [CrossRef]
14. Pravilnik o Prometnim Znakovima Signalizaciji i Opremi na Cestama (NN 92/19) (2019). [In Croatian; National Regulations on Traffic Signs, Equipment and Signalization (NN 92/19) (2019)]. Available online: https://narodne-novine.nn.hr/clanci/sluzbeni/2019_09_92_1823.html (accessed on 17 June 2021).
15. European Parliament and the Council of the European Union. *General Safety Regulation*; Official Journal of the European Union: Brussels, Belgium, 2019.
16. Khalilikhah, M.; Heaslip, K. Prediction of traffic sign vandalism that obstructs critical messages to drivers. *Transport* **2016**, *33*, 399–407. [CrossRef]
17. Babić, D.; Fiolić, M.; Babić, D.; Ščukanec, A. Testing the quality of traffic signs and road markings: Republic of Croatia case study. In Proceedings of the Transport Research Arena 2018, Vienna, Austria, 16–19 April 2018.
18. Babić, D.; Babić, D.; Ščukanec, A. The impact of road familiarity on the perception of traffic signs—Eye tracking case study. In Proceedings of the 10th International Conference: "Environmental Engineering", Vilnius, Lithuania, 27–28 April 2017.
19. Macdonald, W.A.; Hoffmann, E.R. Drivers' awareness of traffic sign information. *Ergonomics* **1991**, *34*, 585–612. [CrossRef]
20. Milosevic, S.; Gajic, R. Presentation factors and driver characteristics affecting road-sign registration. *Ergonomics* **1986**, *29*, 807–815. [CrossRef]
21. European Union Road Federation. *Improved Signage for Better Roads*; European Union Road Federation: Brussels, Belgium, 2015.
22. EuroRap. *Roads That Cars Can Read*; EuroRap: Basingstoke, UK, 2013.
23. Fundation for Road Safety. *Understanding the Impact of Technology: Do Advanced Driver Assistance and Semi-automated Vehicle Systems Lead to Improper Driving Behavior? Technical report*; Fundation for Road Safety: Basingstoke, UK, 2019.
24. Cades, D.; Crump, C.; Lester, B.; Douglas, Y. Driver distraction and advanced vehicle assistive systems (ADAS): Investigating effects on driver behavior. In *Advances in Human Aspects of Transportation*; Springer: Berlin/Heidelberg, Germany, 2017.

Article

Towards Cooperative Perception Services for ITS: Digital Twin in the Automotive Edge Cloud

Viktor Tihanyi, András Rövid, Viktor Remeli, Zsolt Vincze, Mihály Csonthó, Zsombor Pethő, Mátyás Szalai, Balázs Varga, Aws Khalil and Zsolt Szalay *

Department of Automotive Technologies, Faculty of Transportation Engineering and Vehicle Engineering, Budapest University of Technology and Economics, 1111 Budapest, Hungary; tihanyi.viktor@kjk.bme.hu (V.T.); rovid.andras@kjk.bme.hu (A.R.); remeli.viktor@kjk.bme.hu (V.R.); vincze.zsolt@kjk.bme.hu (Z.V.); csontho.mihaly@kjk.bme.hu (M.C.); petho.zsombor@kjk.bme.hu (Z.P.); szalai.matyas@kjk.bme.hu (M.S.); varga.balazs@kjk.bme.hu (B.V.); aws.khalil@bme.edu.hu (A.K.)
* Correspondence: szalay.zsolt@kjk.bme.hu; Tel.: +36-1-463-3226

Abstract: We demonstrate a working functional prototype of a cooperative perception system that maintains a real-time digital twin of the traffic environment, providing a more accurate and more reliable model than any of the participant subsystems—in this case, smart vehicles and infrastructure stations—would manage individually. The importance of such technology is that it can facilitate a spectrum of new derivative services, including cloud-assisted and cloud-controlled ADAS functions, dynamic map generation with analytics for traffic control and road infrastructure monitoring, a digital framework for operating vehicle testing grounds, logistics facilities, etc. In this paper, we constrain our discussion on the viability of the core concept and implement a system that provides a single service: the live visualization of our digital twin in a 3D simulation, which instantly and reliably matches the state of the real-world environment and showcases the advantages of real-time fusion of sensory data from various traffic participants. We envision this prototype system as part of a larger network of local information processing and integration nodes, i.e., the logically centralized digital twin is maintained in a physically distributed edge cloud.

Keywords: cooperative perception; ITS; digital twin; sensor fusion; edge cloud

Citation: Tihanyi, V.; Rövid, A.; Remeli, V.; Vincze, Z.; Csonthó, M.; Pethő, Z.; Szalai, M.; Varga, B.; Khalil, A.; Szalay, Z. Towards Cooperative Perception Services for ITS: Digital Twin in the Automotive Edge Cloud. *Energies* **2021**, *14*, 5930. https://doi.org/10.3390/en14185930

Academic Editor: Marco Pau

Received: 13 July 2021
Accepted: 4 September 2021
Published: 18 September 2021

1. Introduction

1.1. Scope and Significance

The future of connected and automated vehicles (CAVs) and the development of intelligent transportation systems (ITSs) are actively researched topics which open up a multitude of possibilities. With the progress in computation and communication technology in the last decade, some formerly unrealistic constructs are becoming more practically viable and demand proof-of-concept implementations. We envision a future where traffic participants and observers like CAVs and ITSs share their information resources in real-time for a safer and more efficient transportation and traveling experience. Herein, we outline a so-called *Central System* architecture that enables such information sharing and integration. We use the term *central* in a strictly logical sense to denote the emergence of a single, fully integrated and logically consistent environment and decision model in the cloud (the *digital twin*), while the physical implementation itself remains highly *distributed*, i.e., computation and communication loads are delegated to a spatially localized edge processing nodes hierarchy, as well as a network of third-party partners such as trusted data and algorithm providers. Understanding the wide ranging general applicability of building a well-integrated smart road and vehicle IoT, we made a dedicated effort to design the *Central System* as an easily extensible integration framework using industry standard interfaces. In the coming months we expect to be able to connect a part of the fast-developing Hungarian ITS facilities into the *Central System* and to provide non-stop real-time integration of real traffic data and thus demonstrate the first large-scale industrial application. In the current paper, we report the positive results of small-scale experiments conducted in late 2020.

Arguably, the most useful kind of information for traffic participants might very well be the establishment of a so-called *digital twin*: a dynamic and real-time model of the environment, which is the focus of our current work and provides the major information source to build other services upon. The functional prototype we built and present in this paper therefore realizes only the fundamental real-time environment perception function of the *Central System*, which for clarity we will call *Central Perception*—distinguishing it from an envisioned suite of other dependent functionalities like traffic analytics and planning, road infrastructure monitoring and management, cloud-based traffic control and vehicle control, specific applications on CAV testing grounds and logistic grounds, etc. Of course, such derivative services must be introduced in at least some detail to highlight the expected practical significance of our initial efforts. A schematic overview of the planned *Central System*, its participants and functionalities is represented in Figure 1.

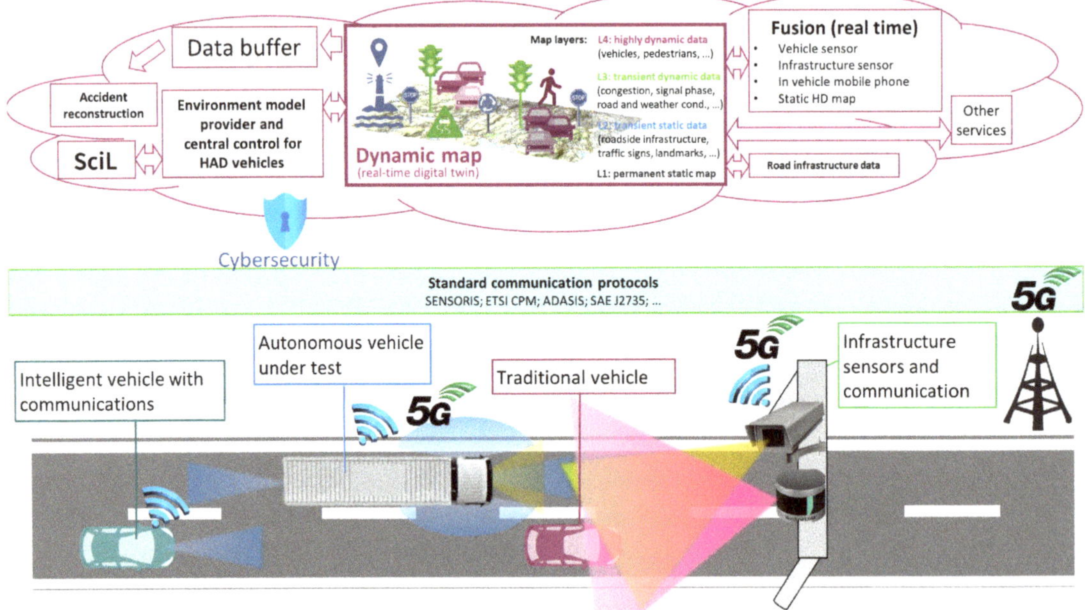

Figure 1. *Central System* overview.

The premise of *Central Perception* is the following: assuming smart roads and vehicles in the near future, the same physical traffic environment will typically be perceived through many sensors and platforms of highly differing setups and capabilities at the same time, and these platforms may or may not be affiliated. Depending on application requirements like expected response time and reliability, the integration of environment perception information from various sources will pose certain challenges including communication, compatibility, synchronization, calibration, fusion, tracking, and end-to-end latency. Such an ad-hoc collective and collaborative perception system also requires certain sophistication in its architecture which must allow for scaling in a dynamically changing environment, especially considering CAVs that constantly change their location and whose sensor data must therefore be integrated with different stationary platforms at different points in time.

1.2. Prior Work

In the last two years, several connected automotive information system prototypes were realized, all of them demonstrating some degree of cooperative perception. Krämmer et al. from TU Munich and Fortiss built "Providentia" [1], an intelligent infrastructure system consisting of several gantries equipped with cameras and radars, serving as a support system for intelligent vehicles and aiding them in perceiving blind spots and objects behind cover. They process and fuse the sensory data both locally at the individual measurement stations and also centrally at designated edge computing nodes before communicating it to the consumer vehicles via 5G. Gabb et al. from Bosch and Karlsruhe showed theoretical guidelines (supported by experiments) for developing a similar system around the same time, at the Intelligent Vehicles conference in Paris [2]. The University of Tokyo released open source software for cooperative perception [3] while the University of California together with Toyota demonstrated a use-case of actually backpropagating the already integrated data from the cloud for driving assistance purposes [4]. Toyota also pursues similar topics with other Universities and independently, focusing on various settings like V2V communication [5] or camera and digital twin integration for visual guidance systems [6]. In Australia, researchers implemented the ETSI CPM messaging standard in an I2V setting, allowing sensor-less vehicles to perceive and autonomously react to pedestrians [7]. Recently, Chinese authorities announced to launch "world's first high-level cloud-controlled autonomous driving demonstration zone" (http://m. news.cctv.com/2020/09/11/ARTIeJEug9svYwuLazxQFzO3200911.shtml (accessed on 14 September 2021)) to be constructed in Beijing with similar long term targets as our *Central System* project.

As prior work we have carried out a measurement campaign (together with international industrial and academic partners) on a real-world motorway section in Hungary, which resulted sensory data useful for future automotive R&D activities due to the available ground truth for static as well as for dynamic content [8].

A possible first industrial application of *Central System* technology is likely to occur where experimental CAVs and ITSs are introduced earliest: on automotive testing grounds. In particular, development of *Central System* based Scenario-in-the-Loop (SciL) [9] control is already underway on the ZalaZONE [10] CAV proving ground in Hungary.

1.3. Primary Contribution

According to the best of our knowledge, we are the first to demonstrate a real-time cooperative perception platform that has both stationary and moving multi-sensor data sources and that combines several levels of data integration such as inter-sensor raw fusion, on-platform tracking and inter-platform local area fusion to finally create and visualize a simultaneous, centrally consistent model (*digital twin*) of all objects of interest in the area covered by the sensors' field of views.

We emphasize that our primary contribution lies in the demonstration of system-level possibilities as they were not demonstrated beforehand, i.e., the maintenance of a *digital twin* in a complex and heterogeneous environment. In this paper we do not claim any scientific novelty wrt. our individual subsystems, nor are they uniquely necessary for the development of our technological demonstration (we could have chosen alternative technological approaches to demonstrate the same concept).

To clarify, we define the following:

- **Digital twin:** a logically centralized, dynamic digital model of the traffic environment that integrates data from heterogeneous sources including both intelligent infrastructure and traffic participants in the cloud real-time.
- **Derivative services:** novel services that are expected to become available via a digital twin. These services fall into following major categories:
    - *Cooperative perception services:* more reliable centralized perception via sensor fusion in the cloud (*Central Perception*).

- *Dataset generation services:* more reliable perception enables more sophisticated scenario extraction and data referencing on multi-sensor datasets, as well as cross-validation of perception algorithms.
- *Accident reconstruction related services*
- *Cloud Control:* certain control functionalities (e.g., emergency braking) could be performed centrally based on cooperative perception inputs assuming sufficiently low latencies (a technological possibility in the near future).
- *Proving ground and logistical yard management:* some of the first potential areas of Central Perception and/or Cloud Control deployment.
- *Analytic services:* miscellaneous real-time datastream and historical data analytics.

2. Problem Definition

Arguably, the most crucial element in providing centrally maintained *digital twin* services is achieving a centrally integrated, real-time perception of the traffic environment. This paper presents a solution to the *Central Perception* problem that we specify as follows.

The overall system must acquire an integral and dynamic object-level view of the real-time traffic situation with at most 100 ms latency. The logical core of the system should be responsible for data integration while the peripheral measurement platforms will act as data sources. The overall architecture and the established interfaces must enable the simultaneous participation of connected intelligent equipment including static road-side infrastructure, mobile (vehicle-borne) and third-party measurement systems. Various sensor types and vendors must be supported, as well the precise and reliable detection of traffic participants including pedestrians, vehicles, obstacles, etc. As additional data sources, the system may use static HD maps and various traffic data (e.g., road meteorology) providers.

We specified and built a proof of concept system covering a substantial subset of above requirements, demonstrating the viability of the approach. Our functional sample provides following capabilities:

- a single central server (edge node);
- covering a spatially localized region of overlapping sensor field of views;
- fusing data from three multi-sensor camera-LiDAR platforms, one of them mobile;
- detecting (currently only) pedestrians;
- visualizing the digital twin in a realistic 3D simulation;
- in real-time (with less than 100 ms latency).

3. Overview of the Central Perception System Architecture

In order to support the requirements mentioned in Section 2, numerous design decisions had to be made regarding the prototype development of the measurement systems, the central server, and the communication between them. The following sections will discuss the overall design in some detail, the current section giving only a brief overview.

The realized system consists of one central server in the cloud and three wirelessly connected measurement systems, two of which are stationary and one that is mounted on a vehicle. At present we focus on fusion-based perception since we want to utilize the strengths of different sensor types on a single platform. Each measurement system has a sensory setup consisting of a camera-LiDAR pair, with the exception of the vehicle which has one camera and two LiDARs. For precise distance measurements and large-scale 3D reconstruction in automotive applications, stereo vision is becoming a less and less viable choice simply due to the precision loss at distances that are relevant to driving (compared to the high precision and falling prices of LiDAR technology) [11].

The measurement systems use the RTMaps software framework for data acquisition, synchronization, and data-flow processing. Detection and tracking is performed locally on the GPU-s and CPU-s of the measurement systems. The 3D pedestrian detection is done using low-level (raw) data fusion on the local system, i.e., the camera image and the corresponding LiDAR pointcloud are both necessary and are considered together in calculating the 3D position of the detected pedestrian (in contrast to object-level fusion where each sensor arrives at a detection estimate separately, and fusion occurs only afterwards). The fused detections of each sensor cluster are then tracked locally to smooth out any remaining uncertainties or missing datapoints. The tracks are then communicated using standard SENSORIS message formats over 5G or DSRC to the central server, where the inter-systems track fusion occurs. Finally the resulting locally and globally fused tracks are displayed real-time in a digital twin simulation developed in Unity. The overview of the system components and connections is shown in Figure 2.

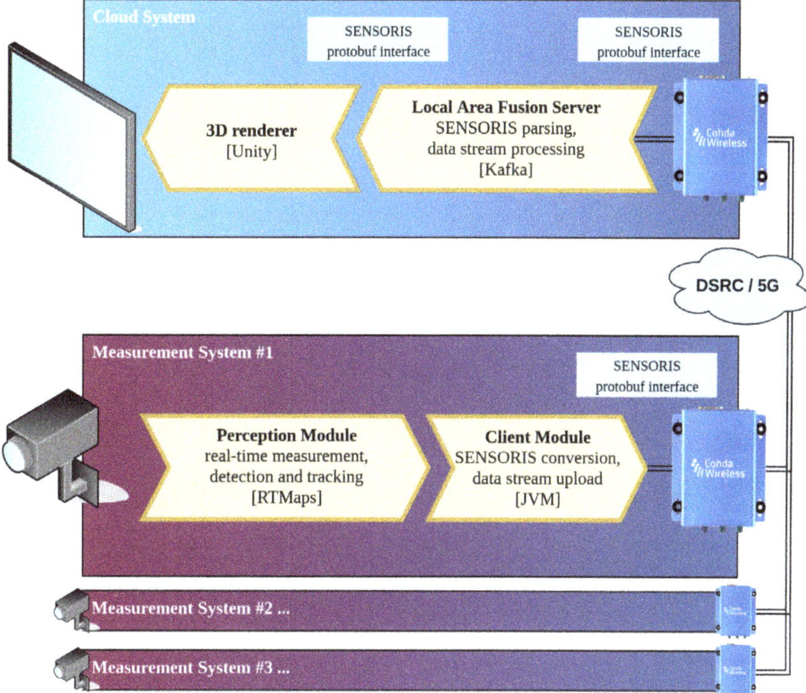

Figure 2. *Central Perception* prototype main components and protocols.

4. Perception Module

4.1. System Calibration

The proposed *Central System* integrates numerous different type of sensors (each having an assigned local coordinate system) such as LiDARs, cameras deployed in the infrastructure or attached to a vehicle. In addition to these sensors—in case of vehicles—the IMU/dGPS stand for an additional key element. For the system to work properly the calibration parameters have to be estimated first, i.e., all the intrinsics and extrinsics. Here, the intrinsics cover the internal parameters of cameras (such as focal length, principal point coordinates, skew, radial and tangential lens distortion) and the extrinsics stand for the transformations between the local coordinate systems of attached sensors as well as transformations from and to the Universal Transverse Mercator (UTM) frame which was selected to represent the global reference frame. For camera calibration the method published in [12] has been applied. The simplified calibration setup is illustrated by Figure 3. Let us briefly introduce the calibration approaches used to calibrate the proposed *Central System*.

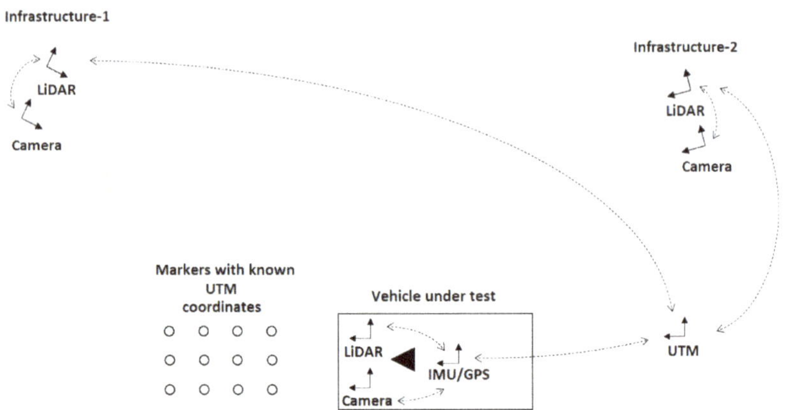

Figure 3. Illustration of the simplified setup of calibration containing two infrastructure stations and a vehicle each equipped with a camera and a LiDAR. In the vehicle there is an IMU/dGPS system, as well.

4.1.1. Chessboard Based Camera-LiDAR Calibration

The estimation of the rotation and translation between the camera-LiDAR pairs is more challenging than that of the camera-camera pairs, since we have to identify 3D points in the LiDAR point cloud and their corresponding image points in camera images. The estimation of LiDAR-camera extrinsics was based on the method proposed by authors in [13], which is a fully automatic extrinsic calibration approach aimed for LiDAR-camera extrinsics calibration by using a printed chessboard attached to a rigid planar surface. The key element of the method is to determine the 3D locations of chessboard corners in LiDAR's coordinate system. A full-scale model of the chessboard (A0 sized) is fitted to the segmented 3D points corresponding to the chessboard in the LiDAR point cloud. The intensities of light rays reflected form black and white patches of the chessboard are different and well distinguishable, thus the model is fitted to a 4D point cloud where the last dimension corresponds to the intensity of the given LiDAR point). The corners of the fitted model are considered to be the 3D corners of the chessboard.

The extrinsic calibration of the camera and LiDAR is performed by minimizing the re-projection error (given the estimated corners $\mathbf{M}_i$ in the LiDAR frame, their measured projections in the camera image $\mathbf{m}_i$ as well as the intrinsics of the camera (camera matrix $\mathbf{K}$, radial and tangential distortion coefficients p_1, p_2, p_3, q_1, q_2). N stands for the number of corner points considered.

$$\min_{\mathbf{R},\mathbf{t}} \sum_{i=1}^{N} \left\| \mathbf{m}_i - \hat{\mathbf{m}}_i(\mathbf{M}_i, \mathbf{K}, \mathbf{R}, \mathbf{t}, p_1, p_2, p_3, q_1, q_2) \right\|^2 \tag{1}$$

Figure 4 shows the process of data acquisition. The blue and yellow colors correspond to different LiDAR point intensities. Figure 5 shows the LiDAR points projected onto the camera image. In the chessboard image we can see both the detected corners and the 3D corners identified in the LiDAR point cloud and projected onto the camera image. The 3D viewer shows the detected corners together with the point cloud of the chessboard colored based upon the black and white patches of the fitted chessboard model. The achieved RMSE in case of five different poses of the chessboard can be followed in Table 1.

Figure 4. Data acquisition for chessboard-based camera-LiDAR calibration.

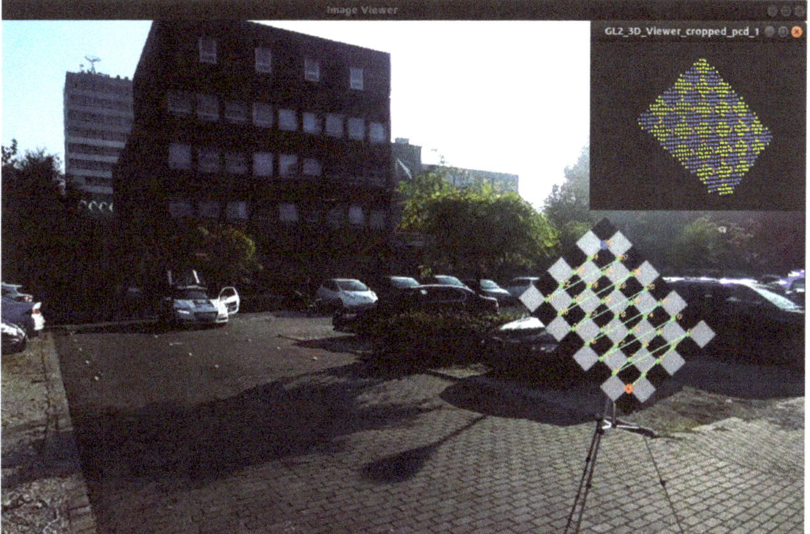

Figure 5. chessboard-based camera-LiDAR calibration results.

Table 1. The achieved RMSE in case of five poses.

Chessboard Pose Index	RMSE [px]	Chessboard Size
1	1.66	The chessboard
2	1.30	pattern is 6×8
3	1.49	with a cell size
4	0.87	of 140×140 mm
5	0.87	
The whole image set	1.28	

4.1.2. Box Based Calibration

In this second approach, the Camera-LiDAR extrinsics calibration relies on box corners, instead of a chessboard. The calibration box is placed at different locations (with known UTM coordinates measured in advance by a portable dGPS modul) of the working area. For each pose of the box the point cloud and the corresponding camera image is acquired. In the LiDAR pointcloud, the box corners were determined by segmenting the points which correspond to the box in the point cloud followed by fitting the box model. Since the box corners are obtained with respect to the LiDAR coordinate frame and the corresponding UTM coordinates are known, the LiDAR pose wrt. the UTM frame can be estimated. Similarly, the camera pose can be determined by minimizing the re-projection error (see Equation (1)) associated with the selected corners with known UTM coordinates. Since the UTM coordinates of box corners as well as the IMU pose wrt. UTM are known, the transformation between the LiDAR and IMU coordinate systems can also be determined. Another well known approach to estimate LiDAR-UTM extrinsics is the hand-eye calibration which requires at least two motions (with non-parallel rotation axes) of the sensors (LiDAR and IMU) [14].

4.2. Data Synchronization

In order to keep the data streams synchronized among infrastructural and vehicular sensors, a common time source as well as a time protocol is needed. As the most commonly used time protocols, the Network Time Protocol (NTP) and the Precision Time Protocol (PTP) might be emphasized. During our experiments the NTP was utilized and the GPS time was used as time source. Each station (including two infrastructural stations and one measurement vehicle) was equipped with an on-board unit having an integrated GPS time source and running the NTP service (see Section 6.1). Each computing node's (PCs, DrivePX2 Tegra A and Tegra B) system clock has been synchronized with the GPS time by relying on the NTP protocol. Prior to testing the *Central Perception* system, the synchronicity of data streams from different sensors have been verified by experiments. To each data frame (independently on what type of sensor it originates from), a timestamp is assigned as it enters the computing framework. In the computing framework the data frames (from different streams) being closest in time are associated and processed afterwards as depicted by Figure 6.

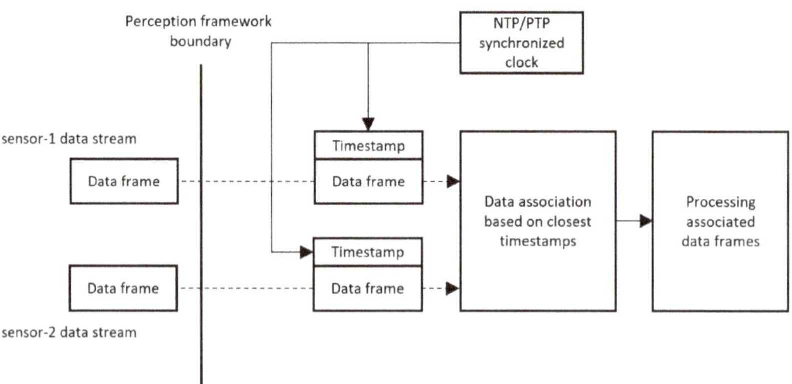

Figure 6. Illustration of the data flow and timestamp based assignment of data frames.

As another alternative for time synchronization the Precision Time Protocol might be used, which instead of millisecond-level synchronization, aims to achieve nanosecond- or even picosecond-level synchronization. In case of the PTP, switches with PTP support are required for each station. For most commercial and industrial applications, NTP is more than accurate enough [15].

4.3. 3D Object Detection

3D object detection plays crucial role in environment perception and understanding. During the development of the *Central System* we payed much attention on the development of robust 3D object detection methods which are considered to be essential from the overall system performance point of view. We have considered two type of approaches:

1. camera-lidar based approach benefits from the high resolution of cameras and the high position accuracy of 3D LiDAR points.
2. a single camera based detection where the YOLO 2D detection [16] algorithm and the homography between the road plane and the image plane was considered. This approach is cost efficient (due to the camera only requirement) and is preferred to be applied in such scenarios where the camera is static.

4.3.1. Yolo and Point Cloud Based Approach

Camera-based systems perform outstandingly well in case of recognition tasks, but when it comes to position estimation they are less accurate than LiDAR-based systems. Depending on the resolution and the number of used cameras, the baseline length, the accuracy of calibration as well as the accuracy of the pixel coordinates of points of interests, the position estimation might be improved; however, by including one or more LiDARs the location estimation of object's might significantly be improved.

The method presented below combines the advantages of the two sensors (i.e., the high resolution of cameras and the localization capabilities of LiDARs). In order to fuse camera images with LiDAR point clouds the sensors have to be calibrated (see Figure 3). Another crucial point here is to guarantee real-time processing which puts additional constraint (depending on the used hardware) on the complexity of applied algorithms. Nevertheless, the data streams of different sensors must be kept synchronized to ensure that data frames closest to each other in time are associated and processed accordingly (see Section 4.2).

As first step the detector receives images on its input and 2D object detection is performed by the YOLOv4 object detector [16]. The speed and accuracy of the algorithm are in line with the requirements defined, which means that the frame rate of the overall system was set to be at least 20 FPS (which currently stands for the upper limit for LiDARs). During the experiment pedestrians and cars have been considered as primary objects of interest, however the algorithm can easily be extended to detect additional classes such as motorcycles, buses, etc. The 2D detection may take several milliseconds even on the most powerful hardware ($\sim$30 ms).

In the next stage, the point cloud is projected onto the camera image and each estimated 2D bounding box gets associated with the LiDAR points which projections are bounded by the given box. As result a set of frustums is obtained (one for each 2D bounding box) containing the 3D points of the objects of interest. Let us denote the set of these frustums by $\mathcal{F}_i$. Given $\mathcal{F}_i$ the 3D bounding box corresponding to the given object might either be estimated on neural basis by a convolutional neural network trained to perform detection in frustums or the position of the object might be determined based on a simple reasoning.

By the reasoning based approach first the false points (foreground, background points) from $\mathcal{F}_i$ are eliminated and a small 2D window inside each bounding box is defined. The size and position of the window is proportional to the size of the original box. The scaling factor and position were set empirically based on the type of object. Since these windows generate significantly narrower frustums, the points falling inside it are more likely to belong to the object of interest. Let us denote the set of these points as $\mathcal{F}_i'$. The location of the object is determined as the mean of the points falling inside the volume bounded by $\mathcal{F}_i$ and satisfying the constraint

$$d_{min} < \|\mathbf{p}_j - \mathbf{c}\| < d_{min} + \delta, \quad \mathbf{p}_j \in \mathcal{F}_i, \quad j = 1..N_i, \tag{2}$$

where $\mathbf{c}$ stands for the camera center and N_i represents the number of points in $\mathcal{F}_i$, $\delta = \max\{object_{width}, object_{height}\}$.

$$d_{min} = \underset{\mathbf{q} \in \mathcal{F}_i'}{\mathrm{argmin}} \|\mathbf{q} - \mathbf{c}\|. \tag{3}$$

This is an extremely simple and therefore very fast way to filter out unnecessary points and localize the object of interest Figure 7. The latency of the detection can be followed in Figure 8.

Figure 7. Detected objects.

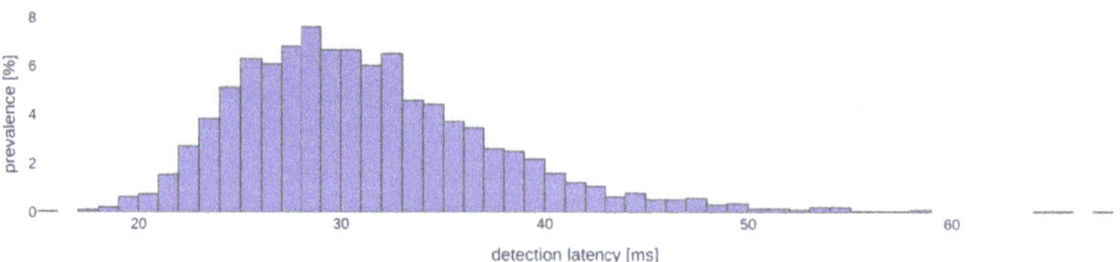

Figure 8. Yolo based object detector latency.

4.3.2. Yolo and Homography Based Approach

The detector described above uses the lidar point cloud to estimate the 3D location of the target, the method introduced in this section focuses on a single camera based 3D localization of targets. Homography and its estimation is well known topics in the literature, but let us briefly summarize it: Let us denote a world point by $\mathbf{M}$ and its image coordinates by $\mathbf{m}$. Let us consider the scenario when the world points of interest are lying on the XY plane, thus their Z coordinate is zero. These points are projected onto the image plane of the camera as follows:

$$\mathbf{m} = \mathbf{PM} = \mathbf{K}[\mathbf{R} \mid \mathbf{t}] \begin{bmatrix} X \\ Y \\ 0 \\ 1 \end{bmatrix} = \mathbf{K}[\mathbf{r}_1 \quad \mathbf{r}_2 \quad \mathbf{r}_3 \quad \mathbf{t}] \begin{bmatrix} X \\ Y \\ 0 \\ 1 \end{bmatrix} = \underbrace{\mathbf{K}[\mathbf{r}_1 \quad \mathbf{r}_2 \quad \mathbf{t}]}_{\mathbf{H}} \begin{bmatrix} X \\ Y \\ 1 \end{bmatrix}, \tag{4}$$

where r_i denote columns of the rotation matrix $\mathbf{R}$, $\mathbf{t}$ stands for the translation and $\mathbf{K}$ denotes the camera matrix containing the camera intrinsics. In order to estimate the homography

H the following cost function is minimized (measurement error is considered in both the image and world plane):

$$\min_{\mathbf{H},\hat{\mathbf{m}}_i',\hat{\mathbf{m}}_i} \sum_{i=1}^{N} \|\mathbf{m}_i - \hat{\mathbf{m}}_i\|^2 + \|\mathbf{m}_i' - \hat{\mathbf{m}}_i'\|^2, st. \ \hat{\mathbf{m}}_i' = \mathbf{H}\hat{\mathbf{m}}_i, \ \forall \, i, \tag{5}$$

where $\mathbf{m}_i$ and $\mathbf{m}_i'$ stand for the measured point pairs while $\hat{\mathbf{m}}_i'$ and $\hat{\mathbf{m}}_i$ stand for the estimated perfectly matched correspondences, i.e., $\hat{\mathbf{m}}_i' = \mathbf{H}\hat{\mathbf{m}}_i$ [17].

We have used 18 markers $\mathbf{m}_i'$ with known UTM coordinates (measured by a mobile GNSS system in advance with an accuracy of $\sim$20 mm) and their image projections $\mathbf{m}_i$ to estimate the homography. $\mathbf{m}_i$ stand for the undistorted normalized image points. The detection part of the approach uses the YOLO4 [16] neural network to detect targets of various types in images (during our experiment pedestrians were the main objects of interest, however other object types are also supported by the proposed perception system). The point of interest for each detected pedestrian was set to be the center point of the bottom edge of its 2D bounding box. Let us denote these points by $\mathbf{m}_i$. By applying the estimated homography $\mathbf{H}$, the image points m_i can be transformed to the XY plane of the UTM coordinate system as $\mathbf{m}_i' = \mathbf{H}\mathbf{m}_i, \ \forall i$. Here we omit the true altitude, thus it was set to zero for each point. Although this kind of 3D detection is very useful for static cameras (for example cameras installed in the infrastructure), in case of cameras attached to a vehicle, change in pitch or roll of the vehicle (caused for example when accelerating or making a hard turn, etc.), invalidates the estimated homography. In addition, the uncertainty of the measured image points $\mathbf{m}_i$ must also be considered. Given both the uncertainty of $\mathbf{H}$ and $\mathbf{m}_i$, the covariance of the estimated points $\mathbf{m}_i'$ is given by:

$$\Sigma_{\mathbf{m}_i'} = \mathbf{J}_\mathbf{h} \, \Sigma_\mathbf{h} \, \mathbf{J}_\mathbf{h}^T + \mathbf{J}_{\mathbf{m}_i} \, \Sigma_{\mathbf{m}_i} \, \mathbf{J}_{\mathbf{m}_i}', \tag{6}$$

where $\Sigma_{\mathbf{m}_i'}$, $\Sigma_{\mathbf{m}_i}$ and $\Sigma_\mathbf{h}$ stand for the covariance matrix of the estimated road point $\mathbf{m}_i'$, the measured image point $\mathbf{m}_i$ and the estimated homography $\mathbf{h}$, respectively (vector $\mathbf{h}$ is composed from the concatenated rows of $\mathbf{H}$). Furthermore, $\mathbf{J}_{\mathbf{m}_i}$ and $\mathbf{J}_\mathbf{h}$ stand for the Jacobians of $\mathbf{m}_i' = \mathbf{H}\mathbf{m}_i$ wrt. $\mathbf{m}_i$ and $\mathbf{h}$, respectively [17].

Since the vehicle is moving, the detected objects have to be transformed according to the actual pose of the vehicle to a global coordinate system (e.g., UTM). Firstly, the detections should be estimated wrt. a selected reference coordinate system and then based on the actual pose of the vehicle transformed to the UTM frame. The reference coordinate system for the vehicle was set to be the coordinate system of the IMU shifted along the vertical axes to the ground level (road level). Let us refer to this coordinate system as IMU_{gl}. In order to estimate the homography which transforms the image points directly to IMU_{gl}, one needs to estimate the marker coordinates in IMU_{gl}. Since the IMU modul (used during our experiments) includes a differential GPS with a dual antenna system, the UTM coordinates and the heading of the vehicle can be measured with an accuracy of $\sim$20 mm which might be considered to be sufficient for autonomous driving related applications. Based on the measured pose of the vehicle, the UTM to IMU_{gl} rigid transformation can be determined, thus the markers in IMU_{gl} can be calculated. By applying the homography (estimated based upon markers in the IMU_{gl} and the corresponding image plane points) to points m_i, the 3D position of each detected target is obtained directly in IMU_{gl}. Since the pose of the vehicle is continuously measured with a sampling rate of 100 Hz, the detections can directly be transformed to the UTM frame in real-time.

4.4. Object Tracking

4.4.1. Overview

There are several different types and solutions for tracking objects [18]. Kalman filter based methods are widely used for target position tracking.

For tracking objects with high manoeuvring capabilities, utilisation of the Interacting Multiple Model (IMM) filter is a good practice. Although the IMM filter is a well known approach for object tracking, let us briefly point out the basic principle. The IMM filter considers multiple motion models (e.g., constant velocity, constant acceleration, constant turn rate models) each associated with a dedicated Kalman filter. The Kalman filters are running simultaneously in parallel and their outputs are blended to generate the estimated state of the system according to the likelihoods of being in a certain motion mode. The higher the probability of a mode, the higher its contribution to the blended state. The state of a more probable mode is affected slightly by less probable modes [19,20]. During this process, the likelihoods of being in a certain mode (e.g., constant velocity mode) and the likelihoods of transitions between modes are calculated based on the last state. In order to reduce the transient period every filter is reinitialized with the mixed estimate of state and covariance [21].

4.4.2. Implementation

The tracker component receives a description data structure from all recognised objects as input. From that data structure, it pulls the position coordinates and the corresponding timestamp and combines them into a position list for a given frame. The core of the tracker is an IMM filter. The filter consists of three different motion models, which are the constant velocity, constant acceleration and constant turn-rate models. In each step, the tracker gives an estimation of current positions for all the registered tracks. Then the tracker component pairs the tracks with the input positions using Munkres global nearest neighbour assignment algorithm. Then it manages the tracks in the following manner: If no existing track can be paired with a position, it creates a new one. If a track was paired with any positions 5 times within the last 7 frames, then it flags it as confirmed. With this method, any false positive detection can be filtered out. The component deletes a track when it has not been assigned with any positions at least 22 times within the last 25 frames. These settings fit the Yolo and point cloud based approach. Due to behaviour differences, the Yolo and homography based approach requires other settings for the tracker component for best results. Therefore, a tracker with optimised settings has been implemented for each detector solution. After the track management, the component compiles a list with the positions of the confirmed tracks. The output of the component is a data structure that contains the ID and position of the tracks and the position and orientation information of the sensor system. The output data structure has the same format as the input structure. The pseudo code of the tracker is listed below:

The latency histogram of tracking can be followed in Figure 9. First of all it is influenced by the number of current tracks and detections. The average response time of the component is 690.7 µs for the sample sequence.

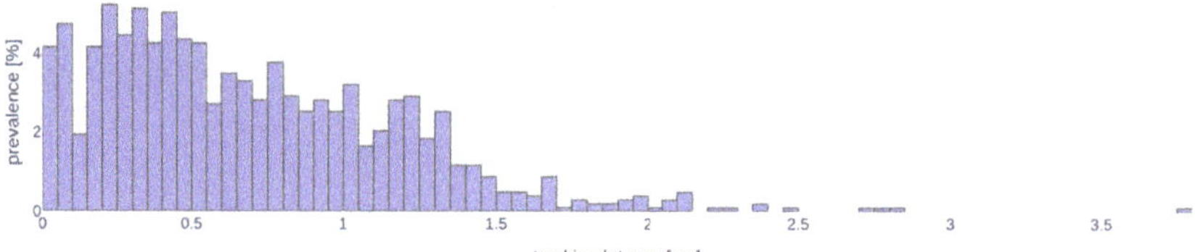

Figure 9. Latency histogram of tracking.

```
READ input
FOR elements in detection data:
READ position
FOR each track:
    LOAD last position
    ESTIMATE new position with IMM Filter
ASSIGN estimations with current detection positions
FOR each track:
    IF track was paired with a detection for n times in the last m
    frames:
        REGISTER track as ,,Confirmed''
    IF track was paired with detection less times than j in the last
    k frames:
        DELETE track
IF there are any detection which were not assigned to a track:
    FOR each unassigned detection:
        CREATE new track based on current detection
ASSEMBLE a list from the positions of ,,Confirmed'' tracks
CREATE output data structure
ADD system origin position and yaw information from input data
WRITE output
```

5. Local Area Fusion Server

5.1. Stream Setup

The local area fusion server we set up for our current demonstration automatically processes and converts the incoming detections streams across five sequential processing steps until we get the fused result in the final stream. The five so-called fusion-processors can be observed in Figure 10.

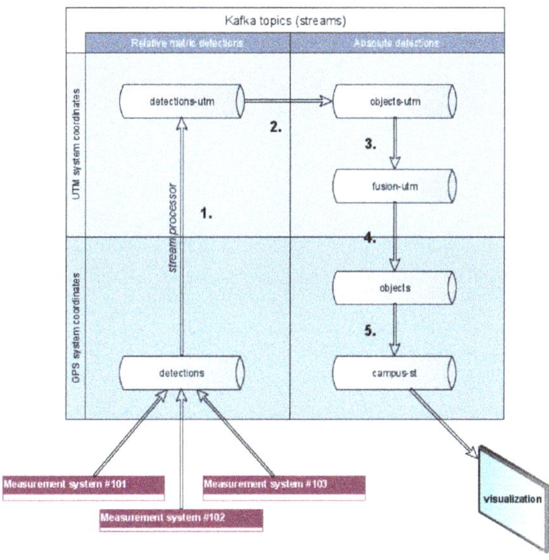

Figure 10. Current stream setup in the *Central Perception* functional sample server (cylinders represent topics/streams, while the numbered arrows represent stream processors).

The sequentially numbered stream processors from Figure 10 have the following responsibilities:

1. Converts system origin coordinates from WGS84 to UTM. Position uncertainty remains the same since the SENSORIS input was in (square) meters already. Rotation and its uncertainty do not change because WGS84 and UTM frames point in the same directions.
2. Recalculates all object coordinates from system-relative to absolute UTM.

    (a) **Position:** The transformation matrix is straightforward to derive from the relative frame (e.g., vehicle IMU): we just calculate the rotation matrix from the platform's current orientation and append its current position as a translation vector.

    (b) **Orientation:** The global heading is obtained by adding the relative (IMU-based) object yaw to the system yaw. Calculating global pitch and roll is more involved and was skipped since this data is not represented in our current environment model. The pitch and roll values are set to zero.

    (c) **Position covariance:** the object position covariance matrix has to be backrotated and added to system position covariance, assuming no cross-covariance between system and object positions since they are independent.

    $$\Sigma_{op} = \Sigma_{pp} + R\,\Sigma_{rp}\,R^{\top} \tag{7}$$

    R denotes the IMU-to-UTM rotation matrix, while $\Sigma_{op}, \Sigma_{pp}, \Sigma_{rp}$ denote the resulting object position covariance, the platform position covariance and the IMU-based relative object position covariance, respectively.

    (d) **Orientation covariance:** the object heading variance is added to the system heading variance, the pitch and roll uncertainties are disregarded (set to identity).

3. Here, the fusion algorithm itself is performed on the objects-utm stream. Exact details will be given in the next subsection. For convenient further utilization a very specific rule for populating the output fusion-utm stream is applied: the chosen fusion input messages and the fused output message are written sequentially to the out-stream. So later reading the messages in offset order will yield an alternating sequence of fusion inputs followed by the corresponding fusion output. Note that not every objects-utm message becomes a fusion input.
4. Converts all objects from UTM coordinates into WGS84 coordinates.
5. Optionally filters certain objects according to position in relation to demonstration area.

5.2. Fusion Algorithm

Assuming no cross-correlation between sources, we employed a Kalman filter and Global Nearest Neighbor (GNN) association based central tracking source-to-track fusion method called trackerGNN (https://www.mathworks.com/help/fusion/ref/trackergnn-system-object.html (accessed on 14 September 2021)), which is an integral part of the Sensor Fusion and Tracking Toolbox of Matlab. TrackerGNN maintains a single hypothesis (set of central tracks) about the environment and it follows the central tracking algorithm template detailed in the following subsection. The theory behind the implementation is based on [22]; notably it solves GNN association using the Kuhn-Munkres [23] algorithm, also known as the Hungarian method [24].

5.2.1. Central Tracking

Central tracking, sensor-to-track or source-to-track (S2T) fusion has detections from multiple sources (usually sensors) as inputs and is expected to produce a single set of central tracks as output. Therefore, the detections have to be integrated across time and across sources. If we first perform the time-integration (tracking) and subsequently perform the source-integration (fusion), we get the equivalent of a track-to-track (T2T) fusion approach.

In contrast, if we perform source-integration (fusion) before time-integration (tracking), we are talking about S2T fusion.

The general S2T fusion framework assumes the maintenance of a single set of central tracks throughout the filtering steps. An S2T fusion step usually follows the template given below:

1. Collect measurements within a time interval between previous and currently queried step time. Each potential source should provide exactly zero or one measurement containing a number of simultaneous detections. Sequential measurements from the same source within the data collection interval can be handled by (a) discarding all but the last, as done in our approach; or (b) keeping all but technically regarding them as different sources with shared measurement model parameters.
2. Assign each detection of each source to exactly one track (either pre-existing or newly-created). Make sure that no two detections from the same source are assigned to the same track. Thus each track is assigned 0 to s detections (s being the number of sources at this step).
   (a) Track lifecycle management is done during this step (trackerGNN uses parametrizable heuristics as detailed in Section 4.4.2).
   (b) The assignment algorithm may handle passage of time. A simple solution like trackerGNN would disregard time and only use spatial data for assignment. A sophisticated solution might have to assign and integrate each measurement individually, ordered by time, iterating between steps 2 and 3, increasing computation costs.
3. Filter (predict and update) each live track with the assigned measurements ordered by time.
4. Output the prediction for all tracks for the same moment in time (which was provided as the fusion query argument).

5.2.2. Integration

In order to make use of trackerGNN as part of an efficient stream processor outside of Matlab, several technical challenges had to be overcome, most notably:

- C code generation and compilation from Matlab,
- generation of a Java wrapper using the SWIG (http://swig.org/ (accessed on 14 September 2021)) framework for integration with Kafka Streams API,
- devising and implementing an appropriate input buffering scheme within the stream processor, and finally
- making sure both real-time and playback fusion options are supported.

5.2.3. Buffering

The issue of input message buffering is not entirely trivial, since the order of message arrival in the topic partition does not necessarily follow any kind of (e.g., timestamp-based) ordering. We have to somehow make sure that the tracker is always fed appropriately chosen inputs and that no inputs are wasted or discarded prematurely. The regular intervals the fusion is queried at (in our case 120 ms) require a flexible buffering method that supports fusion with missing or no data, old and future data, etc.

Our buffering method:

- tries to collect and buffer all available data immediately and continues to collect as long as the stream is accessible;
- discards messages with past timestamps that precede the most recent fusion step;
- preserves far-future data points without running out of memory;

- collects data falling—according to timestamp—into each inter-fusion time window into separate sets;
- when the time for the next fusion step comes, the cluster of messages that falls into the preceding time-window is regarded: a collection of one latest message per sensor is retained as fusion input, the rest discarded.

5.2.4. Real-Time vs. Playback Mode

The behavior of the fusion module should be different when we play back data from the past then when we stream the present data. In both cases fusion is performed at fixed real time (not timestep time) intervals, and fusion time can proceed only forward (strictly greater than previous) (see Table 2).

Table 2. Comparison of real-time and playback fusion modes.

Real-Time	Playback
Pace: always jump forward to the most recent available timestep.	Pace: always read input data with a natural pace: for every second passed in consumed timesteps, a second should pass in reality.
Missing data: our requirement for fused tracks is to disappear when no data is received from any of the sensors, i.e., to artificially advance fusion in time and "wind down" within a dozen simulated steps.	Missing data: when no data is received from any of the sensors, we want to freeze everything as it is and to not step the time forward. Empty (0-detection) data can clear the scene, but no-data should freeze it.
Fusion restart: not required, since real time can flow only forward.	Fusion restart: required when rewinding.

5.2.5. TrackerGNN Parametrization

The method parameters for the trackerGNN fusion component were set to:

```
tracker = trackerGNN( ...
    'TrackerIndex', 0, ...
    'FilterInitializationFcn', @initcvkf, ...
    'Assignment', 'MatchPairs', ...
    'AssignmentThreshold', 15*[1 Inf], ...
    'TrackLogic', 'History', ... \% History|Score
    'ConfirmationThreshold', [2 3], ...
    'DeletionThreshold', [5 5], ...
    'DetectionProbability', 0.9, ...
    'FalseAlarmRate', 1e-6, ...
    'Beta', 1, ...
    'Volume', 1, ...
    'MaxNumTracks', 100, ...
    'MaxNumSensors', 20, ...
    'StateParameters', struct(), ...
    'HasDetectableTrackIDsInput', false, ...
    'HasCostMatrixInput', false ...
);
```

The detection-to-track assignment upper threshold was set to half of the default since pedestrians are smaller and slower than vehicles and are expected to have smaller uncertainty. Track confirmation threshold was set to 2 out of 3 detections, although our detections often come with a nonzero object type meaning instant confirmation. The track deletion threshold was set to 5 out of 5 misses. Besides, a custom rule was also introduced removing all input detections with any position covariance value larger then an experimentally chosen threshold $\epsilon = 5.0$ m^2.

5.2.6. Results

The processing time of the local area fusion component was measured and found sufficiently performant for our requirements. Latencies are on the order of 2–3 ms, counting not only fusion itself, but including also stream (de)serialization and message parsing. There were some acceptably rare outliers: 0.14% of cases required more than 5 ms and none more than 35 ms (see Figure 11).

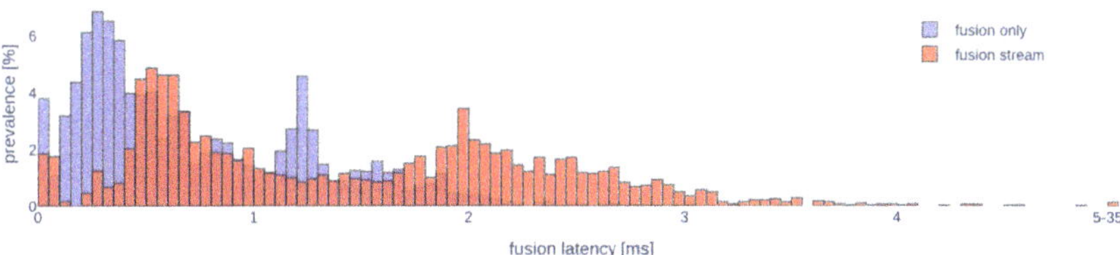

Figure 11. Distribution of processing times of the local area fusion component.

We have developed two distinct tools for visualizing fusion outputs. One is a 3D rendering demonstration tool described in Section 7. The other is a 2D monitoring dashboard for internal use that lets us step through each fusion cycle individually. A screenshot of the fusion results is presented in Figure 12 below.

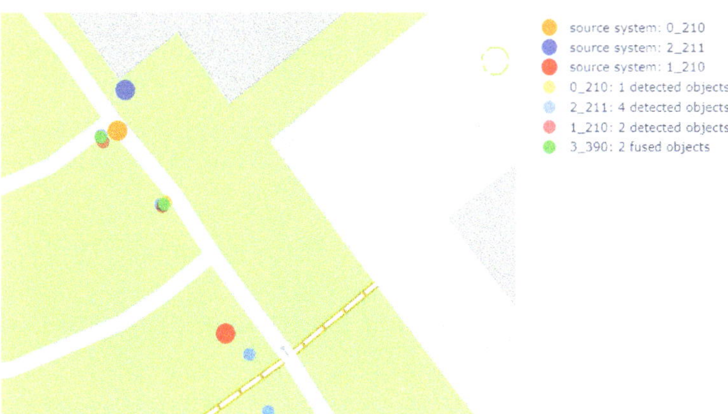

Figure 12. *Central System* Dashboard: our monitoring and analysis tool.

6. Client Module

6.1. DSRC Communication

Real-time communication, such as 5G cellular network or WiFi-based 802.11p (DSRC—Dedicated Short Range Communication), plays a major role in the system architecture. In the *Central Perception* functional sample of the *Central System*, the dedicated DSRC 5.9 GHz radio communication has made it possible for distant system components—such as the infrastructure stations, the vehicle, and the central server—to communicate with each other in real-time via radio frequency (RF).

For DSRC in our functional sample we used Cohda Wireless MK5 OBUs, which use the Software-Defined Radio (SDR) baseband processor SAF5100 and the dual-radio (antenna A, B) multi band RF transceiver TEF5100. These chips offer adjustable parameters for radio wave modulation schemes. The unit includes a dedicated HSM (Hardware Secure Module) for data encryption, compression, decryption and also the keys used for encryption using this chip [25]. The data rate corresponding to the modulation schemes of the device (BPSK, QPSK, QAM, etc.) can be changed from 3 Mbps to a maximum of 27 Mbps. At faster data rates, one of the most critical metrics, the PDR (Packet Delivery Ratio) is less than 100% so a trade-off had to be made and a medium rate, more reliable modulation option than BPSK (Binary Phase Shift Keying) was chosen.

The MK5 OBU complies with the following standards and protocols: IEEE 802.11 (part of the IEEE 802 set, the most widely used wireless networking standard), IEEE 1609 WAVE (Standard for Wireless Access in Vehicular Environments), ETSI ES 202 663 (European profile standard for the physical and medium access control layer for Intelligent Transport System operating in the 5 Ghz frequency band), SAE J2735 (Dedicated Short Range Communications (DSRC) Message Set Dictionary).

The 802.11p protocol compliance grants the following advantages:

- No additional infrastructure requirement: 802.11p does not require any additional infrastructure part, just the receiver and the transceiver units. This is because an ad-hoc network is formed, as soon as two DSRC units come in each other's radio range.
- Low latency: Road experiments have shown the latency at MAC layer to be around 2 ms or less in an optimal setup. The latency value depends on several different factors, such as payload size, vehicle speed (if the unit is mounted in a vehicle), radio interference, line of sight, etc.
- Range: The range is dependent on other variable factors like data rate and environmental factors. According to documentation it offers data exchange among vehicles and roadside infrastructure within a range of 1000 m, with a transmission rate of up to 27 Mbps and a vehicle speed up to 260 km/h.

The OBUs have another key role in the *Central System* architecture because they are also used for time synchronization, using the Chrony module and the GNSS antenna. The MK5 runs a gpsd server to allow applications to access GPS data. Chrony [26] is a versatile implementation of the NTP (Network Time Protocol), and it can synchronize the system clock with the NTP servers and reference clocks. With the help of the Chrony module all of the OBUs can be configured to have a reference time with microsecond accuracy.

Regarding the network topology in the DSRC setup, we define four subnetworks: two infrastructure stations, one vehicle, and one central server. Every subnetwork contains one PC for data acquisition, processing and visualization, one wireless router and one Cohda Wireless MK5 OBU.

The MK5 module has an Ethernet connection interface, which supports Ethernet 100 Base-T with 100 Mbps data rate. For the *Central Perception* functional sample the Cohda OBUs have been configured as IPv4 (Internet Protocol v4) gateways to provide a fully transparent communication between subnetworks. This means that all subnets are seen by each other, so real-time data exchange between nodes can be easily achieved. Figure 13 represents the subnetwork layout of the communication architecture of the *Central Perception* prototype.

6.2. Kafka Streaming Platform

For communication middleware we have chosen to use Kafka, the popular open-source "dumb broker" streaming platform maintained by the Apache Foundation. Judging by its main functionality Kafka can also be considered a distributed commit log, although it is primarily used for messaging. The aim of the project is to provide a real-time, high-throughput, low-latency streaming platform. Kafka provides horizontal scalability via distribution of message topic partitions across respective partition leader brokers while also providing fault tolerance by replicating each partition across non-leader brokers in a way reminiscent of RAID redundancy and fallback mechanisms. The distributed brokers and topic partitions architecture perfectly fits our long-term hierarchical edge computing vision if we assume that message topics should be divided into partitions according to the source area of measurements. We already tested our system in a 3-broker, 3-way-partitioned and triply-replicated (one original and two replicas) setup and experienced no perceptible lag or slowdown. When a broker was deliberately terminated, one of the remaining brokers automatically took up partition leadership; and when the temporarily disabled broker came back to life, the load balancing mechanism automatically reassigned it to partition leadership once again.

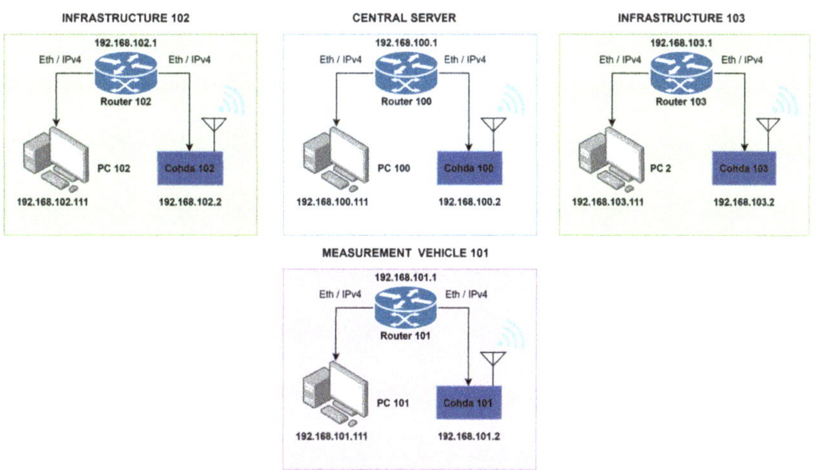

Figure 13. Subnetwork layout of the communication architecture.

In order to connect to the Kafka middleware, we developed a universal and platform-independent client module that runs in the Java Virtual Machine runtime environment in order to create a convenient socket-based API for uploading processed sensor data (detections, tracks, source system positions, etc.) to the distributed Kafka cloud in all the supported standard business domain level formats and protocols. This currently extends to the SENSORIS and ETSI CPM protocols, of which SENSORIS was used in the prototype demonstration. The client module's API encapsulates Kafka specifics and accepts standard SENSORIS messages. For further convenience we also provided a python wrapper API that we can easily call from the RT Maps client-side real-time dataflow-processing framework.

6.3. SENSORIS Message Standard

Exactly one message is sent for each source's each output (after each measurement-detection-tracking cycle). The source is not necessarily a single sensor, it might be, e.g., a raw sensor fusion based untracked detection, or the output of a tracker. The data that we collect via SENSORIS v1.0.0 (https://sensoris.org/ (accessed on 14 September 2021)) messages therefore contains the following elements:

- Message identification
- Source system information
    - identification (platform UUID, sensor UUID, sensor SUID) (UUID stands for universally unique identifier; SUID stands for system-wide unique identifier)
    - GPS PPS synchronized timestamp of originating measurement (event time)
    - localization (position, orientation) and its uncertainty
- Detected objects (i.e., detections or tracks) information [given for each object]
    - Object SUID

- Object existence uncertainty
- Object type and type uncertainty
- Object position and orientation in relative coordinate system and its uncertainty
- Object size and uncertainty

7. 3D Renderer

In order to visually represent the information provided by the individual sensor systems, as well as the central fusion system, it is necessary to use a digital twin rendering module. A properly constructed 3D visualization demonstrates the cooperative perception of scenario participants in scenarios where a single on-site sensor would not ensure proper operation. The visualization system must communicate with the *Central System*, including reading SENSORIS messages, and being able to decode and display this information in real time. It is also assumed that a digitized 3D model of the real environment of the on-site demonstration is available so that the visualized information can be compared with the real-world scenario. In our case, we used Unity 3D software to implement the visualization, which communicates with the *Central System* over a TCP connection. The localization of the measurement stations and their respective object detections (fused or raw) are available on a Kafka topic as encoded Sensoris messages. In order to visualize the measurement vehicle and the surrounding pedestrians, these data are accessed and forwarded to the visualization module in a proper structure.

7.1. Virtual Environment

Virtual imaging of the real environment is most accurate when based on laser measurements. Therefore, testing on the university campus was preceded by a laser measurement that provides a digitized description of the area as a LiDAR point cloud. This point cloud had to be brought from las format to some readable, xyz format to display within Unity software. In addition to the transformation of the format, it is important to place the lateral and longitudinal coordinate pairs relative to some center point in the x-y coordinate system defined by us so that the distances can also be interpreted in the Unity software. This transformation requires the use of the ellipsoid WGS84 as well as the determination of a clearly definable (0, 0) coordinate. This coordinate will later become the center of Unity's virtual world, as well as the basis for the transformation of all information that comes in during testing. The xyz data created in this way can already be read in a csv or txt file, and spheres representing the points can be created for the coordinate points it contains. In this way, it becomes interpretable in the virtual space of Unity, and based on this the various landmarks are clearly outlined (Figures 14 and 15). During the demonstration, the most important thing is that the roads are positioned correctly in the digital world, so we performed additional GPS measurements at their corner points. The origin of the virtual world was also determined during these measurements.

The shape and texture of the buildings surrounding the campus have been modelled according to reality. The shape and location of the vegetation and other components in the parking lot could be modeled based on the point cloud. The vegetation has been designed to vary the colour and density of the foliage according to the seasons. Unity software also provides the ability to model current lighting conditions using various skyboxes. However, for proper running performance, the generation of lights is not done in real-time. Still, a so-called baked lightmap is created, which predetermines illumination with the given settings.

The digital replica of the environment is best presented through cameras that can be matched to each real sensor. For a scenario to be well demonstrated, it is necessary to be able to present the given environment from several perspectives. We placed virtual cameras in the positions corresponding to the two infrastructure cameras as well as the cameras placed on the test vehicle, applying the basic properties of the real sensors.

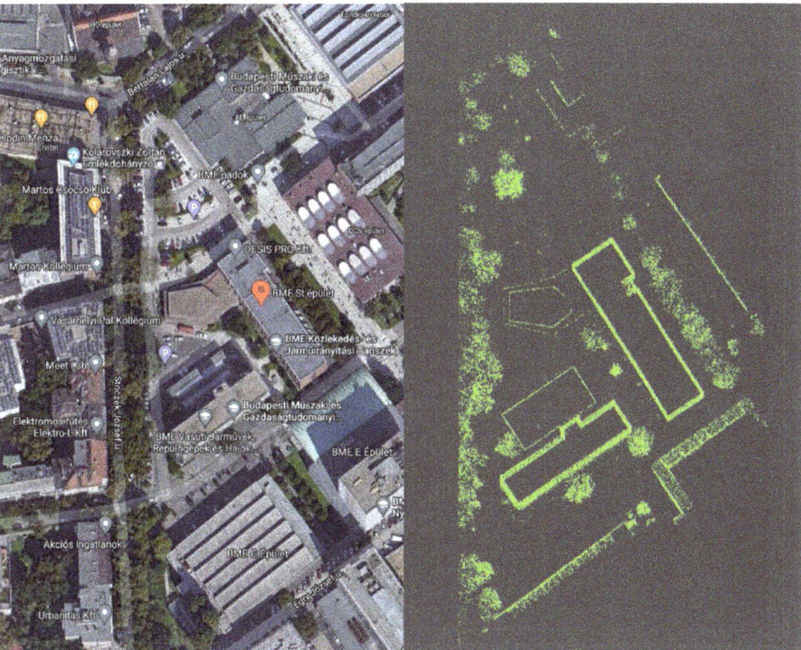

Figure 14. University campus based on LiDAR point cloud (map source: http://maps.google.com (accessed on 14 September 2021)).

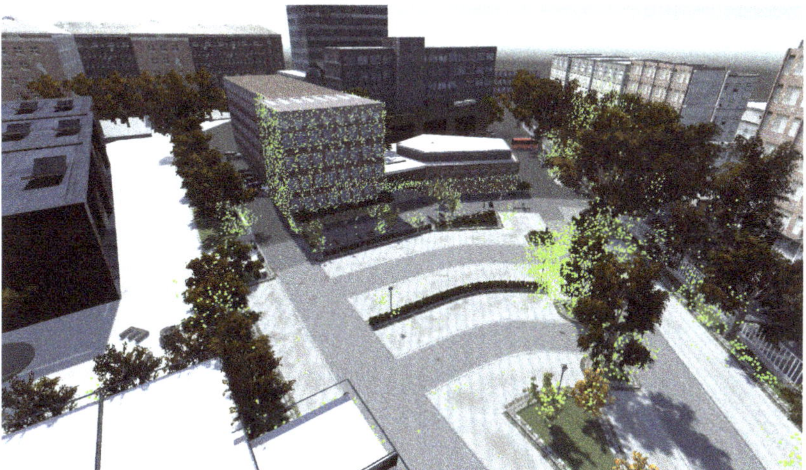

Figure 15. The Unity model of the University campus with the point cloud.

7.2. Sensor Detection Visualization

The test vehicle is displayed according to the method detailed in [27]. High-frequency real-time GPS data is available from the test vehicle that is accurate enough to place the vehicle in the virtual world. In this case, the lateral and longitudinal position of the vehicle and its heading are used. The movement of the vehicle's wheels was not modelled. When handling sensor detections—either from static stations or from the moving vehicle—it is necessary to separate information from different sources, as well as to handle different objects. Although only pedestrians were detected during our current measurements, the system is also prepared to handle all static and dynamic objects defined by SENSORIS. The detections from the sensors always reflect the state closest to real-time, i.e., only the most recent objects are always displayed. This also means that the objects existing in the previous update step must be moved or deleted. We also had to consider that the frequency of messages from different stations and sensor types is not the same in all cases, and may change dynamically. Residual detections—object tracks that get no confirmation within a short time period—are only rendered for the time specified by a parameter, after which they are automatically deleted. In our simulations, this time was set to 0.25 s. With these solutions, the movement of the detected pedestrians is continuous, there is no vibration in the display process, and the objects do not multiply during the movement, they do not draw a strip.

A visual distinction was made between detections of different origin, which helps us to understand the scenario. In addition, different sources also assign different tags to objects, which allows one to treat the objects belonging to that tag as a group, whether it is to turn off the display of objects or even delete objects. Pedestrian objects are generated based on a predefined cylindrical shape, the properties of which, such as size, colour, or permeability, are set based on the data associated with the detection.

The system provides a sufficiently high frequency to ensure that the motion is clearly continuous. There are two ways to test the visualization system, displaying real-time uploaded detections online and playing back data already present on the server offline. During the tests, we had two main expectations for the viewer, the first of which was to display the detection sent to it in real-time, and the second was to position both the environment and the detections accurately in the virtual world. With these conditions fulfilled, we observe a complete and synchronous copy of reality within the simulation. The system also allows one to turn off the display of detections of any given sensor for separate analysis. Figure 16 simultaneously shows the simulated camera FoV areas of all three sending infrastructures and a larger camera FoV overlooking the entire simulation environment. This figure also shows that the vehicle sensor sees a garage door (lower right corner), meaning that the snapshot came from a replay when the test vehicle did not participate in the measurement.

Figure 16. Detections in simulation.

We can best evaluate our digital twin during real-time tests, where we can see the scenario in reality and in its digital version at the same time. The streaming of raw camera images also provides additional checking possibilities. In the offline state, when a recorded stream is played back, the video played can also reveal whether there is a substantial difference in the digitized world compared to reality, as can be observed in Figure 17.

Figure 17. Top-left image: objects sensed by the infrastructure-1; **Bottom-left** image: objects sensed by the infrastructure-2; **Top-right** image: objects sensed by the vehicle; **Bottom-right** image: Real-time digital twin generated by the central system.

8. Conclusions and Future Work

We have proposed a cooperative perception system capable of generating and maintaining the digital twin of the traffic environment in real-time by fusing higher level data of multiple sensors (deployed either in the infrastructure or in intelligent vehicles), thus providing object detections of higher reliability and at the same time extending the sensing range. Besides giving a general idea on cooperative perception we have also introduced the key building blocks of this system including the calibration, 3D detection, tracking, fusion, data synchronization, communication and visualization. In case of time-critical components we have also presented the underlying algorithms and pointed out the relevant implementation details, as well. The functional prototype of the proposed system has also been created and tested under real circumstances on-line. We have demonstrated a single service of the proposed perception system, namely the real-time visualization of the generated digital twin of the environment including pedestrians as dynamic objects of interest communicated using standard SENSORIS message formats over 5G or DSRC to the central server. The system can further be extended to support other type of objects, as well such as cars, bicyclist, etc. Besides the digital twin generation a broad range of new derivative services can be facilitated, as well, including cloud-assisted and cloud-controlled ADAS functions, various analytics for traffic control, etc., which are subjects of further research. We have also shown that the proposed perception system is able to operate in real-time, meaning that an overall latency of less than 100ms has been achieved. As already stated, we envision this prototype system as part of a larger network of local information processing and integration nodes, where the logically centralized digital twin is maintained in a physically distributed edge cloud in real-time.

We have encountered several noteworthy practical questions and lessons during the implementation which led to the establishment of certain best practices that can not be treated adequately within the bounds of this paper, but can be at least mentioned. Some of them include considering sensor latencies, triggering simultaneous snapshots and associating data from sensors with different frequencies. There are effects related to vehicle movement during a full LiDAR rotation. There are problems with creating a perfectly flat and orthogonal calibration points layout in the field. As already mentioned in some detail, GPS time based inter-platform synchronization was a cardinal issue. Detection can suffer from all the problems inherent in deep learning systems: unfamiliar lighting, background, or anything that takes the input image beyond the domain and distribution the neural network was trained for can influence the algorithm adversely. Of course, deep learning models are also susceptible to deliberate adversarial attacks like "invisibility T-shirts" [28] on pedestrians, etc. Foreground clutter in chest height (even a stretched out hand) can destabilize our LiDAR-reliant raw fusion method. DSRC communication tends to break down in the presence of obscuring objects: the installation height and placement of on-board/road-side communication units is crucial, communication hand-off between moving vehicle platforms and stationary road side units has to be solved. On the server side, managing the spatially distributed digital twin across several edge computing nodes and their overlapping areas of responsibility is a theoretical problem we are currently investigating. Practical considerations like system security, authentication, authorization and information integrity are undeniably safety critical issues that must be tackled before industrial application. So is the adherence to automotive standards like ASIL D and the use of provably real-time hardware and software systems that come with industrial-grade guarantees. Despite numerous challenges, technological enablers like cheap LiDAR-s, powerful deep learning and ubiquitous 5G are making the road towards cooperative perception services more attainable by the day.

Author Contributions: Conceptualization, Z.S. and V.T.; methodology, V.T., A.R. and V.R.; sensor calibration, A.R., M.C. and A.K.; platform setup: A.R., V.R., Z.V. and M.C.; platform software: data acquisition, object detection, A.R. and M.C.; platform software: tracking, Z.V.; communication: Z.P. and V.R.; central software: stream processing and local area fusion, V.R.; simulation and visualization, M.S. and B.V.; writing—original draft preparation, A.R. and V.R.; writing—eview and editing, A.R. and V.R.; supervision, project administration, V.T.; funding acquisition V.T. and Z.S. All authors have read and agreed to the published version of the manuscript.

Funding: The research reported in this paper and carried out at the Budapest University Technology and Economics has been supported by the National Research Development and Innovation Fund (TKP2020 National Challenges Subprogram, Grant No. BME-NC) based on the charter of bolster issued by the National Research Development and Innovation Office under the auspices of the Ministry for Innovation and Technology. In addition the research was supported by the National Research, Development and Innovation Office through the project "National Laboratory for Autonomous Systems" under Grant NKFIH-869/2020. Funder NKFIH Grant Nos.: TKP2020 BME-NC and NKFIH-869/2020.

Acknowledgments: The research reported in this paper and carried out at the Budapest University of Technology and Economics has been supported by the National Research Development and Innovation Fund (TKP2020 National Challenges Subprogram, Grant No. BME-NC) based on the charter of bolster issued by the National Research Development and Innovation Office under the auspices of the Ministry for Innovation and Technology. In addition the research was supported by the Ministry of Innovation and Technology NRDI Office within the framework of the Autonomous Systems National Laboratory Program.

Conflicts of Interest: The authors declare no conflict of interest. The funders had no role in the design of the study; in the collection, analyses, or interpretation of data; in the writing of the manuscript, or in the decision to publish the results.

Abbreviations

The following abbreviations are used in this manuscript:

4G/5G	Fourth/fifth generation technology standard for broadband cellular networks
ADAS	Advanced Driver-Assistance Systems
API	Application Programming Interface
CAV	Connected and Autonomous Vehicle
CPU	Central Processing Unit
dGPS	Differential Global Positioning System
DSRC	Dedicated Short-Range Communications
FoV	Field of View
GNN	Global Nearest Neighbor
GNSS	Global Navigation Satellite System
GPS	Global Positioning System
GPU	Graphics Processing Unit
HAD	Highly Automated Driving
HD map	High Definition map
I2V	Infrastructure to Vehicle
IMM filter	Interacting Multiple Model filter
IMU	Inertial Measurement Unit
ITS	Intelligent Transportation System
JVM	Java Virtual Machine
LiDAR	Light Detection and Ranging (sensor)
MAC	Medium Access Control (sublayer of the data link layer in the OSI networking model)
NTP	Network Time Protocol
PPS	Pulse Per Second (synchronization signal)
PTP	Precision Time Protocol
R&D	Research and Development
RAID	Redundant Array of Independent Disks
S2T	Central tracking, aka. sensor-to-track fusion or source-to-track fusion
SciL	Scenario in the Loop
SUID	System-wide Unique Identifier
T2T	Track-to-track fusion
TCP	Transmission Control Protocol
UTM	Universal Transverse Mercator
UUID	Universally Unique Identifier
V2V	Vehicle to Vehicle
WGS84	World Geodetic System 1984
WiFi	Wireless Fidelity (network protocol family)
YOLO	You Only Look Once (object detection system)

References

1. Krämmer, A.; Schöller, C.; Gulati, D.; Knoll, A. Providentia—A large scale sensing system for the assistance of autonomous vehicles. *arXiv* **2019**, arXiv:1906.06789.
2. Gabb, M.; Digel, H.; Muller, T.; Henn, R.W. Infrastructure-supported perception and track-level fusion using edge computing. In Proceedings of the IEEE Intelligent Vehicles Symposium, Paris, France, 9–12 June 2019; Institute of Electrical and Electronics Engineers Inc.: Piscataway, NJ, USA, 2019; pp. 1739–1745. [CrossRef]
3. Tsukada, M.; Oi, T.; Kitazawa, M.; Esaki, H. Networked roadside perception units for autonomous driving. *Sensors* **2020**, *20*, 5320. [CrossRef] [PubMed]
4. Wang, Z.; Liao, X.; Zhao, X.; Han, K.; Tiwari, P.; Barth, M.J.; Wu, G. A Digital Twin Paradigm: Vehicle-to-Cloud Based Advanced Driver Assistance Systems. In Proceedings of the IEEE Vehicular Technology Conference, Antwerp, Belgium, 25–28 May 2020; [CrossRef]
5. Kobayashi, H.; Han, K.; Kim, B. Vehicle-to-Vehicle Message Sender Identification for Co-Operative Driver Assistance Systems. In Proceedings of the 2019 IEEE 89th Vehicular Technology Conference (VTC2019-Spring), Kuala Lumpur, Malaysia, 28 April–1 May 2019; pp. 1–5. [CrossRef]
6. Liu, Y.; Wang, Z.; Han, K.; Shou, Z.; Tiwari, P.; Hansen, J.H. Sensor Fusion of Camera and Cloud Digital Twin Information for Intelligent Vehicles. *arXiv* **2020**, arXiv:2007.04350.

7. Shan, M.; Narula, K.; Wong, R.; Worrall, S.; Khan, M.; Alexander, P.; Nebot, E. Demonstrations of cooperative perception: Safety and robustness in connected and automated vehicle operations. *Sensors* **2020**, *21*, 200. [CrossRef] [PubMed]
8. Tihanyi, V.; Tettamanti, T.; Csonthó, M.; Eichberger, A.; Ficzere, D.; Gangel, K.; Hörmann, L.B.; Klaffenböck, M.A.; Knauder, C.; Luley, P.; et al. Motorway Measurement Campaign to Support R&D Activities in the Field of Automated Driving Technologies. *Sensors* **2021**, *21*, 2169. [CrossRef] [PubMed]
9. Szalay, Z.; Szalai, M.; Tóth, B.; Tettamanti, T.; Tihanyi, V. Proof of concept for Scenario-in-the-Loop (SciL) testing for autonomous vehicle technology. In Proceedings of the 2019 IEEE International Conference on Connected Vehicles and Expo (ICCVE), Graz, Austria, 4–8 November 2019; pp. 1–5.
10. Szalay, Z.; Hamar, Z.; Simon, P. A multi-layer autonomous vehicle and simulation validation ecosystem axis: Zalazone. In *International Conference on Intelligent Autonomous Systems*; Springer: Berlin/Heidelberg, Germany, 2018, pp. 954–963.
11. Zhao, M.; Mammeri, A.; Boukerche, A. Distance measurement system for smart vehicles. In Proceedings of the 2015 7th International Conference on New Technologies, Mobility and Security (NTMS), Paris, France, 27–29 July 2015; pp. 1–5. [CrossRef]
12. Zhang, Z. A flexible new technique for camera calibration. *IEEE Trans. Pattern Anal. Mach. Intell.* **2000**, *22*, 1330–1334. [CrossRef]
13. Wang, W.; Sakurada, K.; Kawaguchi, N. Reflectance Intensity Assisted Automatic and Accurate Extrinsic Calibration of 3D LiDAR and Panoramic Camera Using a Printed Chessboard. *Remote Sens.* **2017**, *9*, 851. [CrossRef]
14. Daniilidis, K. Hand-Eye Calibration Using Dual Quaternions. *Int. J. Robot. Res.* **1999**, *18*, 286–298. [CrossRef]
15. Vyacheslav, I.V.; Illya, E.K.; Irina, P.C. Accurate Time Synchronization for Digital Communication Network. In Proceedings of the 2007 17th International Crimean Conference—Microwave Telecommunication Technology, Sevastopol, Ukraine, 10–14 September 2007; pp. 259–260. [CrossRef]
16. Bochkovskiy, A.; Wang, C.Y.; Liao, H.Y.M. YOLOv4: Optimal Speed and Accuracy of Object Detection. *arXiv* **2020**, arXiv:2004.10934.
17. Hartley, R.; Zisserman, A. *Multiple View Geometry in Computer Vision*, 2nd ed.; Cambridge University Press: Cambridge, UK, 2004; [CrossRef]
18. Yilmaz, A.; Javed, O.; Shah, M. Object tracking: A survey'ACM computing surveys (CSUR). *ACM Comput. Surv.* **2006**, *38*, 13. [CrossRef]
19. Labbe, R. Kalman and bayesian filters in python. *Chap* **2014**, *7*, 246.
20. Chong, C.Y. Tracking and data fusion: A handbook of algorithms (bar-shalom, y. et al; 2011)[bookshelf]. *IEEE Control. Syst. Mag.* **2012**, *32*, 114–116.
21. Blom, H.; Bar-Shalom, Y. The interacting multiple model algorithm for systems with Markovian switching coefficients. *IEEE Trans. Autom. Control* **1988**, *33*, 780–783. [CrossRef]
22. Blackman, S.S.; Popoli, R. *Design and Analysis of Modern Tracking Systems*; Artech House Radar Library: Boston, MA, USA; London, UK, 1999.
23. Munkres, J. Algorithms for the assignment and transportation problems. *J. Soc. Ind. Appl. Math.* **1957**, *5*, 32–38. [CrossRef]
24. Kuhn, H.W. The Hungarian method for the assignment problem. *Nav. Res. Logist. Q.* **1955**, *2*, 83–97. [CrossRef]
25. Abd El-Gawad, M.A.; Elsharief, M.; Kim, H. A comparative experimental analysis of channel access protocols in vehicular networks. *IEEE Access* **2019**, *7*, 149433–149443. [CrossRef]
26. Dinar, A.E.; Merabet, B.; Ghouali, S. NTP Server Clock Adjustment with Chrony. In *Applications of Internet of Things*; Springer: Berlin/Heidelberg, Germany, 2021; pp. 177–185.
27. Szalai, M.; Varga, B.; Tettamanti, T.; Tihanyi, V. Mixed reality test environment for autonomous cars using Unity 3D and SUMO. In Proceedings of the 2020 IEEE 18th World Symposium on Applied Machine Intelligence and Informatics (SAMI), Herlany, Slovakia, 23–25 January 2020; pp. 73–78. [CrossRef]
28. Xu, K.; Zhang, G.; Liu, S.; Fan, Q.; Sun, M.; Chen, H.; Chen, P.Y.; Wang, Y.; Lin, X. Adversarial T-shirt! Evading Person Detectors in A Physical World. *arXiv* **2020**, arXiv:1910.11099.

Article

Model Predictive Controller Design for Vehicle Motion Control at Handling Limits in Multiple Equilibria on Varying Road Surfaces

Szilárd Czibere, Ádám Domina, Ádám Bárdos * and Zsolt Szalay

Department of Automotive Technologies, Budapest University of Technology and Economics, 6 Stoczek St., Building J, H-1111 Budapest, Hungary; szilard.czibere@edu.bme.hu (Sz.Cz.); domina.adam@edu.bme.hu (Á.D.); szalay.zsolt@kjk.bme.hu (Zs.Sz.)
* Correspondence: bardos.adam@kjk.bme.hu

Abstract: Electronic vehicle dynamics systems are expected to evolve in the future as more and more automobile manufacturers mark fully automated vehicles as their main path of development. State-of-the-art electronic stability control programs aim to limit the vehicle motion within the stable region of the vehicle dynamics, thereby preventing drifting. On the contrary, in this paper, the authors suggest its use as an optimal cornering technique in emergency situations and on certain road conditions. Achieving the automated initiation and stabilization of vehicle drift motion (also known as powerslide) on varying road surfaces means a high level of controllability over the vehicle. This article proposes a novel approach to realize automated vehicle drifting in multiple operation points on different road surfaces. A three-state nonlinear vehicle and tire model was selected for control-oriented purposes. Model predictive control (MPC) was chosen with an online updating strategy to initiate and maintain the drift even in changing conditions. Parameter identification was conducted on a test vehicle. Equilibrium analysis was a key tool to identify steady-state drift states, and successive linearization was used as an updating strategy. The authors show that the proposed controller is capable of initiating and maintaining steady-state drifting. In the first test scenario, the reaching of a single drifting equilibrium point with $-27.5°$ sideslip angle and 10 m/s longitudinal speed is presented, which resulted in $-20°$ roadwheel angle. In the second demonstration, the setpoints were altered across three different operating points with sideslip angles ranging from $-27.5°$ to $-35°$. In the third test case, a wet to dry road transition is presented with 0.8 and 0.95 road grip values, respectively.

Keywords: autonomous drifting; model predictive control (MPC); successive linearization; adaptive control; vehicle motion control; varying road surfaces; vehicle dynamics

Citation: Czibere, S.; Domina, Á.; Bárdos, Á.; Szalay, Z. Model Predictive Controller Design for Vehicle Motion Control at Handling Limits in Multiple Equilibria on Varying Road Surfaces. *Energies* **2021**, *14*, 6667. https://doi.org/10.3390/en14206667

Academic Editor: Sheldon Williamson

Received: 6 September 2021
Accepted: 5 October 2021
Published: 14 October 2021

Publisher's Note: MDPI stays neutral with regard to jurisdictional claims in published maps and institutional affiliations.

1. Introduction

To achieve the goal of automated driving, classical electronic vehicle dynamics systems are essential to prevent the vehicle from going beyond the limit of handling. Systems such as ABS or ESP can make interventions to keep the vehicle in a safe condition that improves road safety significantly. As stated in [1–3], road traffic crashes are a considerable concern in motorized countries because of their impact on society and the economy. As an evolution, in this paper, the authors introduce a new breed of electronic vehicle dynamics system which, instead of preventing the drift scenario in the case of an emergency situation, it uses it as an optimal cornering method to stabilize the vehicle. Moreover, the proposed method also provides a solution for high-sideslip cornering on varying road surfaces. The performance of autonomous vehicles should be as good as human drivers or better to gain social trust and acknowledgement. Controlling drift scenarios on varying road surfaces proves a high level of controllability even in situations that can be challenging for professional human drivers. The control and better understanding of these kinds of motions can support the more widespread usage of fully automated cars of the future.

In the last decade, the development of fully automated vehicles seems to have been one of the most primary focuses of development for many automobile manufacturers. On the other hand, the control of normal driving scenarios such as in [4–7] have been at the center of their research in most cases. One drawback of these methods is that they are unable to stabilize the vehicle beyond the limit of handling. To improve road safety, automated vehicles should also be capable of performing high-sideslip maneuvers such as drifts.

This manuscript describes drift maneuvers using equilibrium points as stated in [8,9]. It was shown in [10] that a three-state bicycle model with a nonlinear tire model could be used for steady-state drift control design. The three-state bicycle model describes the motion of the vehicle by three state variables, which are the longitudinal velocity, the lateral velocity, and the yaw-rate. Measurements on a real vehicle were taken for adequate model parametrization.

In [11], the advantages, such as more accurate tracking of the desired operating point and a better disturbance rejection capability, of MIMO (multiple-input multiple-output) controllers were stated over SIMO (single-input single-output) approaches. The authors used optimal control to achieve drift maneuvers. An LQR (linear quadratic regulator) controller was designed for this purpose.

A study was conducted in [12], where the authors performed path control at the limit of handling using the state-dependent Riccati equation technique. Their objective was to minimize lateral path-tracking error while tires operate in limit handling. The proposed controller showed robust path-tracking performance even when the rear wheels were operated beyond their friction limits, and large body sideslip prevailed.

A robust, state-feedback control approach was introduced in [13] with uncertain disturbances to maintain drift maneuvers. A 4-DOF nonlinear vehicle dynamics model was established with the so-called UniTire model (for more information, see [14]). The authors performed drift both in steady and in transient states with this approach. It was also concluded that vehicles, which can drive in drift conditions, are considerably safer.

The model predictive control (MPC) scheme is an optimizing control theory that is becoming popular owing to abilities such as handling constraints directly or future prediction in the design process. Nonetheless, MPC is not a common choice in the field of autonomous drifting but more favored in path tracking, as shown in [15]. In that paper, a so-called multilayer MPC was designed. Three path-tracking controllers were used with fixed velocities and with a velocity decision controller. The author pointed out the outstanding performance of the multilayer MPC as an appropriate choice for real-time application in the field of vehicle motion control.

Adaptive cruise control for cut-in scenarios based on MPC was introduced in [16]. The author defined a finite state machine to manage vehicle control in different scenarios. In addition, MPC was used to realize coordinate control of the host vehicle and the cut-in vehicle.

Since the number of actuators that directly control the motion of a vehicle is expected to multiply [17–19], this will open up new possibilities for controlling maneuvers such as drift more precisely. Having proper control of these actuators will allow performing maneuvers that are impossible even for a professional human driver.

The aim of the current article is to perform steady-state drifting in multiple equilibria on varying road surfaces conditions, using the adaptive MPC topology. Whilst multilayer MPC is a cluster of predefined linear controllers with a decision controller, the adaptive MPC proposed by the authors is a more general approach of one controller with a predefined model with an updating strategy, for which successive linearization was chosen.

Controller performance was investigated in a simulation environment in different test cases for which a three-state, one-track model was defined with a nonlinear tire model. It is shown that the developed adaptive MPC is a practical solution for motion control and automated drifting that can handle changing road conditions, model nonlinearities, and actuator constraints. The computational requirement of a nonlinear MPC is high, and the nonlinear optimization problem might not be solved in real time. In this article, the authors use a linear MPC coupled with a quadratic cost function, of which the computational needs can be handled by currently available hardware. The contribution of this paper is twofold. The first contribution is the application of a multilayer MPC for vehicle drift control, as there was previously no research in which MPC was applied. The second contribution is the ability of the proposed controller to handle the change in friction during vehicle drifting. In the other research presented above, the authors applied a fixed constant friction coefficient during vehicle drift in both simulation and measurement. In this paper, the proposed controller can handle the changes in friction without aborting the drift motion, which results in excellent adaptability of the proposed method, and which is also crucial for real-life driving scenarios.

2. Vehicle Modeling and Simulation Setup

In this section, the vehicle model which served as a basis for the controller design and implementation, as well as for software-in-the-loop (SIL) evaluation of the proposed MPC framework, is presented. For model identification purposes, measurements with a series production coupé sports car were used. A single-track model was built up for both simulations and controller design with a nonlinear tire model.

2.1. Vehicle Modeling

First and foremost, a vehicle model must be defined for the appropriate controller design and SIL testing. Considering the desired application, high fidelity is required to capture the dynamical behavior of the vehicle; moreover, low computational demand is also necessary to allow embedded application. In this paper, a three-state, two-wheel bicycle model with a nonlinear tire model was chosen for this purpose. It was shown in [10,11,20] that a two-wheel bicycle model can be sufficient for optimal control methods to achieve steady-state self-drifting.

2.1.1. Two-Wheel Planar Vehicle Dynamics

The equations of motion in longitudinal and lateral directions, and the yaw dynamics of the vehicle are presented in differential Equations (1)–(3) based on the Newtonian laws, as described in detail in [21]. Please refer to Appendix A for a better understanding of the applied nomenclature.

$$\dot{V}_x = \frac{1}{m}(F_x - F_A) + rV_y. \tag{1}$$

$$\dot{V}_y = \frac{1}{m}F_y - rV_x. \tag{2}$$

$$\dot{r} = \frac{1}{I_z}M_z. \tag{3}$$

The longitudinal (Equation (4)) and lateral (Equation (5)) forces, and the yaw torque (Equation (7)) can be derived from the tire forces as described below, along with the air drag (Equation (6)). The yaw torque (7) is applied at the center of gravity (CoG) of the vehicle.

$$F_x = F_{xF}\cos\delta + F_{xR} - F_{yF}\sin\delta. \tag{4}$$

$$F_y = F_{yF}\cos\delta + F_{yR} + F_{xF}\sin\delta. \tag{5}$$

$$F_A = \frac{1}{2}A\,\rho\,C_A\,V_x^2. \tag{6}$$

$$M_z = aF_{yF}\cos\delta - bF_{yR} + aF_{xF}\sin\delta. \tag{7}$$

The sideslip angle of the front (Equation (8)) and rear (Equation (9)) wheels, and the sideslip angle of the vehicle, which is applied at the center of gravity (Equation (10)), can be calculated in the following way:

$$\alpha_f = arctan\left(\frac{V_y + ar}{V_x}\right) - \delta, \tag{8}$$

$$\alpha_r = arctan\left(\frac{V_y - br}{V_x}\right), \tag{9}$$

$$\beta = arctan\left(\frac{V_y}{V_x}\right). \tag{10}$$

The air drag force F_A is considered zero in the above equations since the speed of drift results in neglectable air drag. Due to small values, rolling and pitching dynamics are also neglected. Since the test vehicle is rear-wheel-driven, and there is no braking during this specific drift maneuver, F_{xF} is considered zero throughout this paper.

2.1.2. Nonlinear Tire Model

The forces in the above-defined equations awaken in the contact patches of the tires. Owing to the complex, heterogeneous structure of pneumatic rubber tires [22], tire models vary in a wide-scaled spectrum regarding their specific application. According to complexity levels, finite element models [23] and empirical models [24] are also widespread in the literature. For achieving a simple, but meaningful tire model for a control-oriented approach, the so-called brush tire model was selected for calculating the lateral tire force.

The equations below can be used for describing the lateral force of the tires in both saturated and unsaturated phases. Saturation refers to the operating range of the tire, where the change in the lateral slip does not affect the generated force. Physically meaningful consideration was derived in [25,26].

For the front tire where no longitudinal forces are considered, the lateral tire force can be calculated on the basis of the following equations, where α_{sl} denotes tire sideslip limit where the tires become saturated:

$$F_{yF} = \begin{cases} -C_{\alpha F}tan(\alpha) + \frac{C_{\alpha F}^2}{3\mu F_{zF}}tan(\alpha)|tan(\alpha)| - \frac{C_{\alpha F}^3}{27\mu^2 F_{zF}^2}tan(\alpha)^3, & |\alpha| \leq \alpha_{sl} \\ -\mu F_{zF}sgn(\alpha), & |\alpha| > \alpha_{sl} \end{cases}, \tag{11}$$

$$\alpha_{sl\_F} = arctan\left(\frac{3\mu F_{zF}}{C_{\alpha F}}\right). \tag{12}$$

In the case of the rear tire, a solution must be found to also add the longitudinal forces to the model. A similar approach could be applied that was also used for deriving the lateral forces, although the wheel speeds would appear as additional states in the model. To overcome this matter, the connection between the longitudinal and lateral forces was determined with the friction circle and described by a coupling factor, as suggested in [10]. As stated below, ζ determines the amount of lateral force that can be reached directly from the friction.

$$F_y^{max} = \zeta\mu F_{zR}. \tag{13}$$

$$\zeta = \frac{\sqrt{(\mu F_{zR})^2 - F_{xR}^2}}{\mu F_{zR}}. \tag{14}$$

Accordingly, Equation (11) can be augmented to compute both the lateral and the longitudinal forces for the rear tire.

$$F_{yR} = \begin{cases} -C_{\alpha R}tan(\alpha) + \frac{C_{\alpha R}^2}{3\zeta\mu F_{zR}}tan(\alpha)|tan(\alpha)| - \frac{C_{\alpha R}^3}{27\zeta^2\mu^2 F_{zR}^2}tan(\alpha)^3, & |\alpha| \leq \alpha_{sl} \\ -\zeta\mu F_{zR}sgn(\alpha), & |\alpha| > \alpha_{sl} \end{cases}. \tag{15}$$

$$\alpha_{sl\_R} = arctan\left(\frac{3\xi\mu F_{zR}}{C_{\alpha R}}\right).\tag{16}$$

2.2. Model Parameter Identification Measurements

Model parameter identification was conducted by the authors. The test vehicle, which was used as a source of model parameter identification measurements, possesses a 3.0 L twin-turbocharged straight-six engine. As for the characteristics of the engine, it is able to produce 550 N·m torque between 2350 and 5230 RPM and 411 HP (302 kW) between 5230 and 7000 RPM. The test vehicle has a rear-wheel drive and a seven-speed dual-clutch automatic transmission.

The position of the center of gravity, depicted in Figure 1, was determined by measuring the axle weights. As a result, 925 kg was measured on the first axle, and 895 kg was measured on the second axle. Taking the total length of the wheelbase, which is 2.69, the position of the center of gravity is situated 1.37 m far away in front of the second axle.

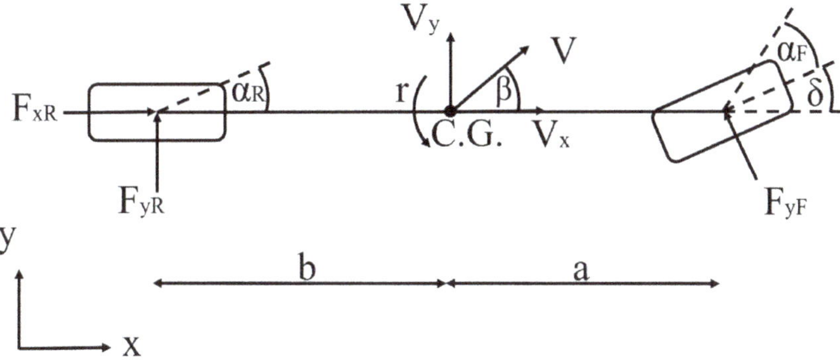

Figure 1. Bicycle model notations used in the equations.

Parameters of the above-defined brush tire model, such as friction coefficients or cornering stiffness, were necessary to identify in advance. For this purpose, the ISO 13674-2 standard was used with ramp steering maneuvers. The measurement was performed from an initial straight line until the vehicle reached a constant velocity of 100 km/h. In the next step, the neutral gear was selected, and a slow, constant velocity ramp was given as an input to the steering wheel. The ZalaZone Dynamical Platform [27] served as a testing ground for the parameter identification. The friction coefficient of its special, dry asphalt was close to 1 in the case of both tires. As a result of the measurements, the cornering stiffness turned out to be 300,000 N/rad for the front tires and 500,000 N/rad for the rear tires [28]. The measured data points and the characteristics of the tires can be seen in Figure 2.

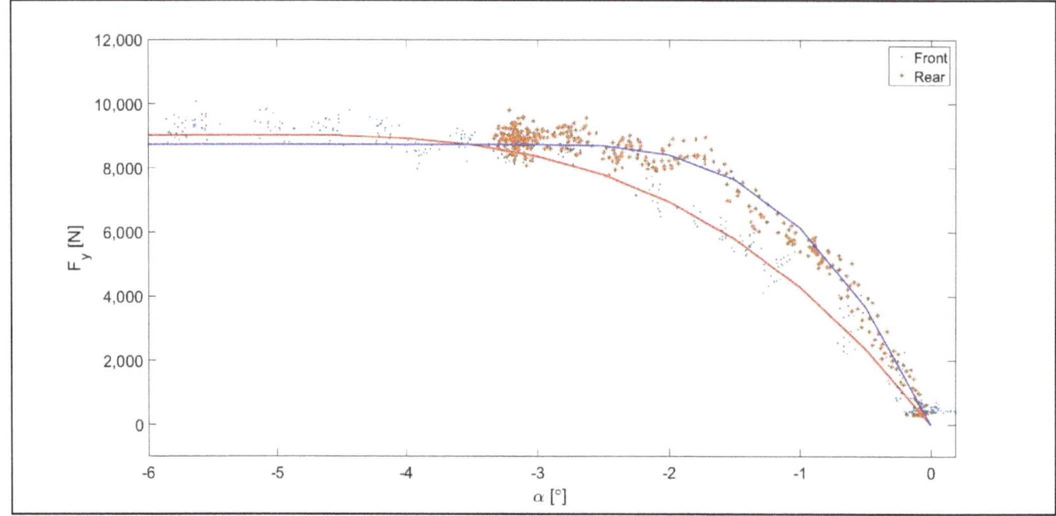

Figure 2. The measured tire curves for the front and rear tires.

3. Equilibrium Analysis

In this study, steady-state drifting cornering is targeted. It occurs in special locations of the state space, where all the state derivatives are equal to zero, called equilibrium points. With this idea in mind, the equilibria of the system must be determined and analyzed for the desired operation.

According to the three-state bicycle model, the state matrix of the system (Equations (1)–(3)) can be written in the form of

$$\mathbf{x} = \begin{bmatrix} V_y \\ r \\ V_x \end{bmatrix}, \tag{17}$$

where the states were chosen as the lateral vehicle speed, yaw rate, and longitudinal vehicle speed.

The control inputs were defined as the roadwheel angle and the rear traction force.

$$\mathbf{u} = \begin{bmatrix} \delta \\ F_{xR} \end{bmatrix}. \tag{18}$$

To locate the equilibria of the system, the following algebraic equation system must be satisfied:

$$\dot{\mathbf{x}} = F(\mathbf{x}^{\text{eq}}, \mathbf{u}^{\text{eq}}) = 0. \tag{19}$$

It can be deduced that our system is an underactuated system since we have one more state than control input. This yields that not all the state derivatives can be driven to zero from an arbitrary point of the state space. Furthermore, considering all the equations, one can see that we have fewer equations than variables; therefore, we need to choose some variables to be fixed in advance. This observation indicates that only specific locations of the state space are solutions of the equation system. To solve this problem, the velocity is set to 10 $\frac{m}{s}$ as a known variable, and the steering angle ranged between $-35°$ and $+35°$ rad with a step size of $2.86°$. Figures 3–5 show the calculated results of Equation (19) with the described parameters.

For controller performance evaluation, three equilibrium points were chosen to be used in the subsequent sections. Parameter values for the test cases are summarized in Table 1.

Table 1. Equilibrium points used in the controller evaluation simulations.

Case No.	δ (°)	F_{xR} (N)	V_y (m/s)	r (rad/s)	V_x (m/s)
1	−20.05	4753	−5.21	0.776	10
2	−28.65	5500	−6.99	0.713	10
3	−22.92	5254	−6.36	0.735	10

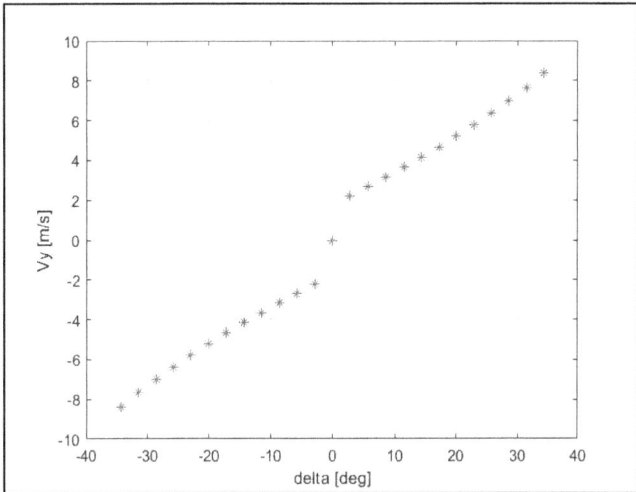

Figure 3. Equilibrium lateral speeds as a function of roadwheel angle.

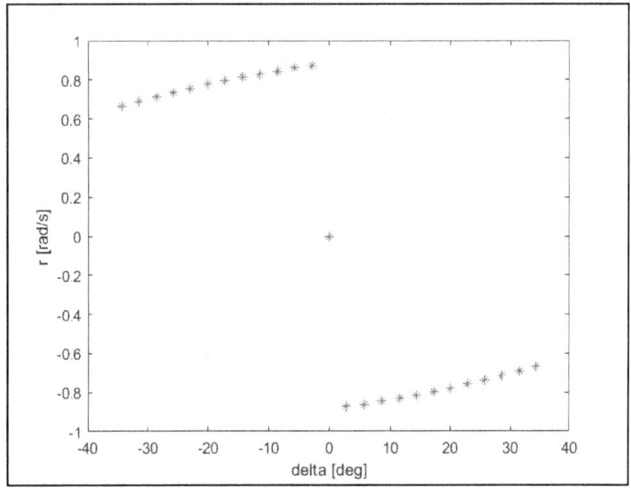

Figure 4. Equilibrium yaw rates as a function of roadwheel angle.

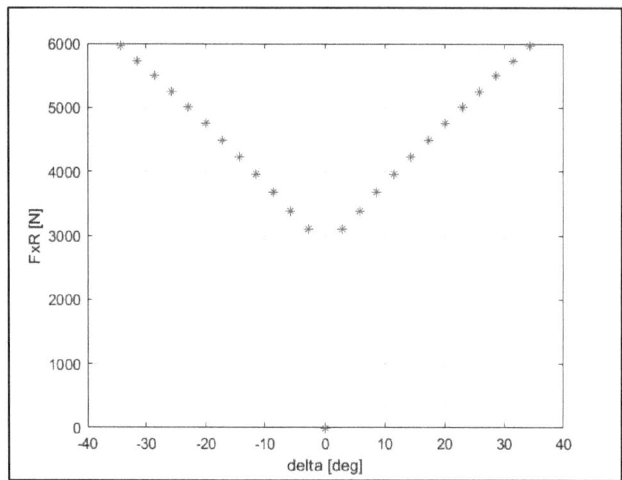

Figure 5. Equilibrium traction forces versus roadwheel angle.

4. Controller Design

In this section, the formulation of the control problem is stated. As previously discussed in Section 1, a MIMO controller is necessary to achieve the desired performance. In this manuscript, an adaptive MPC was designed for this purpose. The state matrices are updated for solving the quadratic problem (QP) at every time step. Updating the state matrices can provide a better control performance during varying environmental conditions due to the utilization of actual model parameter estimates such as the road grip. Moreover, it allows using our controller in multiple equilibria with only one controller.

Below, a system with N_x states and N_u inputs is considered. The system can be described by ordinary differential equations as presented in Section 2.1 and written in the form of Equation (20).

$$\dot{x} = F(x, u), \quad y = z(x, u). \tag{20}$$

4.1. MPC Formulation

Model predictive control is an optimizing control theory that uses the discrete model of the system to predict its behavior in the future. MPC calculates a series of inputs to minimize the predefined cost function at every time step (T_s). When the cost function is minimized, the first element of the calculated control inputs is applied. Crucial parameters are the prediction horizon N_p and the control horizon N_c. N_p defines the capability of the system in terms of how far it can see into the future. The time horizon $N_{p,t}$, which corresponds to the prediction horizon N_p can be written in the following form:

$$N_{p,t} = N_p T_s, \tag{21}$$

where m is a positive integer, and $T_s = 10$ ms throughout this paper. N_c defines the number of control inputs that can be optimized in one time step. One should carefully set these parameters to see the transient of the system in advance, thereby avoiding unnecessarily high computational needs. The MPC uses full state feedback that minimizes the cost function below.

$$J(k) = \frac{1}{2} \left(\hat{x}^T \, \bar{\bar{Q}}_x \hat{x} + \bar{u}^T \bar{\bar{R}} \, \bar{u} + \hat{z}^T \, \bar{\bar{Q}}_z \hat{z} \right), \tag{22}$$

where $\hat{x}$ is the stacked vector of the predicted states, $\hat{z}$ denotes the performance, and $\bar{u}$ is the control inputs at every time step k. $Q_x = Q_x{}^T \geq 0 \in R^{3\times3}$, $Q_x = Q_x{}^T \geq 0 \in R^{3\times3}$, and $R = R^T > 0 \in R^1$ are weight matrices. Based on this, $\bar{\bar{Q}}_x =$

$$diag(Q_x, Q_x, \ldots Q_x) \in R^{N_x N_p \times N_x N_p}, \quad \overline{\overline{Q}}_x = diag(Q_z, Q_z, \ldots Q_z) \in R^{N_x N_p \times N_x N_p}, \text{ and}$$
$$\overline{\overline{R}} = diag(R, R, \ldots R) \in R^{N_u N_p \times N_u N_p}.$$

A further advantage of MPC, for instance, over LQ control, is that it handles the constraints of the system directly. Constraints were set in case of the control inputs.

$$-u_{min} < u < u_{max}. \tag{23}$$

During the simulation, the physical constraints of the test vehicle were taken into consideration. This means that the steering angle δ could range between $-35°$ and $+35°$, and the driving force F_{xR} at the rear wheel was between 0 N and 7000 N.

The evolution matrices in Equation (24) can be determined according to [29]. The D selection matrix was zero throughout in this paper. The system state $x(k)$ is measured at the time step k.

$$\overbrace{\begin{bmatrix} x(k+1) \\ x(k+2) \\ \vdots \\ x(k+N_p) \end{bmatrix}}^{\hat{x}} = \overbrace{\begin{bmatrix} A \\ A^2 \\ \vdots \\ A^{N_p} \end{bmatrix}}^{\overline{A}} x(k) + \overbrace{\begin{bmatrix} B & 0 & \cdots & 0 \\ AB & B & \cdots & 0 \\ \vdots & & \ddots & \vdots \\ A^{N_p-1}B & \cdots & \cdots & B \end{bmatrix}}^{\overline{\overline{B}}} \overbrace{\begin{bmatrix} u(k) \\ u(k+1) \\ \vdots \\ u(k+N_p-1) \end{bmatrix}}^{\overline{u}} \tag{24}$$

The cost function must penalize both the magnitude of the applied control inputs, steering angle, traction force, and deviation of the states from the reference trajectory in order to formulate the optimization problem, the solution of which is the optimal control inputs. The error vector z is the following:

$$\overbrace{\begin{bmatrix} z(k+1) \\ z(k+2) \\ \vdots \\ z(k+Np) \end{bmatrix}}^{\hat{z}} = \overbrace{\begin{bmatrix} C & 0 & \cdots & 0 \\ 0 & C & \cdots & 0 \\ \vdots & & \ddots & \vdots \\ 0 & \cdots & \cdots & C \end{bmatrix}}^{\overline{\overline{C}}} \overbrace{\begin{bmatrix} x(k+1) \\ x(k+2) \\ \vdots \\ x(k+Np) \end{bmatrix}}^{\hat{x}}. \tag{25}$$

The cost function is calculated by substituting Equations (24) and (25) into Equation (22).

$$J = \frac{1}{2}\overline{u}^T H \overline{u} + f^T \overline{u}. \tag{26}$$

For a more detailed derivation, see [28]. The QP problem is formed in the following equation:

$$\min_u \left[\frac{1}{2}\overline{u}^T H \overline{u} + f^T \overline{u} \right], \tag{27}$$

subject to

$$-0.6 < u_1 < 0.6, \tag{28}$$

$$0 < u_2 < 7000. \tag{29}$$

4.2. Successive Linearization

Traditional MPC works on linear and discrete models. Nonetheless, the vast majority of dynamical systems are nonlinear, which is the case of the presented vehicle model introduced in Section 2.1. To be able to create an MPC control scheme, it is reasonable to linearize the nonlinear system and discretize the continuous system afterward. Successive linearization was selected according to [30]. Successive linearization is a key tool to make our model adaptive to environmental and conditional changes in real time. Successive linearization is a linearization method followed by discretization. The idea is to create a linear MPC for every time step and make it adaptive to the changing conditions by updating the model that the MPC uses to formulate the QP problem, instead of using a more complex control scheme like nonlinear MPC in [31].

The system described in Equation (20) can be linearized in specific operating points by knowing its gradient. One can interpret it as Jacobian linearization. According to the aforementioned specifications, the state matrices of the linearized system can be derived as follows:

$$A_c(i,j) = \frac{\partial F_i}{\partial x_j}, \; B_c(i,j) = \frac{\partial F_i}{\partial u_j}, \tag{30}$$

$$C_c(i,j) = \frac{\partial z_i}{\partial x_j}, \; D_c(i,j) = \frac{\partial z_i}{\partial u_j}. \tag{31}$$

By knowing the operating points at every time step, the matrices in Equations (30) and (31) can be determined in numerical form. Since MPC works on discrete systems, discretization is required.

The linear continuous system of Equations (30) and (31) can be discretized as defined in Equations (32)–(34).

$$A_d = e^{A_c T_s}. \tag{32}$$

$$B_d = A_c^{-1}\left(e^{A_c T_s} - I\right)B_c. \tag{33}$$

$$C_d = C_c, \; D_d = D_c. \tag{34}$$

Equations (32)–(34) can be substituted into Equations (24) and (25) at every time step.

A MATLAB and Simulink environment was used for model and controller implementation (the implemented MPC controller and the nonlinear vehicle model are marked in blue in Figure 6), as well as for software-in-loop testing. In order to appropriately tune the weight matrices of Equation (22), a simple linear MPC was created for one equilibrium. Since the aim is to control a nonlinear system with a linear controller around an equilibrium, one should take coordinate system transformation into consideration as a result of Jacobian linearization. Coordinate system transformation was applied according to [32]. The control scheme is depicted in Figure 6. Blocks responsible for coordinate transformation between the controller and the vehicle model are marked in yellow.

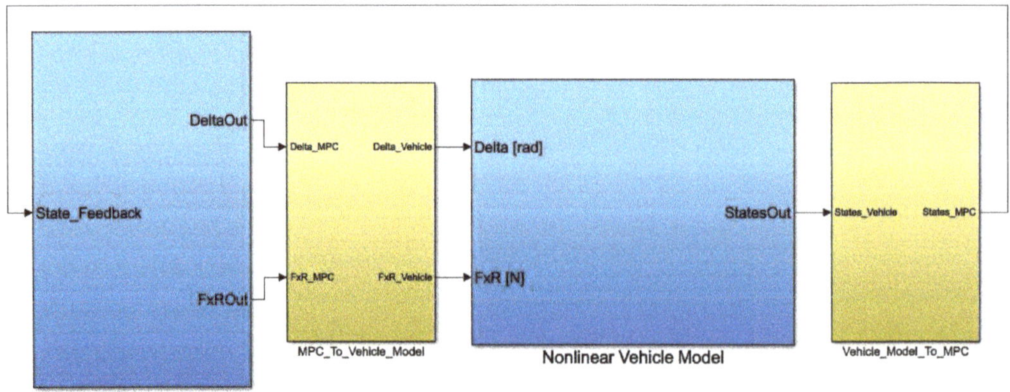

Figure 6. The control scheme.

5. Simulation Results

In this section, the simulation results are revealed. The first test case was created to tune the parameters of the controller and test its performance. In the second test case, the vehicle was driven through multiple equilibria. Lastly, the alteration of the road surface was considered while driving through multiple equilibria.

5.1. Steady-State Drift in a Single Equilibrium

Firstly, the control scheme was tested only using Case No. 1 of Table 1 (Equation (1)). This was an important step in order to tune the weight matrices. A PI (proportional–integral) cruise controller was used to accelerate the vehicle from 0 to 8 m/s, after which the control was given to the MPC drift controller. The controller was able to drive the car into a drift state and keep it there simultaneously. Figure 7 shows the simulation results after 8 m/s was reached. The controller needed to cope with the task of accelerating the vehicle by 2 m/s to reach the desired drifting speed. It can be deduced from the results that the designed linear controller was capable of controlling the defined nonlinear system around the desired equilibrium after the proper tuning. This was proven by the simulations where the vehicle performed a high-sideslip cornering motion, as depicted in Figure 8. The movement of the vehicle can also be tracked in this figure. Firstly, the drift controller took over the control at (0,0), after which there was an acceleration period. When all the states were driven to the operating points, the steady-state drift was maintained.

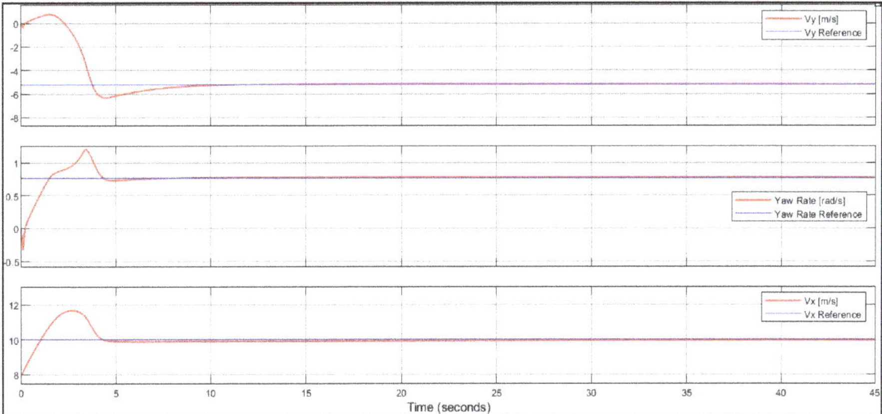

Figure 7. Drift scenario around equilibrium Case No. 1.

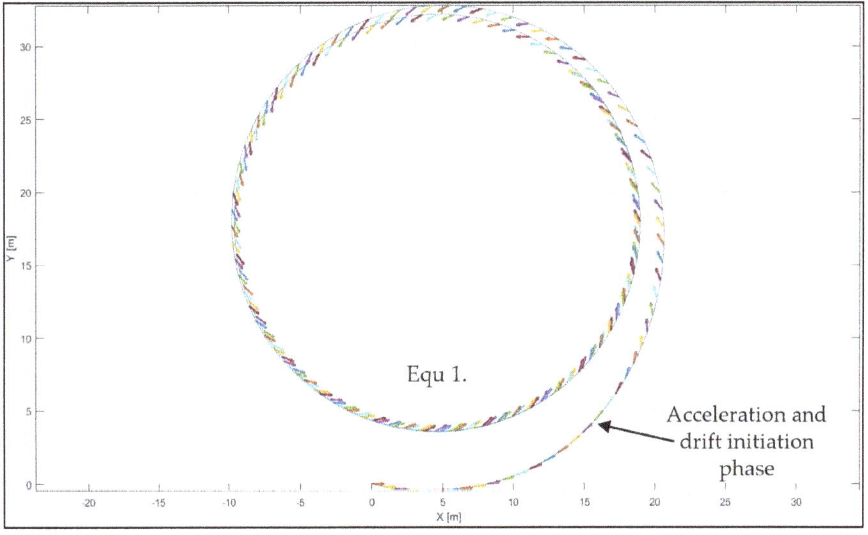

Figure 8. The heading of the vehicle during steady-state drift in equilibrium Case No. 1.

5.2. Steady-State Drift in Multiple Equilibria

The controller introduced in Section 5.1 was able to maintain the system at the desired operating point. However, it performed poorly in other segments of the state space. The adaptive MPC structure can be a solution for this limitation. This section points out the advantages of adaptive MPC over linear MPC for multi-operating point purposes.

The second step after Section 5.1 was to drift around three different equilibria. In this case, the system was updated, and a new MPC was formed online at every time step. The MPC could not only keep the vehicle at the limit of handling, but it was also able to drive the system from one equilibrium to another. The order of the steady points was based on the idea to test the robustness of the controller. First of all, the controller needed to accelerate the system from 8 m/s to 10 m/s while driving it into a drift state. Furthermore, after drifting in the first circle, the controller needed to drive the vehicle into a circle with a smaller radius and into a considerably bigger one after that. The transition in the first case from Equation (3) to Equation (1) was relatively smaller (these were closer operating points) than that from Equation (1) to Equation (2), as can be seen in Figure 9 by examining the overshoots. The controller was able to cope with both acceleration and deceleration tasks. Figure 10 depicts the control inputs during steady-state drift at the operating points. As can be seen, the control inputs were the same in the steady-state cases as the precalculated values in Table 1.

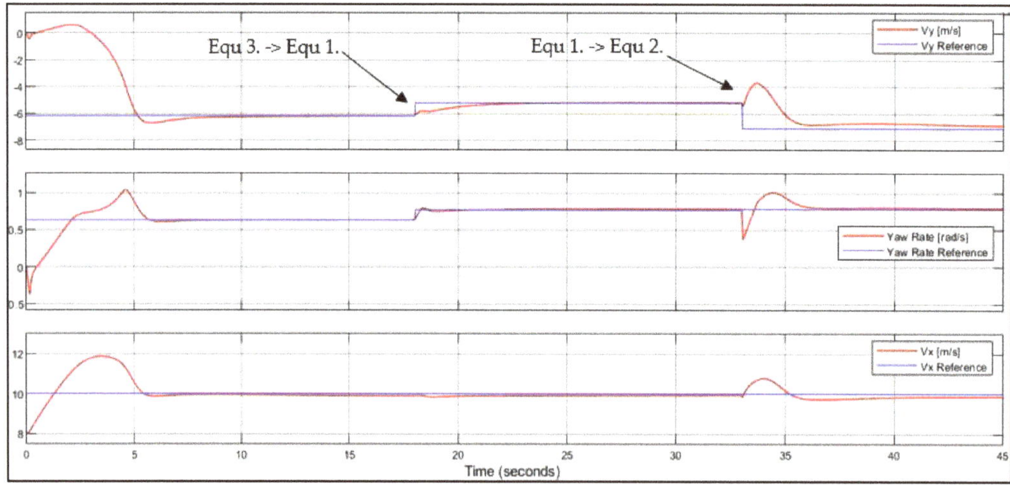

Figure 9. Evolution of states while settling in the three different equilibria.

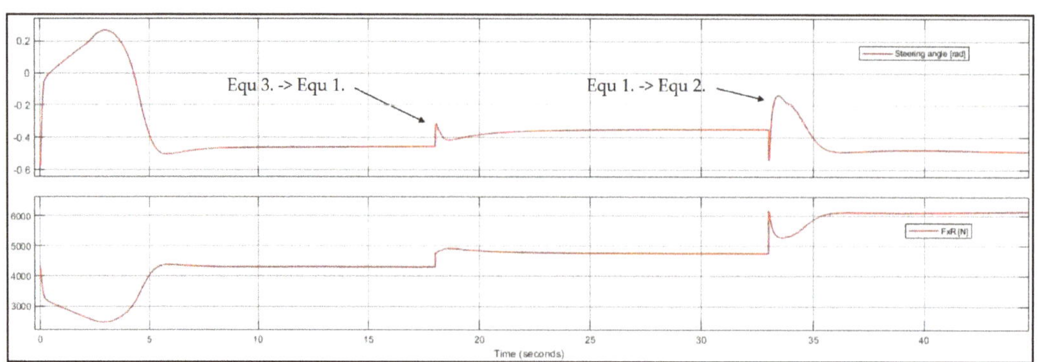

Figure 10. Control inputs during the test while reaching the targeted three different equilibria.

5.3. Drifting on Varying Road Surfaces

In real-life scenarios, changing road surfaces is a common occurrence. Moreover, during drifting, the vehicle can also go off-road or across wet asphalt areas. The varying friction of road surfaces raises many challenges for autonomous vehicles. Road surface estimation is an active topic in the field of autonomous vehicle development. Machine learning is a promising methodology for road surface estimation, as shown in [33–35]. A new idea was proposed in [35] based on a machine-learning approach to estimate road surface. The authors introduced a methodology that relies on large datasets that can be collected using onboard sensors or dynamic simulations. Online road surface estimation is an essential extension of the proposed method introduced in Section 5.2. Nevertheless, it is beyond the scope of the present work. Hence, preset friction coefficients were used for testing the adaptability of the software.

Simulations confirmed the fact that the tolerance of the linear controller was limited toward road changes. Setting the friction coefficient of the road surface to 0.8 from 0.95, for instance, resulted in poor performance or even instability. Adaptive topology can also be used to eliminate this problem. By having an estimation of the road surface, the friction coefficient can be updated along with the other parameters to provide a more accurate model for the MPC to work on. Road friction is considered a variable in the state matrices updated at every time step before passing them to the MPC for prediction.

The friction coefficient was changed as a step, assuming a sudden alteration of the road surface. The change was made at the same time when the controller attempted to drive the system into a new equilibrium. The grip suddenly changed from 0.8 to 0.95, which is roughly equivalent to a wet to dry road transition. Overall, the alteration of the road friction could be interpreted as a special type of transition between equilibria. Figure 11 shows the evolution of the states during the test case, while Figure 12 depicts the trajectory traveled. Despite the sudden change in the operating conditions, the proposed adaptive MPC controller was able to precisely stabilize both targeted drift equilibria by utilizing the available information of the road grip.

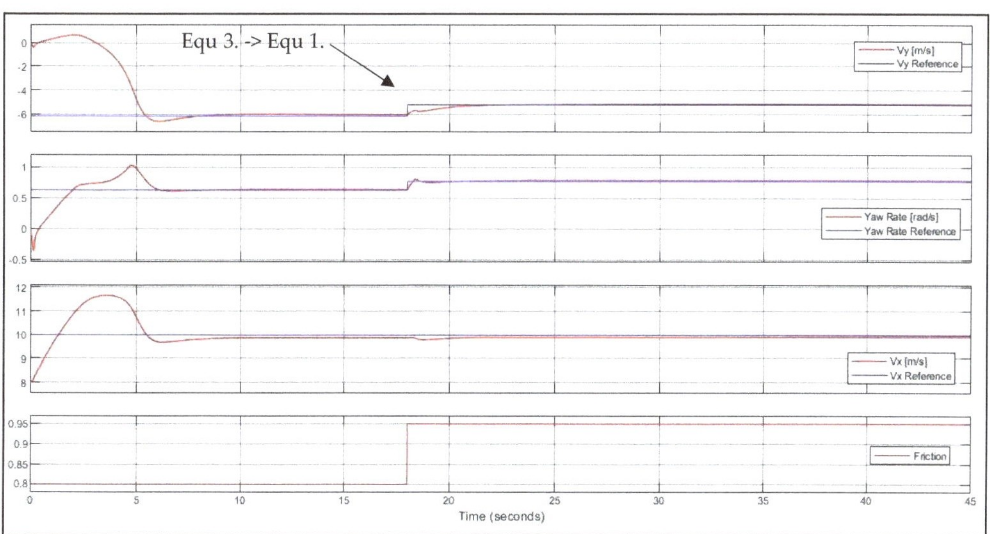

Figure 11. Vehicle states during the test case showing the handling of rapid change of road surface grip.

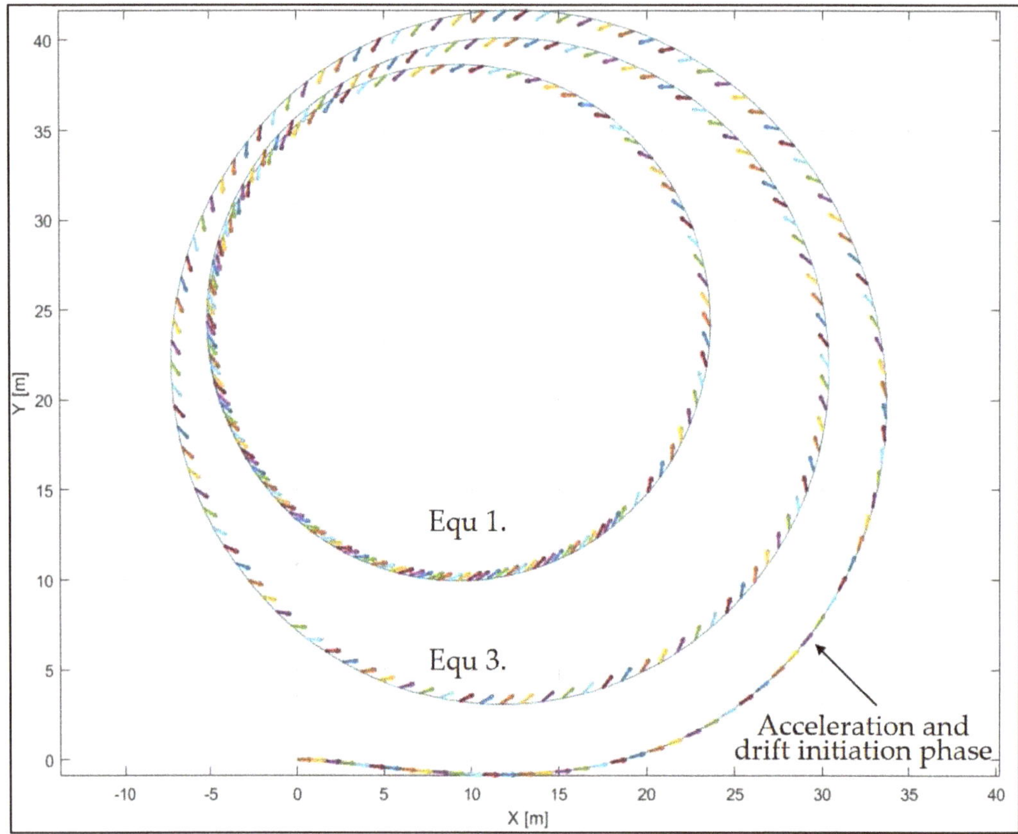

Figure 12. Path and heading of the vehicle in the changing road grip test case.

6. Conclusions

In this article, the authors highlighted the importance of vehicle motion control at handling limits and automated drifting as one of the critical developments of fully autonomous vehicles of the future, aiming at improved road safety. Vehicle automated drifting capabilities were demonstrated during different operating conditions with a systematic controller development process, as outlined below. The handling of changing road conditions and vehicle model nonlinearities is among the key advantages of the proposed adaptive MPC control framework. In Section 2, a nonlinear dynamic vehicle model and a nonlinear tire model were introduced for control-oriented purposes, and the parameter identification was carried out. In Sections 3 and 4, the control scheme and the updating strategy were explained in detail.

In Section 5.1, the weight matrices were determined using a single equilibrium for testing. It was shown that a linear MPC is capable of driving the nonlinear system into the desired drifting equilibrium, keeping it there with the appropriate tuning.

Linear MPC is a simpler control methodology compared to nonlinear MPC, but it performs poorly if the change in vehicle dynamic behavior is far from the design point due to nonlinearities or when it faces parameter variations (road surface changes). According to the proposed idea, these drawbacks can be eliminated by updating the discrete model of the vehicle, reformulating the MPC in an adaptive way, and utilizing measured or estimated model parameter values. This allows using the same MPC structure by updating it instead of using a more different linear MPC and a decision controller for drifting in multiple equilibria as a gain scheduling.

Another advantage of this method in vehicle motion control is directly handling changes in the road surface. A slight or abrupt variation in friction is a common occurrence in real-life driving scenarios. This paper showed that the controller could handle not only various transitions between multiple equilibria but also a simultaneous change in the road grip.

A reasonable next step of the work is to extend the proposed framework with trajectory-planning and -tracking capabilities and possibly online road surface estimation to demonstrate advantages in critical road safety scenarios.

Author Contributions: Conceptualization, Á.B., Sz.Cz., Á.D. and Zs.Sz.; methodology, Á.B., Sz.Cz., Á.D. and Zs.Sz.; software, Á.B., Sz.Cz., Á.D. and Zs.Sz.; validation, Á.B., Sz.Cz., Á.D. and Zs.Sz.; formal analysis, Á.B., Sz.Cz., Á.D. and Zs.Sz.; investigation, Á.B., Sz.Cz., Á.D. and Zs.Sz.; resources, Á.B., Sz.Cz., Á.D. and Zs.Sz.; data curation, Á.B., Sz.Cz., Á.D. and Zs.Sz.; writing—original draft preparation, Á.B., Sz.Cz., Á.D. and Zs.Sz.; writing—review and editing, Á.B., Sz.Cz., Á.D. and Zs.Sz.; visualization, Á.B., Sz.Cz., Á.D. and Zs.Sz.; supervision, Á.B., Sz.Cz., Á.D. and Zs.Sz.; project administration, Á.B., Sz.Cz., Á.D. and Zs.Sz.; funding acquisition, Á.B., Sz.Cz., Á.D. and Zs.Sz. All authors have read and agreed to the published version of the manuscript.

Funding: TKP2020 Institution Excellence Subprogram, Grant No. BME-IE-MIFM, Autonomous Systems National Laboratory Program Ministry of Innovation and Technology NRDI Office.

Institutional Review Board Statement: Not applicable.

Informed Consent Statement: Not applicable.

Data Availability Statement: Not applicable.

Acknowledgments: The research reported in this paper and carried out at the Budapest University of Technology and Economics was supported by the National Research Development and Innovation Fund (TKP2020 Institution Excellence Subprogram, Grant No. BME-IE-MIFM) based on the charter of bolster issued by the National Research Development and Innovation Office under the auspices of the Ministry for Innovation and Technology. In addition, the research was supported by the Ministry of Innovation and Technology NRDI Office within the framework of the Autonomous Systems National Laboratory Program.

Conflicts of Interest: The authors declare no conflict of interest.

Appendix A

Table A1. Nomenclature: list of variables.

Name	Sign	Dimension
mass of the vehicle	m	kg
vehicle moment of inertia around z-axis	I_z	kg·m^2
lateral force	F_y	N
lateral force at front wheel	F_{yF}	N
lateral force at rear wheel	F_{yR}	N
normal load on the front wheels	F_{zF}	N
normal load on the rear wheels	F_{zR}	N
drive force at the rear wheel	F_{xR}	N
drive force at the front wheel	F_{xF}	N
air drag	F_A	N
applied torque at the center of gravity	M_z	Nm
yaw-rate	r	rad/s

Table A1. *Cont.*

Name	Sign	Dimension
longitudinal velocity	V_x	m/s
lateral velocity	V_y	m/s
distance between center of gravity and front axle	a	m
distance between center of gravity and rear axle	b	m
derating factor	ζ	-
cornering stiffness of the front tire	$C_{\alpha F}$	N/rad
cornering stiffness of the rear tire	$C_{\alpha R}$	N/rad
air drag coefficient	C_A	-
frontal cross-section surface of the vehicle	A	m^2
air density	ρ	kg/m^3
front tire sideslip angle	α_F	rad
rear tire sideslip angle	α_R	rad
sideslip boundary angle of the front tire	α_{sl_F}	rad
sideslip boundary angle of the front rear	α_{sl_R}	rad
friction coefficient	μ	-
vehicle sideslip angle at center of gravity	β	rad
steering angle	δ	rad
state vector	$\mathbf{x}$	-
input vector	$\mathbf{u}$	-
continuous time state matrices	A_c, B_c, C_c, D_c	-
discrete time state matrices	A_d, B_d, C_d, D_d	-

References

1. Ágoston, G.; Madleňák, R. Road Safety Macro Assessment Model: Case Study for Hungary. *Period. Polytech. Transp. Eng.* **2021**, *49*, 89–92. [CrossRef]
2. Ghadi, M.Q.; Török, Á. Evaluation of the Impact of Spatial and Environmental Accident Factors on Severity Patterns of Road Segments. *Period. Polytech. Transp. Eng.* **2020**, *49*, 146–155. [CrossRef]
3. Hegedűs, T.; Németh, B.; Gáspár, P. Challenges and Possibilities of Overtaking Strategies for Autonomous Vehicles. *Period. Polytech. Transp. Eng.* **2020**, *48*, 320–326. [CrossRef]
4. Paden, B.; Cap, M.; Yong, S.Z.; Yershov, D.; Frazzoli, E. A Survey of Motion Planning and Control Techniques for Self-Driving Urban Vehicles. *IEEE Trans. Intell. Veh.* **2016**, *1*, 33–55. [CrossRef]
5. Wu, N.; Huang, W.; Song, Z.; Wu, X.; Zhang, Q.; Yao, S. Adaptive dynamic preview control for autonomous vehicle trajectory following with ddp based path planner. In Proceedings of the 2015 IEEE Intelligent Vehicles Symposium (IV), Seoul, Korea, 28 June–1 July 2015; pp. 1012–1017.
6. Bacha, S.; Saadi, R.; Ayad, M.Y.; Aboubou, A.; Bahri, M. A review on vehicle modeling and control technics used for autonomous vehicle path following. In Proceedings of the 2017 International Conference on Green Energy Conversion Systems (GECS), Hammamet, Tunisia, 23–25 March 2017; pp. 1–6.
7. Zhang, B.; Chen, G.; Zong, C. Path tracking of full drive-by-wire electric vehicle based on model prediction control. In Proceedings of the 2018 IEEE Intelligent Vehicles Symposium (IV), Changshu, China, 26–30 June 2018; pp. 868–873.
8. Velenis, E.; Frazzoli, E.; Tsiotras, P. On steady-state cornering equilibria for wheeled vehicles with drift. In Proceedings of the 48h IEEE Conference on Decision and Control (CDC) Held Jointly with 2009 28th Chinese Control Conference, Shanghai, China, 15–18 December 2009; pp. 3545–3550.
9. Velenis, E.; Frazzoli, E.; Tsiotras, P. Steady-state cornering equilibria and stabilisation for a vehicle during extreme operating conditions. *Int. J. Veh. Auton. Syst.* **2010**, *8*, 10. [CrossRef]
10. Hindiyeh, R.Y. Dynamics and Control of Drifting in Automobiles. Ph.D. Thesis, Stanford University, Stanford, CA, USA, 2013.

11. Bardos, A.; Domina, A.; Szalay, Z.; Tihanyi, V.; Palkovics, L. MIMO controller design for stabilizing vehicle drifting. In Proceedings of the IEEE Joint 19th International Symposium on Computational Intelligence and Informatics and 7th IEEE International Conference on Recent Achievements in Mechatronics, Automation, Computer Sciences and Robotics, Szeged, Hungary, 14–16 November 2019.
12. Wachter, E.; Alirezaei, M.; Bruzelius, F.; Schmeitz, A. Path control in limit handling and drifting conditions using State Dependent Riccati Equation technique. *Proc. Inst. Mech. Eng. Part D J. Automob. Eng.* **2019**, *234*, 783–791. [CrossRef]
13. Xu, D.; Wang, G.; Qu, L.; Ge, C. Robust Control with Uncertain Disturbances for Vehicle Drift Motions. *Appl. Sci.* **2021**, *11*, 4917. [CrossRef]
14. Guo, K.; Lu, D. UniTire: Unified tire model for vehicle dynamic simulation. *Veh. Syst. Dyn.* **2007**, *45*, 79–99. [CrossRef]
15. Bai, G.; Meng, Y.; Liu, L.; Luo, W.; Gu, Q.; Li, K. A New Path Tracking Method Based on Multilayer Model Predictive Control. *Appl. Sci.* **2019**, *9*, 2649. [CrossRef]
16. Chen, C.; Guo, J.; Guo, C.; Chen, C.; Zhang, Y.; Wang, J. Adaptive Cruise Control for Cut-In Scenarios Based on Model Predictive Control Algorithm. *Appl. Sci.* **2021**, *11*, 5293. [CrossRef]
17. Zhang, X.; Göhlich, D. Integrated Traction Control Strategy for Distributed Drive Electric Vehicles with Improvement of Economy and Longitudinal Driving Stability. *Energies* **2017**, *10*, 126. [CrossRef]
18. Zhai, L.; Hou, R.; Sun, T.; Kavuma, S. Continuous Steering Stability Control Based on an Energy-Saving Torque Distribution Algorithm for a Four in-Wheel-Motor Independent-Drive Electric Vehicle. *Energies* **2018**, *11*, 350. [CrossRef]
19. Park, J.; Jang, I.G.; Hwang, S.-H. Torque Distribution Algorithm for an Independently Driven Electric Vehicle Using a Fuzzy Control Method: Driving Stability and Efficiency. *Energies* **2018**, *11*, 3479. [CrossRef]
20. Joa, E.; Cha, H.; Hyun, Y.; Koh, Y.; Yi, K.; Park, J. A new control approach for automated drifting in consideration of the driving characteristics of an expert human driver. *Control. Eng. Pr.* **2020**, *96*, 104293. [CrossRef]
21. Reza, N.J. *Advanced Vehicle Dynamics*; Springer: Cham, Switzerland, 2019; pp. 143–144, ISBN 978-3-030-13060-2.
22. Lugaro, C.; Alirezaei, M.; Konstantinou, I.; Behera, A. *A Study on the Effect of Tire Temperature and Rolling Speed on the Vehicle Handling Response*; SAE Technical Paper 2020-01-1235; SAE International: Warrendale, PA, USA, 14 April 2020.
23. Helnwein, P.; Liu, C.; Meschke, G.; Mang, H. A new 3-d finite element model for cord-reinforced rubber composites application to analysis of automobile tires. *Finite Elem. Anal. Des.* **1993**, *14*, 1–16. [CrossRef]
24. Pacejka, H.B.; Bakker, E. The magic formula tyre model. *Veh. Syst. Dyn.* **1992**, *21* (Suppl. S1), 1–18. [CrossRef]
25. Fiala, E. Lateral forces on rolling pneumatic tires. *Z. VDI* **1954**, *96*, 973–979.
26. Hadekel, R. The mechanical characteristics of pneumatic tyres. In *Sand T Memo TPA3/TIB*; National Association of Housing & Redeve: Washington, DC, USA, 1952.
27. Szalay, Z.; Hamar, Z.; Nyerges, Á. Novel design concept for an automotive proving ground supporting multilevel CAV development. *Int. J. Veh. Des.* **2019**, *80*, 1–22. [CrossRef]
28. Bardos, A.; Domina, A.; Tihanyi, V.; Szalay, Z.; Palkovics, L. Implementation and experimental evaluation of a MIMO drifting controller on a test vehicle. In Proceedings of the 2020 IEEE Intelligent Vehicles Symposium (IV), Las Vegas, NV, USA, 19 October–13 November 2020; pp. 1472–1478.
29. Varga, B.; Tettamanti, T.; Kulcsár, B. Optimally combined headway and timetable reliable public transport system. *Transp. Res. Part C Emerg. Technol.* **2018**, *92*, 1–26. [CrossRef]
30. Zhakatayev, A.; Rakhim, B.; Adiyatov, O.; Baimyshev, A.; Varol, H.A. Successive linearization based model predictive control of variable stiffness actuated robots. In Proceedings of the 2017 IEEE International Conference on Advanced Intelligent Mechatronics (AIM), Munich, Germany, 3–7 July 2017; pp. 1774–1779.
31. Grüne, L.; Pannek, J. Nonlinear Model Predictive Control. In *Nonlinear Model Predictive Control: Theory and Algorithms*; Springer: Cham, Switzerland, 2017; pp. 45–69.
32. Roubal, J.; Husek, P.; Stecha, J. Linearization: Students Forget the Operating Point. *IEEE Trans. Educ.* **2009**, *53*, 413–418. [CrossRef]
33. Panahandeh, G.; Ek, E.; Mohammadiha, N. Road friction estimation for connected vehicles using supervised machine learning. In Proceedings of the 2017 IEEE Intelligent Vehicles Symposium (IV), Los Angeles, CA, USA, 11–14 June 2017; pp. 1262–1267.
34. Roychowdhury, S.; Zhao, M.; Wallin, A.; Ohlsson, N.; Jonasson, M. Machine Learning Models for Road Surface and Friction Estimation using Front-Camera Images. In Proceedings of the 2018 International Joint Conference on Neural Networks (IJCNN), Rio de Janeiro, Brazil, 8–13 July 2018; pp. 1–8.
35. Fényes, D.; Németh, B.; Gáspár, P.; Szabó, Z. Road surface estimation based LPV control design for autonomous vehicles. *IFAC-PapersOnLine* **2019**, *52*, 120–125. [CrossRef]

Article

Evaluation of Non-Classical Decision-Making Methods in Self Driving Cars: Pedestrian Detection Testing on Cluster of Images with Different Luminance Conditions

Mohammad Junaid [†], Zsolt Szalay and Árpád Török *

Faculty of Transportation Engineering and Vehicle Engineering, Budapest University of Technology and Economics, Sztoczek Str. 6, J. Building, V. Floor, 1111 Budapest, Hungary; junaidalidocs1607@gmail.com (M.J.); szalay.zsolt@kjk.bme.hu (Z.S.)
* Correspondence: torok.arpad@kjk.bme.hu
† Current affiliation: Bosch Magyarország, 1103 Budapest, Hungary. Research Performed while at the Budapest University of Technology and Economics.

Abstract: Self-driving cars, i.e., fully automated cars, will spread in the upcoming two decades, according to the representatives of automotive industries; owing to technological breakthroughs in the fourth industrial revolution, as the introduction of deep learning has completely changed the concept of automation. There is considerable research being conducted regarding object detection systems, for instance, lane, pedestrian, or signal detection. This paper specifically focuses on pedestrian detection while the car is moving on the road, where speed and environmental conditions affect visibility. To explore the environmental conditions, a pedestrian custom dataset based on Common Object in Context (COCO) is used. The images are manipulated with the inverse gamma correction method, in which pixel values are changed to make a sequence of bright and dark images. The gamma correction method is directly related to luminance intensity. This paper presents a flexible, simple detection system called Mask R-CNN, which works on top of the Faster R-CNN (Region Based Convolutional Neural Network) model. Mask R-CNN uses one extra feature instance segmentation in addition to two available features in the Faster R-CNN, called object recognition. The performance of the Mask R-CNN models is checked by using different Convolutional Neural Network (CNN) models as a backbone. This approach might help future work, especially when dealing with different lighting conditions.

Keywords: Mask R-CNN; transfer learning; inverse gamma correction; illumination; instance segmentation; pedestrian custom dataset

Citation: Junaid, M.; Szalay, Z.; Török, Á. Evaluation of Non-Classical Decision-Making Methods in Self Driving Cars: Pedestrian Detection Testing on Cluster of Images with Different Luminance Conditions. *Energies* **2021**, *14*, 7172. https://doi.org/10.3390/en14217172

Academic Editor: Wiseman Yair

Received: 20 September 2021
Accepted: 22 October 2021
Published: 1 November 2021

Publisher's Note: MDPI stays neutral with regard to jurisdictional claims in published maps and institutional affiliations.

1. Introduction

Previous studies presented that energy minimization is a critical area of autonomous transport system development, where advanced longitudinal and lateral vehicle control methods will play a key role in achieving expected results [1–7]. Conversely, numerous research papers propose to improve the efficiency of the vehicle control process through the development of sensor systems and image detection methods [8–11]. Based on this, we understand that image detection approaches can directly affect the efficiency of highly automated transport systems. In light of this, our paper discusses the comparison of different models influencing the efficiency of image detection processes.

The recent trends in self-driving cars have encouraged researchers to use several object detection algorithms that include various areas in self-driving cars, such as pedestrian detection (see Figure 1) [12–15], lane detection, traffic signal detection [16], and many more. Due to the recent development in CNN and its outstanding performance in these state-of-the-art visual recognition solutions, these processes have become increasingly intensive. CNN is basically used for image classifying tasks, but it cannot detect objects.

For this purpose, many object detection algorithms are available, for instance, SSD (Single Shot Multibox Detector) [14–17] YOLO (You Only Look Once) [18–21], R-CNN [10], and Fast R-CNN [22,23]. All object detection models localize objects by using bounding boxes and classifying them by labels.

This paper focuses on pedestrian detection, especially in different visibility conditions. The developed dataset is manipulated with the inverse gamma correction method to create images representing the different lighting conditions. After this development phase, the Mask R-CNN model [24–28] is trained with this dataset by transfer learning and fine-tuning techniques [29].

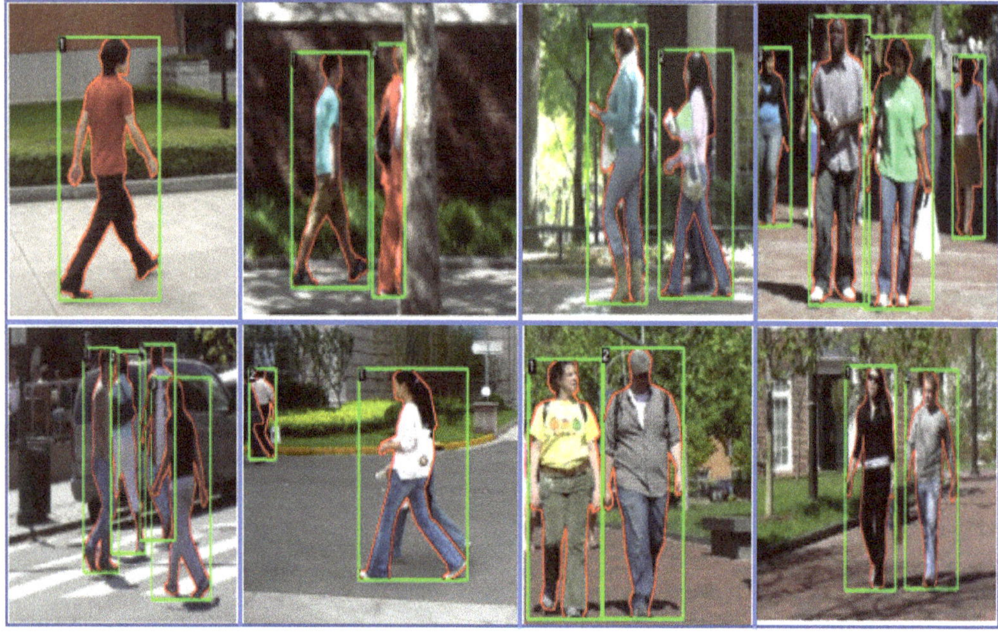

Figure 1. Pedestrian detection using Mask R-CNN with Resnet50 as a backbone, which runs at epoch10. Developed from [30].

There are relevant difficulties related to pedestrian detection [31] from an automated driving point of view, especially when we consider the experiences of highly automated vehicles' accidents. For instance, one of the most famous accidents related to highly automated driving is the well-known Arizona-Uber accident [32,33], where the failure of the detection was seriously affected by lighting conditions. Visibility is the most important factor, such as darkness, brightness, and glaring.

2. Literature Review

This paper gives a review of R-CNN models and their variations. The localization process starts with the coarse scan of the whole image and concentrates on the region of interest, where the sliding window method is used to predict the bounding boxes.

Ross Girshick proposed the R-CNN model in 2014 [23]. He developed a selective search method to create 2000 regions for each image called region proposal. It makes the quality of the bounding box better and helps the CNN model extract high-level features. Thus, R-CNN models take the image as the input, and then a 2k region is proposed by a selective search method. After this, it is cropped to a fixed size, called a warped region. Finally, with the CNN model's help, objects are localized and classified within the region of interest. The CNN model uses the Linear Support Vector Machine (SVM) method [33] to classify the classes of objects as well as non-max suppression methods [16] to suppress the bounding boxes that have a value of less than the critical value.

In other words, R-CNN consists of four processes. First, regions are proposed in the image with the selective method, and then it is warped in a fixed size. After that, the warped region is fed in the CNN model with a fixed size of 227×227 pixels to classify and predict bounding boxes. It extracts 4096 feature vectors from each region proposal. The image contains objects with different sizes and aspect ratios; thus, the region proposal feature comes in different sizes. Before feeding into the CNN model, it is cropped and warped.

The R-CNN method has reasonable limitations in terms of training time, since it takes a huge amount of time to classify the 2 k region proposals in each image. At the same time, it must be mentioned that the selective search algorithm is not a self-learning approach. Thus, to solve this problem, Girshick proposed the fast R-CNN model.

The Fast R-CNN model is nine times faster than the R-CNN model [22,23], in which the VGG16 (Visual Geometry Groups 16) [23] approach is used as the backbone. The architecture is the same as the previous model. However, the input image is fed first into the CNN, and then region proposals are applied to the proposed region. After that, region features are warped with the help of the RoI pooling layer. Then, it is reshaped in fixed size to feed into fully connected layers. Similar to the R-CNN, the 2k region is proposed to CNN every time, but in fast R-CNN, it is fed at once.

A high computational complexity can characterize R-CNN and Fast R-CNN models because both use selective search methods to propose the region. Thus, Shaoqing Ren and his team [22,34] created the idea of a Region Proposal Network (RPN) that replaces the selective search region proposal method. In faster R-CNN, the image is fed into the CNN model first to provide a feature map. A separate Region Proposal Network is then used to predict region proposal, which is further reshaped by using RoI pooling. At last, it is classified and labeled in the Region of Interest.

3. Mask R-CNN

The Mask R-CNN concept [24,25,27,28] is the extended version of the fast R-CNN model. It is used to predict a mask that works parallel to the existing branch of classification and bounding box detection in each region of interest. Because of its simplicity, flexibility, and robustness, Kaiming He and his team won the COCO challenge in 2016. This detection system uses one extra feature called RoI Align, which removes the harsh quantization in RoI Pool.

Mask R-CNN has a similar structure to Fast R-CNN. One additional feature is added, called segmentation masks, that work parallel to each region of interest (RoI) to predict the mask, pixel by pixel. Thus, Mask R-CNN gives one extra output, namely masking, including two existing output: class labelling and bounding box. Mask is quite different from the output mentioned above because it extracts the feature pixel by pixel alignment. Thus, it places a colourful layer (mask) on the object, which is the same in size as the object. At the same time, the bound box has a different aspect ratio that predicts the object through the rectangular box, which is always bigger in size than the instances available in the images.

The Mask R-CNN model is a two-stage detection model. The first stage is designed to provide a proposal for the availability of the object with the help of the Region proposal Network (RPN) [22,35], which is similar to what is used in Fast R-CNN. In the second stage, masking is applied in parallel with the class and bounding box, and it gives a binary mask as an output for each RoI, as shown in Figure 2.

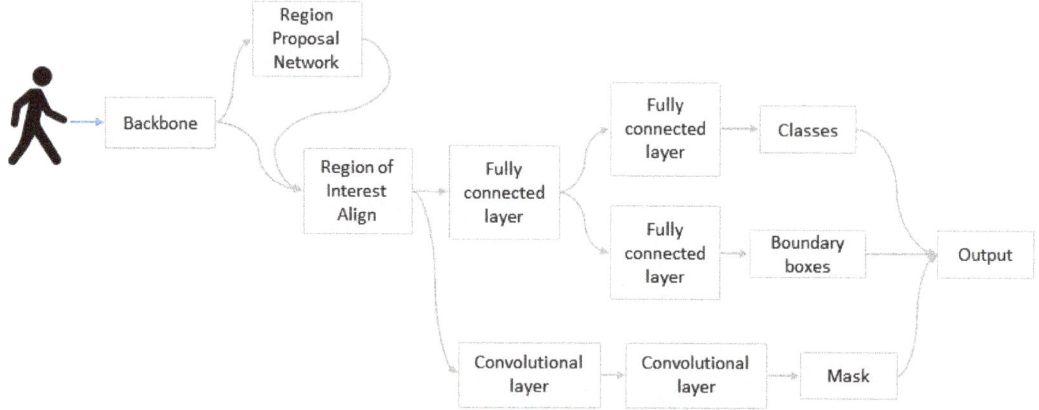

Figure 2. Architecture diagram of Mask R-CNN detection model.

Figure 2 shows that the input image is fed into the convolution neural network to extract the object features. Mask R-CNN uses a new feature called Region of Interest (Roi) Align [32]. This new feature removes the harsh quantization of the RoI Pool. Then, further convolution layers are used to predict instance segmentation, which works in parallel with the classification and localization of objects in each region of interest.

4. Methodology

This section presents the applied methodological approaches related to improving the efficiency of neural network-based detection models. We first describe the concepts of transfer learning and fine-tuning, as these methods are fundamental for improving the efficiency of an existing detection network. In light of the above, by comparing the backbone network types described below, we have the opportunity to determine the network structure that best supports our goals.

4.1. Transfer Learning and Fine Tuning

In the case of transfer learning [36–39], the pre-trained models are applied in the solution of various problems by manipulating relevant layers of the network according to the new application's requirements. In this methodology, some layers are placed in freeze conditions. Fine-tuning is different from transfer learning, where all the layers are used and trained again according to the new application requirements.

This paper uses both techniques to detect the object using Mask R-CNN, where transfer learning techniques replace the backbone. Two classes replace the output of the Mask R-CNN, because the dataset contains two classes, background and masking (foreground), and it is trained again with the help of fine tuning [33,36,40].

4.2. Backbones

As we explained above, the backbone [41] of the Mask R-CNN is a convolution neural network. We tested six different backbone models in the feature extracting and the bounding box identification process. Each Mask R-CNN model with different backbones is trained in different lighting conditions. Since it was not possible to generate the applied dataset during the research, the training and the test procedure was based on a previously developed image dataset (such as images with day or night conditions). In accordance with this, we used a limited number of images from an external database, and we applied the inverse gamma correction method to transform the images into the required lighting conditions. Inverse Gamma Correction Method (IGCM) changes the pixel values to make the picture bright or dark. Each convolutional neural network takes an image with 559×536 pixels as an input and provides a 256-channel output connected to the region proposal network. Accordingly, RPN takes 256 channels as input. Thus, all backbone models are modified according to the input channel of the RPN. In this case, transfer learning and fine-tuning methods are used. Accordingly, we briefly describe the different feature extracting models below.

4.2.1. Alex Net

Alex Net was developed by Alex Krizhevsky and his team in 2012 [42–46]. That year, they won the ImageNet Challenge in visual object recognition. In their approach, recognition refers to the prediction of the bounding boxes and the labelling process of the identified objects in the image. It contains five convolution layers and three fully connected layers to extract the features. In the present paper, we made modifications; for instance, we removed all the fully connected layers. After that, we changed the fifth convolution layer's output, whose channels are equal to the RPN convolution layer's input.

4.2.2. Mobile Net V2

Mobile Net V2 is the extended version of the Mobile Net V1 method [14], which uses an extra layer called a 1×1 expansion layer in each block as compared to Mobile Net V1. Mobile Net V2 [14,17,33,47] replaces the large convolution layer with a depth-wise separable convolution block, and each block contains a 3×3 depth wise kernel to filter the output. Further, it is followed by a 1×1 point wise convolution layer. Thus, it combines the filters and gives new features. Overall, Mobile Net V1 uses 13 depth-wise separable convolution blocks, preceded by a 3×3 regular convolution layer.

At the same time, Mobile Net V2 uses a 1×1 expansion layer in each block in addition to the depth wise and pointwise convolution layer. The pointwise convolution layer is also known as the projection layer because it connects a high number of channels with a low number of channels. Furthermore, the 1×1 expansion layer expands the channel number before going into the depth-wise convolution layer. This model uses new features called residual connection that help in following the gradient through the neural network. Each block contains batch normalization and ReLU6 activation function, but the projection layer does not use the activation function as an output. This model contains 17 residual blocks, and each block contains depth-wise, pointwise, and 1×1 expansion layers. The depth-wise convolution layer is followed by Batch Normalization and Relu6 activation function.

4.2.3. ResNet50

The ResNet model [38,45–47] is based on learning the residual instead of learning the low- or high-level features, i.e., residual network. It is used to go deeper and solve complex problems. Thus, ResNet 50 uses 50 residual blocks [48–50].

4.2.4. VGG11

Karen Simonyan and Andrew Zisserman introduced this model in 2014 [51]. Their team secured first and second place in the localization and classification problems. This model has eight convolution layers and three fully connected layers. However, in our case, we used only the first four layers of this network.

4.2.5. VGG13

One year later, Simonyan and Zisserman, in 2015 [51], investigated the effect of increasing the layers' depth. This model contains eleven convolution layers and three fully connected layers, where a 3×3 kernel is applied on each convolution layer with a stride 1×1 followed by a max pool layer after every two convolution layers.

4.2.6. VGG16

This network [52–54] consists of thirteen convolution layers and three fully connected layers, where 3×3 filters are used in each convolution layer with a stride size of 1×1 and the same padding. Thus, the first two convolution layers contain 64 3×3 kernels.

The input image fed into the first layer has a size of $224 \times 224 \times 64$. It passes through the second layer, and then max pooling is applied to make the channel double. Thus, the third and fourth layers contain 128 3×3 kernels.

Again, the max pool layer is attached to make the channel double. This process is repeated through thirteen layers. The following layers are fully connected that contain 4096 units. These are followed by SoftMax to different 1000 classes. However, we must mention that our investigation considers only convolution layers. It removes the fully connected layers.

As mentioned above, for the three different VGG models, the model's accuracy increases with the depth of the model. The error rate of these three VGG models is introduced in Table 1 below.

Table 1. Error rate of variants of VGG models taken from the paper of VGG 11, 13, and 16 models.

VGG Variant	VGG11	VGG13	VGG16
Error Rate	10.4%	9.9%	9.4%

5. Dataset

This paper uses the Penn-Fudan Database for pedestrian detection as well as segmentation (see Figure 3), which is available on the website (https://www.kaggle.com/jiweiliu/pennfudanpe, accessed on 1 February 2021). It contains 170 images with 345 pedestrian objects, and it is compatible with both COCO [55–57] and Pascal VOC format [54]. We used the dataset during our research in COCO format.

Figure 3. Pedestrian dataset. Developed from [30].

The database consists of three subfiles, namely Annotation, PedMasks, and PNGImages, where annotation files are in text format, and both PNGImages & PedMasks are in png format. Before applying a Mask R-CNN model, the dataset is pre-processed. Each image is normalized and resized to equal sizes, as shown in Tables 2 and 3 below, where the normalization process transforms the pixel value of the images into the range of 0 to 1.

Table 2. The data that is shown in table (**a**) and (**b**) are used to modify the images before importing into the models. (**a**) Normalization of the dataset before importing into the model; (**b**) resizing of all the images in the dataset. Developed from [58].

(a)		
	Mean	Standard Deviation
Normalize	0.485	0.229
	0.456	0.224
	0.406	0.225

(b)		
	Minimum Size	Maximum Size
Resize	800	1333

Table 3. Here, Mask R-CNN with different backbones is trained with pedestrian dataset at epoch 10. In a Mask R-CNN, one extra loss called mask loss is added in addition to the losses in faster R-CNN model, where λ_c = loss of classifier, λ_b = loss of box regression, λ_m = loss of mask, λ_0 = objectiveness loss, λ_r = loss of RPN box, and λ_T = overall loss (the minimum values of the columns are denoted by *).

Backbone	λ_c	λ_b	λ_m	λ_0	λ_r	λ_T
Alex Net	0.0569	0.1345	0.3612	0.1672	1.8658	2.5856
Mobile Net V2	0.0603	0.1248	0.4285	0.17	1.5136	2.2972
ResNet50	0.0199 *	0.0279 *	0.1115 *	0.0002 *	0.0022 *	0.1617 *
VGG11	0.2872	0.4664	0.2734	0.2229	2.5641	3.814
VGG13	0.3089	0.4694	0.2839	0.2462	2.6469	3.9553
VGG16	0.4191	0.6803	0.3671	0.2607	2.8196	4.5468

The table below (Table 3) introduces the results, where the overall loss λ_T [24] indicates the sum of all losses.

$$\lambda_T = \lambda_c + \lambda_b + \lambda_m + \lambda_0 + \lambda_r \tag{1}$$

Equation (1): Total loss (λ_T) is equal to the sum of all losses.

6. Inverse Gamma Correction

The modification of the luminance characteristics can cause reduced visibility of an object and decrease the detection capability of the system [59]. However, the effect of the lighting conditions depends on many other factors, such as the distance of the given object. Beyond this, the lighting contrast between the object and background can also significantly influence detection efficiency. Accordingly, the system can capture sometimes darker or sometimes brighter images depending on the related factors.

Many different algorithms can be used to adjust the contrast and increase or decrease the brightness of the image. For instance, Histogram Equalization (HE) [60] or Bi-Histogram Equalization (BBHE) [61] can be applied to modify the lighting-related characteristics of the investigated images.

This paper uses the inverse gamma correction method to modify the brightness and darkness of the images. Thus, inverse gamma correction transforms the lighting characteristics of the input signal by applying a nonlinear power function. The power coefficient (gamma) represents the nonlinear nature of the human perception process related to the lighting conditions. Accordingly, the inverse gamma correction transformation is given by Equation (1) below.

$$I_0 = I_1^{1/\gamma} \tag{2}$$

Equation (2): Equation of Gamma Inverse Method, where I_0 is the output intensity and I_1 is the input intensity.

The value of I_0 is between 0 and 1, following the introduced model, and I_1 is the transformed intensity. This formula is applied when gamma's value is known, and it is commonly determined experimentally.

In accordance with the blind inverse gamma correction techniques [61–63], gamma is varied between 0.1 and 1.5 with a step size of 0.1, as shown in Figure 4 below. Following this, the gamma value of this image is one. The brightness of the image increases as the gamma value becomes larger, and the image becomes darker as the gamma value decreases.

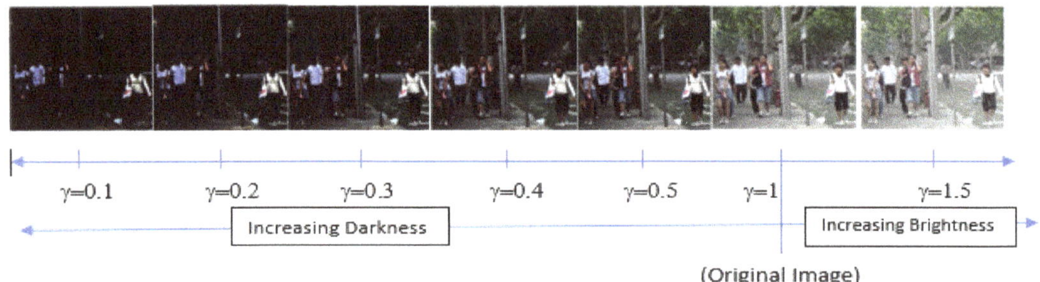

Figure 4. The brightness of the image increases when increasing the gamma value (γ), and it decreases when decreasing the gamma value. Here Gamma Inverse Method is applied to change the luminance intensity of the image. Developed from [30].

7. Instance Segmentation

The instance segmentation [35,58] process involves two main steps. First, it detects and indicates the object by bounding boxes within defined categories, and in the second step, segmentation prediction is performed pixel-wise. Instance segmentation (see Figure 5) is different from semantic segmentation since, beyond the object detection phase, instance segmentation labels the objects, according to the investigated categories' sub-classes. In contrast, semantic segmentation performs the detection and then classifies the objects. We used the method of instance segmentation with Mask R-CNN in our research. This paper uses instance segmentation with Mask R-CNN.

Figure 5. Instance segmentation of the images where GIMP tool is used to segment the images.

8. Results

The gamma value of the used dataset is assumed to be 1 and is in accordance with the observed good, day-light conditions of the included images. The dataset was augmented by the Torch vision 0.3 package's inbuilt processing methods of the PyTorch Framework.

First, the dataset was converted into a tensor since PyTorch accepts this structure during the pre-processing phase. In the next step, the dataset was loaded in the framework with a batch size number 2. After this, the Mask R-CNN model is applied, which is an inbuilt module of the Torch Vision packages. The Mask R-CNN model works on top of the faster R-CNN detection model. It uses one extra feature called mask prediction that is applied parallel to the object recognition system in each region of interest.

Here, the mask R-CNN model's backbone is changed with different CNN pre-trained models through the transfer learning technique. In the figure below (see Figure 6), it can be observed that ResNet50 has the lowest loss as compared to other models, whereas VGG16 has the highest loss. In this Mask R-CNN model, anchor boxes are used with a size (32, 64, 128, 256, 512) where region proposal network generates three different aspect ratios, namely 0.5, 1.0, and 2.0. Apart from this, the number of epochs was 10 during the training, and it is optimized with the Stochastic Gradient Descent Method. Parameter values related to the learning method, the momentum, and the weight decay were 0.005, 0.9, and 0.0005, respectively.

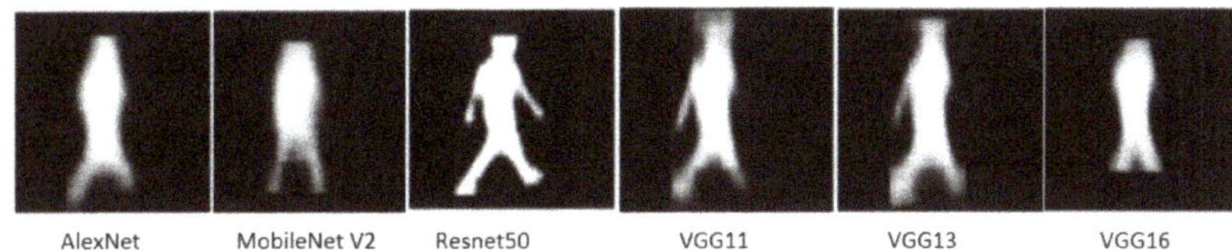

Figure 6. Mask detection of the image FudanPed00035.png with different Mask R-CNN models, where the gamma value (γ) is one.

The Mask R-CNN model detects objects by predicting bounding boxes, which can result in uncertainties due to the process of segmentation prediction, where the images are decomposed to pixels, the number of which is proportional to the size of the instance.

In the figure above, AP parameter indicates the average precision [11,20,26], and AR represents the average recall for both bounding boxes and segmentation. Accordingly, average precision defines how accurate the prediction is. On the contrary, average recall defines how well identified the proper classes are. The table below (see Table 4) shows that Mask R-CNN with the ResNet 50 backbone has the highest AP and AR value compared to the other models because it uses the residual network with a deeper layer. In other words, it contains 17 residual blocks.

Table 4. Values of Average Precision and Average Recall for both bounding boxes and segmentation tabulated while using different backbones (the maximum values of the columns are denoted with *).

Backbone	AP (Bbox)	AR (Bbox)	AP (Segm)	AR (Segm)
Alex Net	0.213	0.409	0.173	0.357
Mobile Net V2	0.175	0.38	0.105	0.23
Resnet50	0.844 *	0.883 *	0.774 *	0.813 *
VGG11	0.233	0.413	0.298	0.427
VGG13	0.237	0.42	0.2878	0.416
VGG16	0.148	0.359	0.188	0.339

At the same time, VGG16 takes the longest time to train the model. Moreover, ResNet needs 0.003 s to evaluate it (see Table 5).

Table 5. Training time and evaluating times are calculated for different backbones (the minimum values of the columns are denoted with *).

Backbone	Model Time	Evaluator Time
Alex Net	0.0452	0.0165
Mobile Net V2	0.0606	0.0111
ResNet50	0.075	0.003 *
VGG11	0.046 *	0.0094
VGG13	0.0699	0.0069
VGG16	0.0827	0.0115

In general, Mask R-CNN with ResNet backbone performed well in all aspects, including AP and AR indicators (see Tables 6–11). Accordingly, we can conclude that ResNet based Mask R-CNN model is robust and flexible.

9. Evaluation

In this section, we introduce the evaluation of the investigated networks by processing 10 images with different gamma values. Images are indicated in the tables below with numbers from 0 to 9. We use score values to compare the different networks to indicate the probability of the proper classification. Their score value is shown in the contingency tables. Besides this, the bottom row contains the average score value related to the different images.

Mask R-CNN with AlexNet Backbone.

Table 6. Top score of the 10 images represented by numbers 0 to 9 with different gamma values (γ) while using Mask R-CNN with backbone Alex Net.

		Top Score Value						
	γ	0.1	0.2	0.3	0.4	0.5	1	1.5
	0	0.1092	0.0587	0	0.0648	0.1715	0.5374	0.1118
	1	0.0854	0	0	0.0685	0.158	0.5369	0.069
	2	0.1225	0.0594	0.0603	0.0698	0.1518	0.5276	0.0852
	3	0.1541	0.0917	0	0.0731	0.1479	0.5605	0.1314
Test	4	0.1415	0.0585	0.0639	0.0824	0.1341	0.5486	0.0953
Images	5	0.1737	0.121	0	0.0913	0.1385	0.5462	0.0887
	6	0.1359	0.08	0.0624	0.0841	0.1049	0.5342	0.1157
	7	0.1526	0.0743	0	0.0525	0.1144	0.5403	0.1174
	8	0.1327	0.0533	0.0509	0.0872	0.1341	0.5252	0.0732
	9	0.1198	0	0	0.0851	0.178	0.5135	0.1021
	Average	0.13274	0.05969	0.02375	0.07588	0.14332	0.53704	0.09898

Mask R-CNN with MobileNet V2 Backbone.

Table 7. Top score of the 10 images with different gamma values (γ) while using Mask R-CNN with backbone Mobile Net V2.

		Top Score Value						
	γ	0.1	0.2	0.3	0.4	0.5	1	1.5
	0	0.1037	0.0935	0.0874	0.0826	0.122	0.4899	0.1032
	1	0.1021	0	0	0	0.0781	0.6261	0.0756
	2	0.109	0.0975	0.0661	0.0771	0.1038	0.6609	0.1351
	3	0.114	0.0896	0.0813	0.077	0.1073	0.6643	0.1171
Test	4	0.1365	0.112	0.0945	0.0606	0.1363	0.6215	0.0946
Images	5	0.1389	0.1277	0.059	0.0858	0.1089	0.6422	0.0807
	6	0.0878	0.0907	0.0746	0.0616	0.0927	0.6987	0.0856
	7	0.0985	0.0954	0.0774	0.0676	0.1007	0.6773	0.1011
	8	0.1076	0.1087	0.0947	0.0699	0.127	0.6693	0.1194
	9	0.1045	0.0549	0.0725	0.071	0.0851	0.5985	0.0569
	Average	0.11026	0.087	0.07075	0.06532	0.10619	0.63487	0.09693

Mask R-CNN with ResNet50 Backbone.

Table 8. Top score of the 10 images with different gamma values (γ) while using Mask R-CNN with backbone ResNet 50.

		Top Score Value						
	γ	0.1	0.2	0.3	0.4	0.5	1	1.5
	0	0.677	0.8755	0.8625	0.8822	0.8686	0.9988	0.88
	1	0.6161	0.9058	0.9036	0.9032	0.9071	0.9976	0.92
	2	0.6315	0.8889	0.88	0.8846	0.8716	0.9958	0.895
	3	0.6486	0.8687	0.8721	0.8778	0.8729	0.9966	0.8862
Test	4	0.613	0.8566	0.8305	0.8427	0.8383	0.9967	0.8588
Images	5	0.6082	0.8319	0.8023	0.7977	0.7933	0.9904	0.8418
	6	0.6438	0.8465	0.8541	0.8556	0.856	0.9972	0.8872
	7	0.6061	0.8737	0.8579	0.8532	0.8507	0.9982	0.8841
	8	0.5955	0.8308	0.8229	0.8121	0.8116	0.998	0.8189
	9	0.5269	0.8534	0.8821	0.8685	0.866	0.9985	0.8851
	Average	0.61667	0.86318	0.8568	0.85776	0.85361	0.99678	0.87571

Mask R-CNN with VGG11 Backbone.

Table 9. Top score of the 10 images with different gamma values (γ) while using Mask R-CNN with backbone VGG11.

		Top Score Value						
	γ	0.1	0.2	0.3	0.4	0.5	1	1.5
	0	0.1889	0.1629	0.1914	0.1337	0.1733	0.7792	0.175
	1	0.0851	0.0993	0.086	0.1407	0.1965	0.8253	0.1482
	2	0.2229	0.1983	0.2062	0.133	0.2092	0.7748	0.1773
	3	0.2128	0.2215	0.2034	0.1219	0.2051	0.7706	0.1749
Test	4	0.1816	0.1708	0.1674	0.1421	0.2152	0.75	0.1775
Images	5	0.2326	0.231	0.2264	0.1227	0.1889	0.7653	0.1853
	6	0.1875	0.2072	0.1849	0.1176	0.1843	0.819	0.1767
	7	0.2115	0.1955	0.1959	0.139	0.1741	0.7952	0.1643
	8	0.2199	0.1863	0.2001	0.1452	0.2207	0.7579	0.1791
	9	0.1191	0.1097	0.1057	0.1298	0.1668	0.5576	0.2018
	Average	0.18619	0.17825	0.17674	0.13257	0.19341	0.75949	0.17601

Mask R-CNN with VGG13 Backbone.

Table 10. Top score of the 10 images with different gamma values (γ) while using Mask R-CNN with backbone VGG13.

		Top Score Value						
	γ	0.1	0.2	0.3	0.4	0.5	1	1.5
	0	0.0706	0.1952	0.1302	0.165	0.1275	0.7111	0.1958
	1	0.0685	0.08	0.1694	0.1439	0.1375	0.7227	0.1742
	2	0.0671	0.1873	0.1622	0.1585	0.1297	0.7334	0.2018
	3	0.064	0.1768	0.14	0.1602	0.1176	0.7446	0.2033
Test	4	0.0703	0.1739	0.1605	0.1666	0.1199	0.6729	0.2156
Images	5	0.0854	0.1767	0.1481	0.1617	0.1222	0.7071	0.193
	6	0.1682	0.1718	0.1436	0.1728	0.1383	0.7209	0.2092
	7	0.0644	0.1727	0.1269	0.1692	0.1232	0.7559	0.2012
	8	0.0714	0.18	0.1202	0.1314	0.1218	0.7607	0.2041
	9	0.0676	0.0995	0.1205	0.1205	0.1181	0.688	0.2397
	Average	0.07975	0.16139	0.14216	0.15498	0.12558	0.72173	0.20379

Mask R-CNN with VGG16 Backbone.

As shown in the heat map below, Mask R-CNN with Resnet 50 had the best performance in all scenarios. If the gamma was 1, the average score value was 99.68%.

As the intensity of the image changes from dark to bright section, the scores of the ResNet increases until gamma 1. Generally, ResNet50 based Mask R-CNN model performs well in all scenarios. Even if the images become brighter, the score of the ResNet 50 decreases much slower than the other models (see Table 12).

Table 11. Top score of the 10 images with different gamma values (γ) while using Mask R-CNN with backbone VGG16.

		Top Score Value						
	γ	0.1	0.2	0.3	0.4	0.5	1	1.5
	0	0.0962	0.0939	0.0966	0.1385	0.0848	0.5924	0.1109
	1	0.0922	0.0702	0.0851	0.1592	0.114	0.588	0.109
	2	0.0976	0.0807	0.0973	0.1341	0.0857	0.5485	0.1101
Test	3	0.0923	0.0792	0.0923	0.1414	0.0779	0.5059	0.1163
Images	4	0.0936	0.0792	0.0958	0.16	0.1059	0.5174	0.114
	5	0.0986	0.1023	0.0995	0.1493	0.1453	0.5552	0.1186
	6	0.0891	0.1008	0.0953	0.1225	0.0978	0.5839	0.1133
	7	0.0988	0.0841	0.0939	0.145	0.0757	0.5848	0.1193
	8	0.0922	0.08	0.0998	0.1355	0.0792	0.5436	0.1232
	9	0.0903	0.0732	0.0885	0.1583	0.0941	0.5375	0.1146
	Average	0.09409	0.08436	0.09441	0.14438	0.09604	0.55572	0.11493

Table 12. Heat map of the Mask R-CNN models with respect to different values of gamma (γ). The colour of the cells changes from green to red, where green indicates high values and red indicates low values.

	$\gamma = 0.1$	$\gamma = 0.2$	$\gamma = 0.3$	$\gamma = 0.4$	$\gamma = 0.5$	$\gamma = 1$	$\gamma = 1.5$
Mask R-CNN_AlexNet	0.13274	0.05969	0.02375	0.07588	0.14332	0.53704	0.09898
Mask R-CNN_MobileNet	0.11026	0.087	0.07075	0.06532	0.10619	0.63487	0.09693
Mask R-CNN_ResNet50	0.61667	0.86318	0.8568	0.85776	0.85361	0.99678	0.87571
Mask R-CNN_VGG11	0.18619	0.17825	0.17674	0.13257	0.19341	0.75949	0.17601
Mask R-CNN_VGG13	0.07975	0.16139	0.14216	0.15498	0.12558	0.72173	0.20379
Mask R-CNN_VGG16	0.09409	0.08436	0.09441	0.14438	0.09604	0.55572	0.11493

In our study, we tested the models on a custom dataset. However, in real life, the system must deal with real-time datasets. Accordingly, in the future, we are planning to test the KITTI dataset. It contains 3D data involving Lidar sensor data, images, etc.

We found a robust and flexible detection model (Mask R-CNN) that can perform well in any scenario, whether it is day or night. In future research steps, we are going to investigate images from rainy and smoky conditions.

Furthermore, self-driving cars are expected to be equipped with high-resolution cameras recording gamma value as well. Following this, it seems reasonable to use the automatic gamma correction method to improve the efficiency of the instance detection process in different driving conditions.

10. Conclusions

In a nutshell, ResNet50-based mask R-CNN model performs well in all lighting conditions, whether it is bright or dark. Conversely, the total loss of this model is 16.17%. Summing up, it is found that ResNet50 based Mask R-CNN is better for real-time detection systems, because self-driving cars run on the road with real data that changes in milliseconds. Second, low qualities of images can be automatically corrected with the gamma correction method. However, a brighter environment can also be a challenging factor. In addition to this, many factors can significantly influence image quality, such as fog, rain, smoke, vehicle speed, etc. Thus, from the above results, ResNet-based Mask R-CNN model is robust, flexible, and can efficiently support the driving process in all driving conditions.

Author Contributions: Conceptualization, Á.T.; methodology, Á.T. and Z.S.; software, M.J.; validation, Á.T.; formal analysis, M.J.; investigation, M.J.; resources, Z.S.; data curation, M.J.; writing—original draft preparation, M.J.; writing—review and editing, Á.T.; visualization, M.J.; supervision, Á.T.; project administration, Á.T. and Z.S.; funding acquisition, Z.S. All authors have read and agreed to the published version of the manuscript.

Funding: The research presented in this paper was supported by the NRDI Office, Ministry of Innovation and Technology, Hungary, within the framework of the Autonomous Systems National Laboratory Programme, and the NRDI Fund based on the charter of bolster issued by the NRDI Office. The presented work was carried out within the MASPOV Project (KTI_KVIG_4-1_2021), which was implemented with support provided by the Government of Hungary in the context of the Innovative Mobility Program of KTI.

Institutional Review Board Statement: Not applicable.

Informed Consent Statement: Not applicable.

Data Availability Statement: To explore the environmental conditions, a pedestrian custom dataset based on Common Object in Context (COCO) was used [55–57].

Conflicts of Interest: The authors declare no conflict of interest. The funders had no role in the design of the study; in the collection, analyses, or interpretation of data; in the writing of the manuscript; or in the decision to publish the results.

References

1. Borek, J.; Groelke, B.; Earnhardt, C.; Vermillion, C. Economic Optimal Control for Minimizing Fuel Consumption of Heavy-Duty Trucks in a Highway Environment. *IEEE Trans. Control. Syst. Technol.* **2019**, *28*, 1652–1664. [CrossRef]
2. Torabi, S.; Bellone, M.; Wahde, M. Energy minimization for an electric bus using a genetic algorithm. *Eur. Transp. Res. Rev.* **2020**, *12*, 2. [CrossRef]
3. Fényes, D.; Németh, B.; Gáspár, P. A Novel Data-Driven Modeling and Control Design Method for Autonomous Vehicles. *Energies* **2021**, *14*, 517. [CrossRef]
4. Aymen, F.; Mahmoudi, C. A novel energy optimization approach for electrical vehicles in a smart city. *Energies* **2019**, *12*, 929. [CrossRef]
5. Skrúcaný, T.; Kendra, M.; Stopka, O.; Milojević, S.; Figlus, T.; Csiszár, C. Impact of the electric mobility implementation on the greenhouse gases production in central European countries. *Sustainability* **2019**, *11*, 4948. [CrossRef]
6. Csiszár, C.; Földes, D. System model for autonomous road freight transportation. *Promet-Traffic Transp.* **2018**, *30*, 93–103. [CrossRef]
7. Lu, Q.; Tettamanti, T. Traffic Control Scheme for Social Optimum Traffic Assignment with Dynamic Route Pricing for Automated Vehicles. *Period. Poly. Transp. Eng.* **2021**, *49*, 301–307. [CrossRef]
8. Li, Q.; He, T.; Fu, G. Judgment and optimization of video image recognition in obstacle detection in intelligent vehicle. *Mech. Syst. Signal Process.* **2020**, *136*, 106406. [CrossRef]
9. Yang, L.; Liu, Z.; Wang, X.; Yu, X.; Wang, G.; Shen, L. Image-Based Visual Servo Tracking Control of a Ground Moving Target for a Fixed-Wing Unmanned Aerial Vehicle. *J. Intell. Robot. Syst.* **2021**, *102*, 81. [CrossRef]
10. Junaid, A.B.; Konoiko, A.; Zweiri, Y.; Sahinkaya, M.N.; Seneviratne, L. Autonomous wireless self-charging for multi-rotor unmanned aerial vehicles. *Energies* **2017**, *10*, 803. [CrossRef]
11. Lee, S.H.; Kim, B.J.; Lee, S.B. Study on Image Correction and Optimization of Mounting Positions of Dual Cameras for Vehicle Test. *Energies* **2021**, *14*, 4857. [CrossRef]
12. Tschürtz, H.; Gerstinger, A. The Safety Dilemmas of Autonomous Driving. In Proceedings of the 2021 Zooming Innovation in Consumer Technologies Conference (ZINC), Novi Sad, Serbia, 26–27 May 2021; pp. 54–58.
13. Zhao, Z.Q.; Zheng, P.; Xu, S.T.; Wu, X. Object Detection with Deep Learning: A Review. *IEEE Trans. Neural Netw. Learn. Syst.* **2019**, *30*, 3212–3232. [CrossRef] [PubMed]
14. Kohli, P.; Chadha, A. Enabling pedestrian safety using computer vision techniques: A case study of the 2018 uber inc. self-driving car crash. *Lect. Notes Netw. Syst.* **2020**, *69*, 261–279. [CrossRef]
15. Barba-Guaman, L.; Naranjo, J.E.; Ortiz, A. Deep learning framework for vehicle and pedestrian detection in rural roads on an embedded GPU. *Electron* **2020**, *9*, 589. [CrossRef]
16. Simhambhatla, R.; Okiah, K.; Slater, R. Self-Driving Cars: Evaluation of Deep Learning Techniques for Object Detection in Different Driving Conditions. *SMU Data Sci. Rev.* **2019**, *2*, 23.
17. Cao, C.; Wang, B.; Zhang, W.; Zeng, X.; Yan, X.; Feng, Z.; Liu, Y.; Wu, Z. An Improved Faster R-CNN for Small Object Detection. *IEEE Access* **2019**, *7*, 106838–106846. [CrossRef]
18. Sandler, M.; Howard, A.; Zhu, M.; Zhmoginov, A.; Chen, L.C. MobileNetV2: Inverted Residuals and Linear Bottlenecks. In Proceedings of the IEEE Conference on Computer Vision and Pattern Recognition (CVPR), Salt Lake City, UT, USA, 18–23 June 2018; pp. 4510–4520. [CrossRef]

19. Alganci, U.; Soydas, M.; Sertel, E. Comparative research on deep learning approaches for airplane detection from very high-resolution satellite images. *Remote Sens.* **2020**, *12*, 458. [CrossRef]

20. Huang, R.; Pedoeem, J.; Chen, C. YOLO-LITE: A Real-Time Object Detection Algorithm Optimized for Non-GPU Computers. In Proceedings of the 2018 IEEE International Conference on Big Data, Seattle, WA, USA, 10–13 December 2018; pp. 2503–2510. [CrossRef]

21. Redmon, J.; Farhadi, A. YOLO v.3. Tech Rep. 2018, pp. 1–6. Available online: https://pjreddie.com/media/files/papers/YOLOv3.pdf (accessed on 1 February 2021).

22. Suhail, A.; Jayabalan, M.; Thiruchelvam, V. Convolutional neural network based object detection: A review. *J. Crit. Rev.* **2020**, *7*, 786–792. [CrossRef]

23. Lokanath, M.; Kumar, K.S.; Keerthi, E.S. Accurate object classification and detection by faster-R-CNN. *IOP Conf. Ser. Mater. Sci. Eng.* **2017**, *263*, 052028. [CrossRef]

24. Girshick, R. Fast R-CNN. In Proceedings of the 2015 International Conference on Computer Vision (ICCV), Santiago, Chile, 7–13 December 2015; pp. 1440–1448. [CrossRef]

25. Almubarak, H.; Bazi, Y.; Alajlan, N. Two-stage mask-R-CNN approach for detecting and segmenting the optic nerve head, optic disc, and optic cup in fundus images. *Appl. Sci.* **2020**, *10*, 3833. [CrossRef]

26. Mahmoud, A.S.; Mohamed, S.A.; El-Khoribi, R.A.; AbdelSalam, H.M. Object detection using adaptive mask R-CNN in optical remote sensing images. *Int. J. Intell. Eng. Syst.* **2020**, *13*, 65–76. [CrossRef]

27. Rezvy, S.; Zebin, T.; Braden, B.; Pang, W.; Taylor, S.; Gao, X.W. Transfer learning for endoscopy disease detection and segmentation with Mask-RCNN benchmark architecture. In Proceedings of the 2nd International Workshop and Challenge on Computer Vision in Endoscopy, Iowa City, IA, USA, 3 April 2020; pp. 68–72.

28. Wu, X.; Wen, S.; Xie, Y.A. *Improvement of Mask-R-CNN Object Segmentation Algorithm*; Springer International Publishing: New York, NY, USA, 2019; Volume 11740.

29. Penn-Fudan Database for Pedestrian Detection and Segmentation. Available online: https://www.cis.upenn.edu/~{}jshi/ped_html/index.html (accessed on 1 February 2021).

30. Tv, C.R.T.; Crt, T.; Intensity, L.; Voltage, C.; Lcd, T. How to Match the Color Brightness of Automotive TFT-LCD Panels. pp. 1–6. Available online: https://www.renesas.com/us/en/document/whp/how-match-color-brightness-automotive-tft-lcd-panels (accessed on 1 February 2021).

31. Farid, H. Blind inverse gamma correction. *IEEE Trans. Image Process.* **2001**, *10*, 1428–1433. [CrossRef] [PubMed]

32. DeArman, A. The Wild, Wild West: A Case Study of Self-Driving Vehicle Testing in Arizona. 2010. Available online: https://arizonalawreview.org/the-wild-wild-west-a-case-study-of-self-driving-vehicle-testing-in-arizona/ (accessed on 1 February 2021).

33. Mikusova, M. Crash avoidance systems and collision safety devices for vehicle. In Proceedings of the MATEC Web of Conferences: Dynamics of Civil Engineering and Transport Structures and Wind Engineering—DYN-WIND'2017, Trstená, Slovak Republic, 21–25 May 2017; Volume 107, p. 00024.

34. Ammirato, P.; Berg, A.C. A Mask-R-CNN Baseline for Probabilistic Object Detection. 2019. Available online: http://arxiv.org/abs/1908.03621 (accessed on 1 February 2021).

35. Zhang, Y.; Chu, J.; Leng, L.; Miao, J. Mask-refined R-CNN: A network for refining object details in instance segmentation. *Sensors* **2020**, *20*, 1010. [CrossRef]

36. Michele, A.; Colin, V.; Santika, D.D. Mobilenet convolutional neural networks and support vector machines for palmprint recognition. *Procedia Comput. Sci.* **2019**, *157*, 110–117. [CrossRef]

37. Sharma, V.; Mir, R.N. Saliency guided faster-R-CNN (SGFr-R-CNN) model for object detection and recognition. *J. King Saud. Univ.-Comput. Inf. Sci.* **2019**. [CrossRef]

38. Ren, S.; He, K.; Girshick, R.; Sun, J. Faster R-CNN: Towards Real-Time Object Detection with Region Proposal Networks. *IEEE Trans. Pattern Anal. Mach. Intell.* **2017**, *39*, 1137–1149. [CrossRef]

39. Girshick, R.; Donahue, J.; Darrell, T.; Malik, J. Region-Based Convolutional Networks for Accurate Object Detection and Segmentation. *IEEE Trans. Pattern Anal. Mach. Intell.* **2016**, *38*, 142–158. [CrossRef] [PubMed]

40. Özgenel, F.; Sorguç, A.G. Performance comparison of pretrained convolutional neural networks on crack detection in buildings. In Proceedings of the 35th International Symposium on Automation and Robotics in Construction (ISARC 2018), Berlin, Germany, 20–25 July 2018. [CrossRef]

41. Iglovikov, V.; Seferbekov, S.; Buslaev, A.; Shvets, A. TernausNetV2: Fully convolutional network for instance segmentation. In Proceedings of the IEEE Conference on Computer Vision and Pattern Recognition (CVPR) Workshops, Salt Lake City, UT, USA, 18–23 June 2018; pp. 228–232. [CrossRef]

42. Saha, R. Transfer Learning—A Comparative Analysis. December 2018. Available online: https://www.researchgate.net/publication/329786975_Transfer_Learning_-_A_Comparative_Analysis?channel=doi&linkId=5c1a9767299bf12be38bb098&showFulltext=true (accessed on 1 February 2021).

43. Girshick, R.; Donahue, J.; Darrell, T.; Malik, J.; Berkeley, U.C.; Malik, J. Rich feature hierarchies for accurate object detection and semantic segmentation. In Proceedings of the IEEE Conference on Computer Vision and Pattern Recognition, Columbus, OH, USA, 23–28 June 2014; Volume 1, p. 5000. [CrossRef]

44. Zhao, Y.; Han, R.; Rao, Y. A new feature pyramid network for object detection. In Proceedings of the 2019 International Conference on Virtual Reality and Intelligent Systems (ICVRIS), Jishou, China, 14–15 September 2019; pp. 428–431. [CrossRef]

45. Krizhevsky, A.; Sutskever, I.; Hinton, G.E. ImageNet Classification with Deep Convolutional Neural Networks. *Commun. ACM* **2017**, *60*, 84–90. [CrossRef]
46. Martin, C.H.; Mahoney, M.W. Traditional and heavy tailed self regularization in neural network models. In Proceedings of the 36th International Conference on Machine Learning, Long Beach, CA, USA, 9–15 June 2019; pp. 7563–7572.
47. Alom, M.Z.; Taha, T.M.; Yakopcic, C.; Westberg, S.; Sidike, P.; Nasrin, M.S.; Van Esesn, B.C.; Awwal, A.A.S.; Asari, V.K. The History Began from AlexNet: A Comprehensive Survey on Deep Learning Approaches. 2018. Available online: http://arxiv.org/abs/1803.01164 (accessed on 1 February 2021).
48. Almisreb, A.A.; Jamil, N.; Din, N.M. Utilizing AlexNet Deep Transfer Learning for Ear Recognition. In Proceedings of the 2018 Fourth International Conference on Information Retrieval and Knowledge Management (CAMP), Kota Kinabalu, Malaysia, 26–28 March 2018; pp. 8–12. [CrossRef]
49. Sudha, K.K.; Sujatha, P. A qualitative analysis of googlenet and alexnet for fabric defect detection. *Int. J. Recent Technol. Eng.* **2019**, *8*, 86–92.
50. Howard, A.G.; Zhu, M.; Chen, B.; Kalenichenko, D.; Wang, W.; Weyand, T.; Andreetto, M.; Adam, H. MobileNets: Efficient Convolutional Neural Networks for Mobile Vision Applications. 2017. Available online: http://arxiv.org/abs/1704.04861 (accessed on 1 February 2021).
51. Reddy, A.S.B.; Juliet, D.S. Transfer learning with RESNET-50 for malaria cell-image classification. In Proceedings of the 2019 International Conference on Communication and Signal Processing (ICCSP), Chennai, India, 4–6 April 2019; pp. 945–949. [CrossRef]
52. Rezende, E.; Ruppert, G.; Carvalho, T.; Ramos, F.; de Geus, P. Malicious software classification using transfer learning of ResNet-50 deep neural network. In Proceedings of the 2017 16th IEEE International Conference on Machine Learning and Applications (ICMLA), Cancun, Mexico, 18–21 December 2017; pp. 1011–1014. [CrossRef]
53. Mikami, H.; Suganuma, H.; U-chupala, P.; Tanaka, Y.; Kageyama, Y. ImageNet/ResNet-50 Training in 224 Seconds, No. Table 2. 2018. Available online: http://arxiv.org/abs/1811.05233 (accessed on 1 February 2021).
54. Simonyan, K.; Zisserman, A. Very deep convolutional networks for large-scale image recognition. In Proceedings of the 2014 International Conference on Learning Representations, Banff, AB, Canada, 14–16 April 2014; pp. 1–14.
55. Rostianingsih, S.; Setiawan, A.; Halim, C.I. COCO (Creating Common Object in Context) Dataset for Chemistry Apparatus. *Procedia Comput. Sci.* **2020**, *171*, 2445–2452. [CrossRef]
56. Lin, T.Y.; Maire, M.; Belongie, S.; Hays, J.; Perona, P.; Ramanan, D.; Dollár, P.; Zitnick, C.L. Microsoft COCO: Common Objects in Context. In *Computer Vision—ECCV 2014*; Fleet, D., Pajdla, T., Schiele, B., Tuytelaars, T., Eds.; Springer: Cham, Switzerland, 2017; Volume 8693, pp. 740–755. [CrossRef]
57. Everingham, M.; van Gool, L.; Williams, C.K.I.; Winn, J.; Zisserman, A. The pascal visual object classes (VOC) challenge. *Int. J. Comput. Vis.* **2010**, *88*, 303–338. [CrossRef]
58. Python Deep Learning Cookbook. Available online: https://livrosdeamor.com.br/documents/python-deep-learning-cookbook-indra-den-bakker-5c8c74d0a1136 (accessed on 1 February 2021).
59. Guan, Q.; Wang, Y.; Ping, B.; Li, D.; Du, J.; Qin, Y.; Lu, H.; Wan, X.; Xiang, J. Deep convolutional neural network VGG-16 model for differential diagnosing of papillary thyroid carcinomas in cytological images: A pilot study. *J. Cancer* **2019**, *10*, 4876–4882. [CrossRef] [PubMed]
60. Tammina, S. Transfer learning using VGG-16 with Deep Convolutional Neural Network for Classifying Images. *Int. J. Sci. Res. Publ.* **2019**, *9*, 9420. [CrossRef]
61. Veit, A.; Matera, T.; Neumann, L.; Matas, J.; Belongie, S. COCO-Text: Dataset and Benchmark for Text Detection and Recognition in Natural Images. September 2016. Available online: http://arxiv.org/abs/1601.07140 (accessed on 1 February 2021).
62. Mahamdioua, M.; Benmohammed, M. New Mean-Variance Gamma Method for Automatic Gamma Correction. *Int. J. Image Graph. Signal Process.* **2017**, *9*, 41–54. [CrossRef]
63. Jin, L.; Chen, Z.; Tu, Z. Object Detection Free Instance Segmentation with Labeling Transformations. 2016, p. 1. Available online: http://arxiv.org/abs/1611.08991 (accessed on 1 February 2021).

Deep Learning-Based Prediction of Throttle Value and State for Wheel Loaders

Jianfei Huang, Xinchun Cheng, Yuying Shen , Dewen Kong and Jixin Wang *

Key Laboratory of CNC Equipment Reliability, Ministry of Education, School of Mechanical and Aerospace Engineering, Jilin University, Changchun 130022, China; jfhuang19@mails.jlu.edu.cn (J.H.); chengxc19@mails.jlu.edu.cn (X.C.); yuyingshen1995@163.com (Y.S.); dwkong@jlu.edu.cn (D.K.)
* Correspondence: jxwang@jlu.edu.cn

Abstract: Accurate prediction of the throttle value and state for wheel loaders can help to achieve autonomous operation, thereby reducing the cost and accident rate. However, existing methods based on a physical model cannot accurately reflect the operator's driving habits and the interaction between wheel loaders and the environment. In this paper, a deep-learning-based prediction model is developed to predict the throttle value and state for wheel loaders by learning from driving data. Multiple long–short-term memory (LSTM) networks are used to extract the temporal features of different stages during the operation of the wheel loader. Two backward-propagation neural networks (BPNNs), which use the temporal feature extracted by LSTM as the input, are designed to output the final prediction results of throttle value and state, respectively. The proposed prediction model is trained and tested using the data from two different conditions. The end-to-end LSTM prediction model and BPNNs are used as benchmark models. The results indicate that the proposed prediction model has good prediction accuracy and adaptability. Furthermore, the relationship between the prediction performance and signal sampling frequency is also studied. The proposed prediction method that combines driving data and deep learning can make the throttle action conform to the decisions of an experienced operator, providing technical support for the autonomous operation of construction machinery.

Keywords: deep learning; wheel loaders; throttle prediction; state prediction; automation

Citation: Huang, J.; Cheng, X.; Shen, Y.; Kong, D.; Wang, J. Deep Learning-Based Prediction of Throttle Value and State for Wheel Loaders. *Energies* **2021**, *14*, 7202. https://doi.org/10.3390/en14217202

Academic Editors: Arno Eichberger, Zsolt Szalay, Martin Fellendorf, Henry Liu and Fernando Sánchez Lasheras

Received: 21 July 2021
Accepted: 20 October 2021
Published: 2 November 2021

Publisher's Note: MDPI stays neutral with regard to jurisdictional claims in published maps and institutional affiliations.

1. Introduction

As the most common mobile construction machinery, wheel loaders are widely used in the construction and mining industry, which are the important economic sectors across the world [1], due to their flexibility and adaptability. The main task of wheel loaders is to transport materials, including soil and rock, from a site to nearby dumpsite or trucks in a complex and changing working environment [2]. Control of the throttle is critical to the operation of wheel loaders. Accurately predicting the throttle action of a wheel loader expert operator can better achieve autonomous operation. The predicted throttle action can be used to directly the machine to imitate the expert operator's operations, to help achieve autonomous operation. In addition, predictions on the state of wheel loaders can be applied to model predictive control and energy management to achieve a good performance in terms of efficiency and fuel consumption.

The automation of construction machinery can reduce the cost and improve the safety of construction sites. Based on this, the last three decades have seen a growing trend towards the automation of construction machinery [3–5]. Many researchers have discussed the division method from manual operation to fully autonomous operation [6,7]. Dadhich et al. [8] proposed five steps to achieve the full automation of wheel loaders: manual operation, in-sight tele-operation, tele-remote operation, assisted tele-remote operation, fully autonomous. Despite the extensive research on automating construction machinery [9–11], a commercial system with autonomous construction machinery is still being explored [12].

Remote-operated construction machinery is being tried for commercial purposes [13,14]. However, this has led to a greater reduction in productivity and fuel efficiency [15] than manual operation because there are not enough sensory inputs to the remote operators. Therefore, to increase the fuel efficiency and productivity of construction machinery, it is necessary to improve the degree of automation of the loader to reduce the operator's remote intervention.

Most previous works related to the automation of construction machinery are based on a physical model that requires accurate mathematical representations [16–18]. Meng et al. [10] presented a way of optimizing the bucket trajectory for the autonomous loading of load-haul-dump (LHD) machines by solving the optimal trajectory through optimizing the minimum energy consumption calculated by Coulomb's passive earth pressure theory. Filla et al. [11] analyzed different autonomous scooping trajectories for wheel loaders by developing a simulation model of the uniform gravel pile. Shen and Frank [12,14] introduced dynamic programming into the solvution of the optimal control variable trajectory based on a mathematical model of the machine. However, these physical-model-based approaches have some common limitations. The method requires a dynamic model of construction machinery to be built, but the dynamic model simplifies machinery in the real world, and the dynamic model of machinery may change under the condition of wear during the operation. Additionally, modeling the interaction force between the tool and material is challenging, as the working environment is unpredictable and variable, and the properties of the different media to be excavated or moved are diverse.

The data-driven approach makes it possible to deal with the complex machinery dynamics [19–22] et al. used the data collected from tests to construct a nonlinear, non-parametric statistical model to predict the behavior of soil excavated by an excavator bucket. Heteroscedastic Gaussian process regression is used as the prediction framework. Machine learning, as a significant means of analyzing complicated data, can adjust its weight parameters by learning from data. In recent years, machine learning has made remarkable progress in solving pattern classification or prediction problems, such as image recognition [23], pattern recognition [24,25], and fault diagnosis [26]. Deep learning has been widely used in construction machinery [12,27,28]. Kim et al. [29] proposed a vision-based action recognition framework that considers the sequential working patterns of earthmoving machinery to recognize the operation types. The earthmoving machinery's sequential patterns are modeled and trained with convolutional neural networks and double-layer long–short-term memory (LSTM).

Due to its powerful ability to characterize complicated systems, process big data, and automatically extract features, deep learning has feasibility and superiority in the prediction task. The deep-learning-based prediction has received great attention for the automated operation of machinery. Yao et al. [30] designed a two-stage Convolutional Neural Networks model, including a classifier and some regressors, to automatically extract image features to obtain the piled-up status and payload distribution of the current state. The final prediction result is output via a backward-propagation neural network. Luo et al. [31] proposed a framework to predict the pose of construction machinery based on historical motions and activity attributes. The Gated Recurrent Unit is used to predict future machine poses, considering working patterns and interaction characteristics. Shi et al. [32] constructed a deep long–short-term memory network to predict the brake pedal aperture for different braking types by combining the driving data of experienced drivers in different driving environments with deep learning. Xing et al. [33] proposed an energy-aware personalized joint time-series modeling approach based on a recurrent neural network and LSTM to accurately predict the trajectory and velocity of the vehicle. Dai et al. [34] employed two groups of LSTM networks to predict the trajectory of the target vehicle. One LSTM is used to model the target vehicle and the individual trajectory of the surrounding vehicle, and the other is used to model the interaction between the target vehicle and each of the surrounding vehicles.

In this study, based on driving data of the experienced operator, a deep-learning-based method is proposed to accurately predict the throttle value and states (including lift cylinder, tilt cylinder, engine speed, vehicle velocity) of wheel loaders to help achieve autonomous operation and make predictive control algorithms and energy management strategies work with an acceptable performance. The sensor signals of wheel loaders under different working conditions are used instead of images as an important basis for predicting the throttle value and states of wheel loaders, as images will be inevitably affected by occlusions, deviations in viewpoint and scale, ambient illumination, and other factors [35,36]. Considering the time series characteristics of the working process of wheel loaders, LSTM networks are used to extract features. To reduce the computation load, the prediction of throttle value and state share the same LSTM network structures and weights. Two backward-propagation neural networks (BPNNs) are introduced to output the prediction results, as the throttle is controlled by the driver and the state of the wheel loaders is randomly influenced by the environment. Each working cycle of wheel loaders consists of several working phases, which possess their own unique characteristics, so the prediction results at different stages are output by neural networks with different weights to improve the prediction accuracy. Two different materials are used to study the adaptability of the prediction model. The relationship between the prediction performance and signal sampling frequency is also studied. Compared with the existing works, the method proposed in this study does not require a physical model and can be applied to different working conditions. The method proposed in this study can provide technical support for the autonomous operation of construction machinery and contribute to the intelligent process of the mining and construction industry.

2. Background

2.1. Problem Statement

The task of wheel loaders is to remove materials, including soil and rock, from a material pile to a nearby dumpsite or an adjacent load receiver in the sophisticated and changing working environment. There are many operation modes for wheel-loaders, including I, V, and T-shaped modes, depending on the route taken by wheel loaders during the loading operation. The difference in operation modes increases the difficulty of data analysis. For wheel-loaders, the V-cycle, which is the most common work cycle, is adopted in this experiment, as illustrated in Figure 1. The single V-cycle is divided into six phases, namely, V1 forward with no load (start and approach the pile), V2 bucket filling (penetrates the pile and load), V3 backward with a full load (retract from the pile), V4 forward and hoisting (approach the dumper), V5 dumping, V6 backward with no load (retract from the dumper), as shown in Table 1.

During the entire working cycle of wheel loaders, the operator needs to constantly modulate the throttle to control the movement of the wheel loaders. The throttle greatly determines the productivity and fuel efficiency during the operation of wheel loaders. When the throttle value is too high, wheel slipping will occur, resulting in a loss of traction as the driving force exceeds the adhesion. Wheel slipping damages tires and results in significant increases in operational cost. While the throttle value is too small, the speed of the vehicle will be lower, resulting in a loss of productivity. For the V-cycle of wheel loaders, the road adhesion coefficient is different for different phases. The quality of the vehicle will vary widely due to loading and dumping, which impacts the throttle value. Therefore, in the process of driving, experienced operators are required to perceive the environment information and select the appropriate throttle value. In this paper, the throttle value of the next moment is predicted.

Lift cylinder pressure, tilt cylinder pressure, engine speed and vehicle velocity are all crucial for wheel loaders. Lift cylinder pressure and tilt cylinder pressure can help to identify the working-cycle stages of wheel loaders. Engine speed and vehicle velocity play an important role in energy management. In this paper, the four parameters are called the state of wheel loaders, and every parameter has a corresponding prediction value.

The state prediction is the basis of many control technologies, including model predictive control. The state of wheel loaders is affected by the operator's driving action and the environment, so the state prediction cannot only take their internal dynamics into account. In the operation process, in addition to the current state, operators usually need to consider the past actions and state of wheel loaders. Thus, the throttle value and state prediction of wheel loaders needs to consider the time series characteristics of the working process.

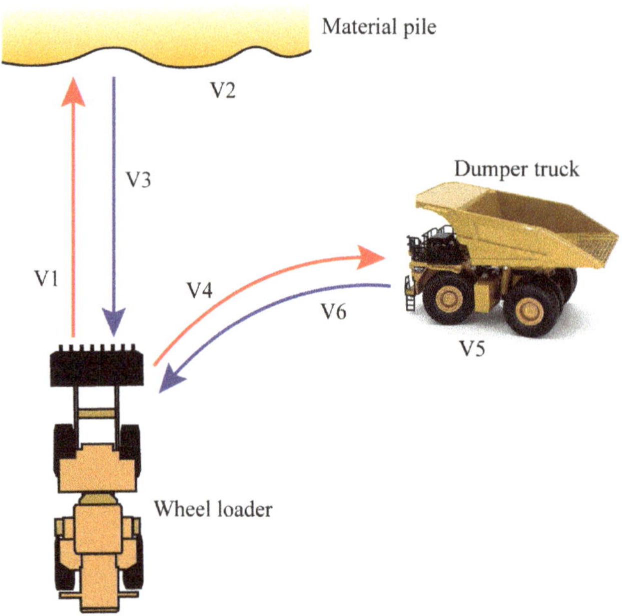

Figure 1. V-cycle of wheel loaders.

Table 1. V-cycle grouping.

Phase	Path
Forward with no load	V1
Bucket filling	V2
Backward with full load	V3
Forward and hoisting	V4
Dumping	V5
Backward with no load	V6

2.2. LSTM Network

Deep learning models automatically learn multiple levels of representations and abstractions from the data [37], which solves the problem that features need to be manually designed in traditional machine learning. Recurrent neural network (RNN) is a type of neural network specialized for the processing of sequence data. However, in practice, a simple RNN cannot cope with the challenge of long-term dependence.

The most efficient sequence model used in practical applications is called gated RNN, which is based on the idea of creating paths through time that have derivatives that neither vanish nor explode. Long short-term memory (LSTM) [38] is a type of gated RNN and a popular solution for processing sequence data. It has been shown to learn long-term dependencies more easily than simple recurrent architectures through gating units. Compared to the gated recurrent unit (GRU), a simpler gated RNN, LSTM is more powerful and more flexible, since it has three gates instead of two. Thus, LSTM is applied in this paper.

The LSTM block diagram is illustrated in Figure 2. The most crucial component of LSTM is the cell state, which is the horizontal line running through the top of the figure, making it easy for information to flow without changing. LSTM's ability to remove or add information to the cell state is controlled by the structure called gates. Gates consisting of a sigmoid neural net layer and a pointwise multiplication operation can selectively let information through. The first step of LSTM is to determine which information is discarded from the cell state. This decision is made by the forget gate, which outputs a number between 0 and 1 for each number in the previous cell state. Second, the new information that will be stored in the cell state needs to be determined. The input gate determines which values will be updated and a tanh layer creates a vector of new candidate values that could be used to update the current cell state. Finally, the current cell state and output gate are used to output the hidden state. The process of LSTM can be expressed as follows:

$$i_t = sigmoid(W_i \cdot [x_t, h_{t-1}] + b_i) \tag{1}$$

$$f_t = sigmoid\left(W_f \cdot [x_t, h_{t-1}] + b_f\right) \tag{2}$$

$$g_t = tanh\left(W_g \cdot [x_t, h_{t-1}] + b_g\right) \tag{3}$$

$$o_t = sigmoid(W_o \cdot [x_t, h_{t-1}] + b_o) \tag{4}$$

$$c_t = f_t * c_{t-1} + i_t * g_t \tag{5}$$

$$h_t = o_t * tanh(c_t) \tag{6}$$

where h_t, c_t and x_t represent the hidden states, cell state and the input sequence of LSTM at time t, respectively. i_t, f_t and o_t represent input gate, forget gate and output gate, respectively. W and b represent the weights and bias, respectively. c_t and g_t represent cell state at time t and candidate cell state at time t. $*$ is the element-wise product.

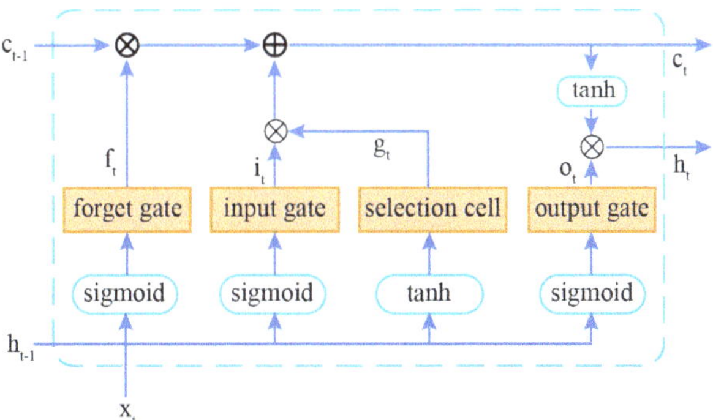

Figure 2. Illustration of Long short-term memory (LSTM).

3. Methodology

3.1. The Overview of the Proposed Deep Learning-Based Framwork

Operators of different proficiency levels can account for great differences in productivity and fuel efficiency. Consequently, deep learning is used to predict the throttle value of wheel loaders based on the driving data of experienced operators so that the driving process of wheel loaders conforms to the driving decisions of experienced operators to meet the vehicle's operational requirements, even in sophisticated driving environments, while ensuring productivity and fuel efficiency. Meanwhile, based on the temporal features extracted by LSTM, the DP neural network is also added to predict the state of the wheel loader, which does not make any assumption regarding its internal behavior and learns the impact of the environment on the state from the data. The flowchart of this proposed framework is shown in Figure 3, which involves three parts.

Part one: Data collection and pre-processing. Neural networks require real driving data from the skilled operator to imitate the experienced operator. For the collection of wheel loader driving data, skilled drivers were required to perform a V-cycle in the actual working environment. To improve the computation speed and prediction accuracy, the driving data are normalized, and the working cycle is divided.

Part two: Sequence model. LSTM, which is capable of extracting temporal features and solving the problem of gradient disappearance in the original RNN, is applied in this paper. Six LSTM networks with the same structure are used for six stages of the working cycle.

Part three: Regression model. In order to output the final results, two BPNNs following the LSTM output of the prediction results of throttle value and state, respectively.

Each part is discussed in detail as follows.

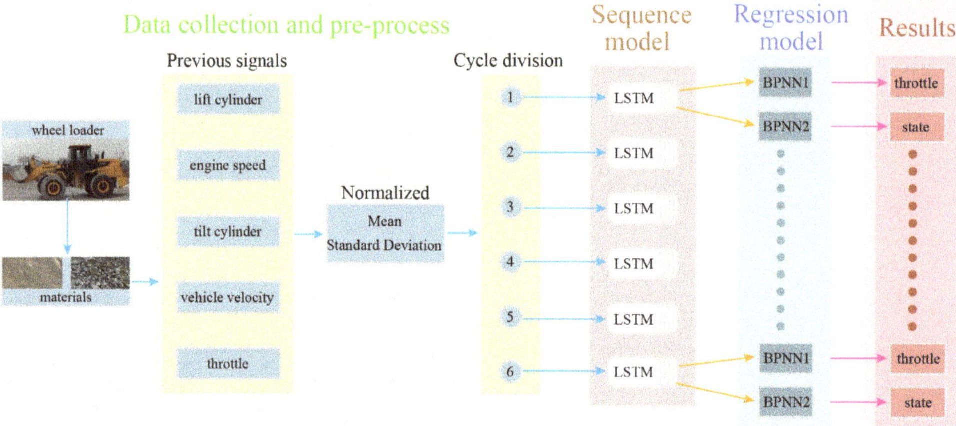

Figure 3. The model presented in this paper.

3.2. Data Collection and Pre-Processing

The data acquisition of the wheel loader is shown in Figure 4, which is equipped with pressure sensors and GPS. Field data were collected in sites with dry ground. To study the adaptability of the proposed prediction approach, it is important to conduct experiments with a variety of materials. Small coarse gravel (SCG) and large coarse gravel (LCG) were used as the operating materials for this experiment, which are shown in Figure 5. Small coarse gravel mainly contains particles with sizes 0–25 mm, while large coarse gravel mainly contains particles with sizes 25–500 mm.

Figure 4. Experimental wheel loader.

<table>
<tr><td>(a) Small coarse gravel</td><td>(b) Large coarse gravel</td></tr>
</table>

Figure 5. Two operating materials.

The V-cycle of wheel loaders consists of six working phases, which possess their own unique characteristics. To improve the prediction precision and computation efficiency, six prediction models were constructed for six phases of the working cycle of wheel loaders. Normalization was used to speed up the training. According to the working characteristics of wheel loaders in the V-cycle, the V-cycle was divided by extracting the working condition features of the actuator and walking device to realize the mapping between the collected data and working state, as shown in Figure 6. For different operating materials, 50 sets of data were collected to train and test the prediction model.

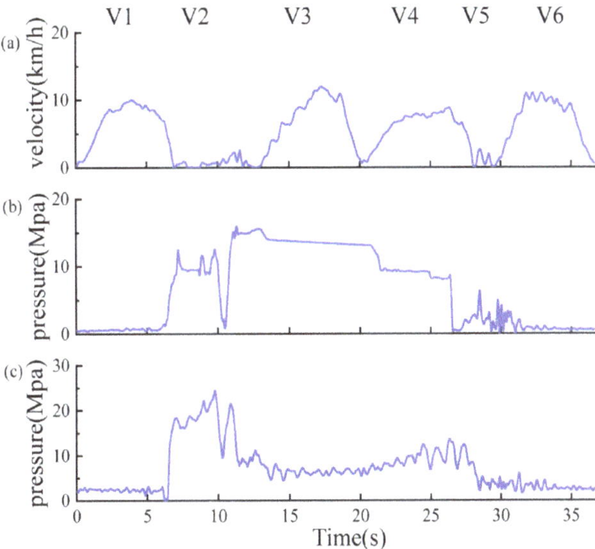

Figure 6. Schematic diagram of working condition division: (**a**) velocity of wheel loader; (**b**) lift cylinder pressure; (**c**) tilt cylinder pressure.

3.3. Construction of LSTM

The proposed deep-learning-based prediction method consists of LSTM and BPNN, as illustrated in Figure 7. LSTM can memorize the temporal relationship in time-series data. The particular gate structure of the LSTM allows the networks to learn when to store and when to forget the relationship. Thus, the temporal information of the driving data is encoded into the LSTM network. In the training process, the high-dimensional temporal information was extracted by the hidden layer from the time-series data.

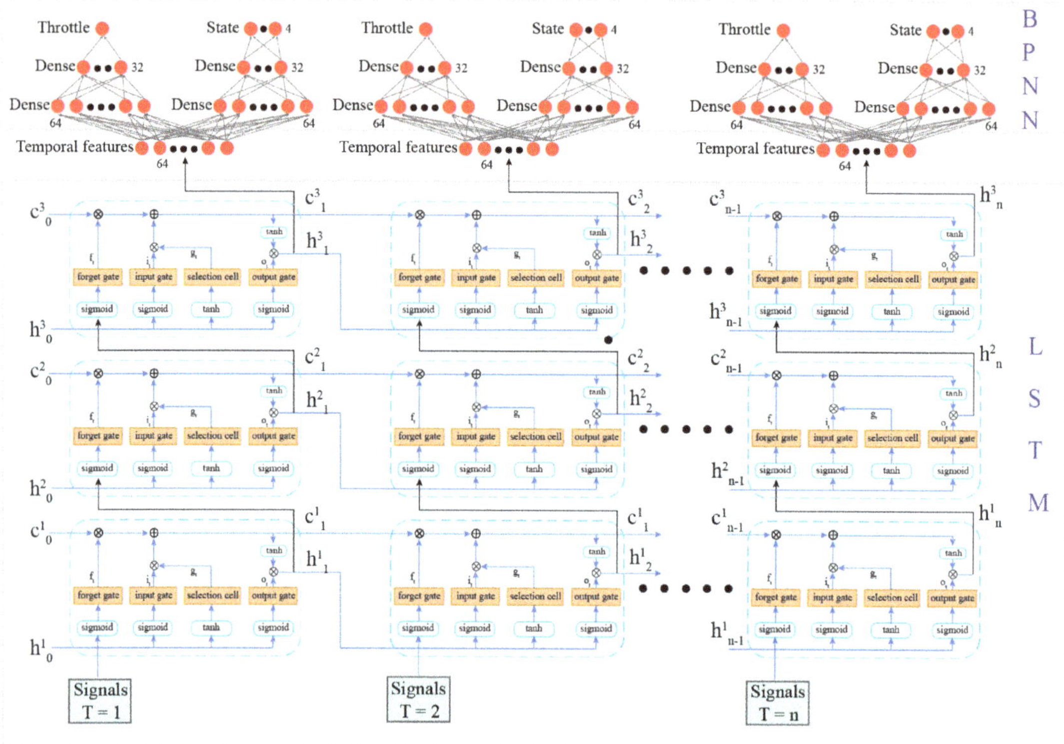

Figure 7. Structure of the proposed LSTM and BPNN.

The LSTM model is developed with triple-stacked LSTM units because this configuration outperformed the double-stacked and the single-stacked LSTM in the training experiment. Meanwhile, compared with quadruple-stacked LSTM, triple-stacked LSTM has similar prediction precision and requires fewer computation resources. This result implies that increasing the structural complexity of the LSTM does not always lead to an improvement in the prediction accuracy.

The prediction of throttle value and state share an LSTM network to extract the temporal features. An alternative option is to use two LSTMs to extract the temporal features and predict the throttle value and state separately. However, two LSTMs introduce the extra burden of training and real-time calculations. In the experiment, both options have a similar effect. A possible explanation for this is that the sequence features required to predict the throttle value and state are similar.

When operators drive wheel loaders to work, the cycle operation time is diverse. Thus, the time-series data have different lengths. For the LSTM network, the time-series data with different sequence lengths need padding to ensure the same length. However, after padding for the time-series data, the prediction ability of the network will be influenced. Therefore, in this paper, the batch-size was set to 1 to ensure prediction accuracy.

The time-series data were taken as the input and the time dimension was $[1,2,\dots t,\dots n]$. each sequence has five parameters: lift cylinder, tilt cylinder, engine speed, vehicle velocity and throttle, respectively. In the training process, all previous throttle values and state values were taken as the inputs to output the corresponding prediction values of the next moment via BPNN, and the real values of the next moment were used as the correct mark values.

For the LSTM, the number of output units is 64. To train the neural networks, the learning rate was 0.001, while the loss function was mean squared error (MSE) and expressed as:

$$MSE = \frac{1}{N} \sum_{i=1}^{N} (\widehat{y_i} - y_i)^2 \tag{7}$$

where $\widehat{y_i}$ and y_i are, respectively, the predicted value and the actual value of the sampling point in the test set, and N is the number of samples in the test set.

The solver was the Adam algorithm [39], which is one of the most common solvers and suitable for training RNN. To assess the quality of training results, the root mean square error (RMSE) is taken as the criterion and expressed as:

$$RMSE = \sqrt{MSE} \tag{8}$$

3.4. Construction of BPNN

Two BPNNs with 64 inputs were used to output the prediction results of throttle value and state. The temporal information extracted from all previous data was taken as the input parameter of BPNN at each moment and the BPNN output the prediction values of the next moment. The two BPNNs have the same structure, with two hidden layers, with 64 and 32 units, respectively. The BPNN structure was proven to be effective and accurate. The BPNN part in Figure 4 depicts the network architecture. The Rectified Linear Units were chosen as the activation function of BPNN because they allow for deep neural networks to be trained with acceptable speed and performance [40].

4. Results and Discussion

TensorFlow was employed for the programming implementation of the benchmark and proposed architectures. The time-series data were imported into Python as a list. The label was placed in the other list. The first 40 elements of the lists were used as the training set, and the last 10 elements formed the test set.

4.1. Performance Analysis of Deep Learning Model for Different Materials

To validate the adaptability of the proposed method on the prediction problem, the experimental wheel loader was required to load different materials with the V-cycle operation mode at two different working sites, and the collected driving data at the two test sites were used as the inputs to train two LSTM network individually. Meanwhile, the throttle value and state of the wheel loader at the next time step were used as the output to train the networks. In addition to the different operating materials, the two different working sites have different driving road surfaces. When loading small coarse gravel, the pavement comprised concrete road surfaces, and when loading large coarse gravel, the pavement comprised native soil road surfaces. For each working material, 50 sets of driving data were collected at a 200 Hz sampling frequency. The data were further divided into training data, consisting of 40 sets, and testing data, consisting of 10 sets.

Figure 8 shows the comparison results of the RMSE of the predicted throttle value and state using small and large coarse gravel as working materials for the six working stages and 10 groups of test data, respectively. Each boxplot represents the quartiles of RMSE, where the current throttle value and state are used as the input, and the prediction results belong to the next time step. It can be seen from Figure 8a that the RMSE of the predicted throttle value for two different materials was less than 1.8 and, compared with the RMSE using small coarse gravel as working materials, the RMSE using large coarse gravel had a higher mean and wider variation range, which indicates worse prediction results. In Figure 8b, the RMSE of the predicted state for two different materials are less than 5, with the same comparison results as the predicted throttle value. A possible explanation for this is that the complexity of working environments has an impact on prediction accuracy. If large coarse gravel is used as the working material, the load of the wheel loader will change drastically during the bucket filling stage (V2) and dumping stage (V5), which increases the difficulty of prediction. In addition to this, native soil pavement is more complicated than concrete pavement, so the interaction between the wheel loader and the environment has stronger randomness. The predicted results and the actual values under two different working conditions are compared in Figure 9. As shown in Figure 9, during the whole working cycle, the LSTM network can predict the throttle value and state with relatively high accuracy under different working conditions, which means that the proposed prediction model has good adaptability.

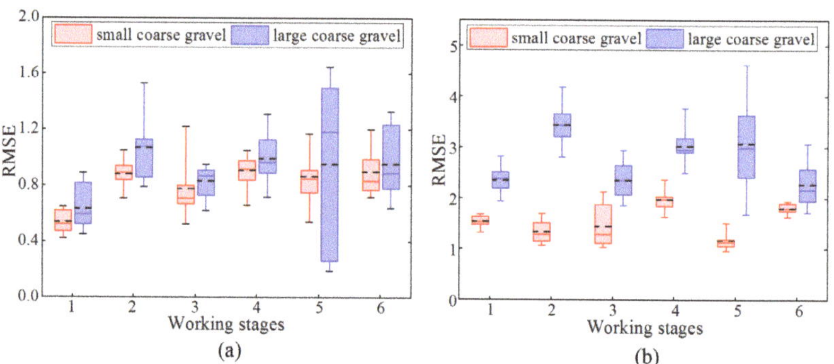

Figure 8. Comparison of RMSE from different materials: (**a**) prediction of throttle value (**b**) prediction of state. mean and median values are shown with '–' and '—' respectively.

4.2. Comparison with Different Deep Learning Models

The single V-cycle of wheel loaders consists of six stages, which have different operation modes and feature data. To more accurately extract unique feature data for each work stage and obtain high prediction accuracy, six LSTM prediction networks are developed for different stages. A single LSTM prediction network can also be used for this work. A single prediction network takes the complete data containing six stages as input and outputs the prediction result, which is end-to-end deep learning. End-to-end deep learning can reduce the hand-designed features and intermediate steps, but requires a considerable amount of data.

Figure 10 compares the RMSE results of the single LSTM prediction network and multiple LSTM prediction networks using small coarse gravel (SCG) as a working material. From Figure 10, it can be seen that the RMSE obtained by the single prediction networks has a higher mean and wider variation range compared with the RMSE obtained by the multiple prediction networks. Particularly for the bucket-filling stage (V2) and dumping stage (V5), the multiple prediction networks significantly outperforms the single prediction network in the prediction effect. The above finding can be further confirmed by Figure 11, which shows the RMSE comparison results using large coarse gravel (LCG) as a working material. There are two possible reasons for this result. The first reason is that there is a change in the load of the wheel loader during the bucket-filling stage and the dumping stage. The lift and tilt of the working device also account for this result. Therefore, in the case of limited data, to obtain accurate prediction results, it is necessary to establish different prediction networks for different working stages. However, it should be noted that the single prediction network may achieve the same performance as the multiple prediction networks with sufficient data.

BPNN is also used as a benchmark model for different stages. Figures 12 and 13 compare the RMSE results of BPNNs and LSTM networks. The result shows that the LSTM network has a better prediction effect. The better prediction result can be ascribed to the fact that LSTM can extract temporal features, which can make the model understand the environment and wheel loader more accurately. The RMSE of throttle value for different operating materials and models is shown in Table 2, and the RMSR of state is shown in Table 3.

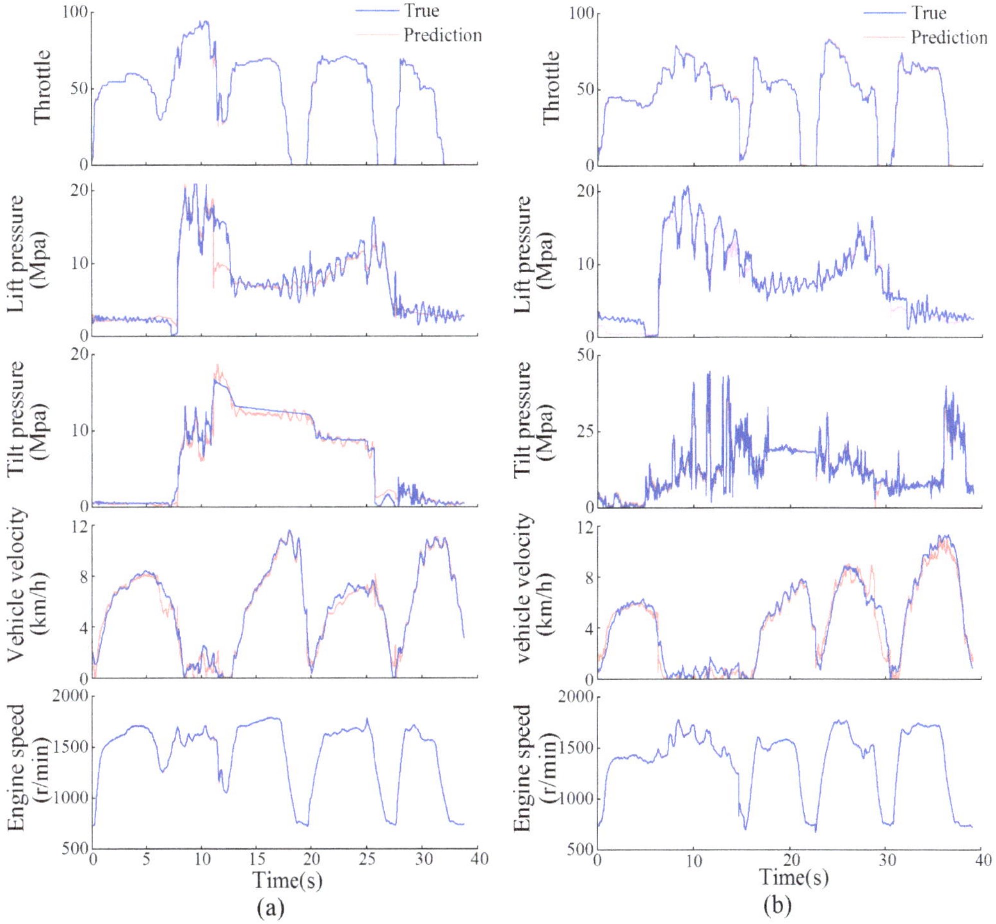

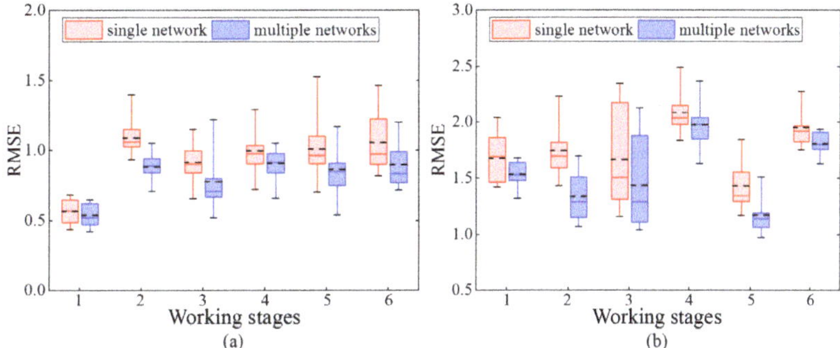

Figure 9. Driving data of experienced drivers and the predicted value from different materials: (**a**) small coarse gravel (**b**) large coarse gravel.

Figure 10. RMSE comparison of different LSTM networks using small coarse gravel: (**a**) prediction of throttle value (**b**) prediction of state.

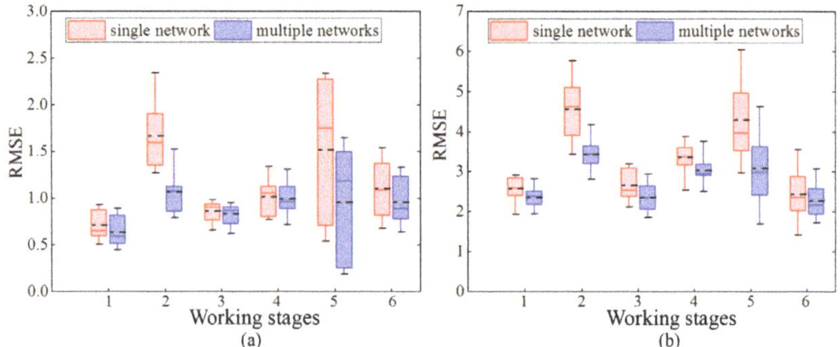

Figure 11. RMSE comparison of different LSTM networks using large coarse gravel: (**a**) prediction of throttle value (**b**) prediction of state.

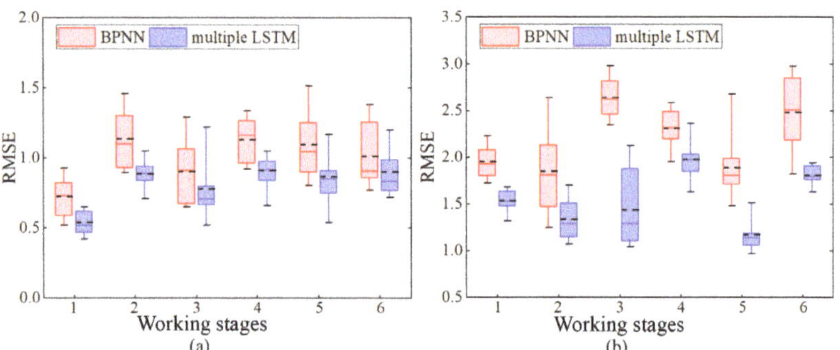

Figure 12. RMSE comparison of BPNNs and LSTM networks using small coarse gravel: (**a**) prediction of throttle value (**b**) prediction of state.

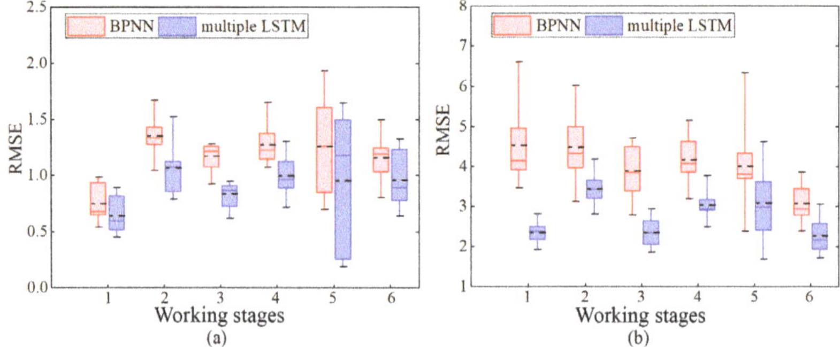

Figure 13. RMSE comparison of BPNNs and LSTM networks using large coarse gravel: (**a**) prediction of throttle value (**b**) prediction of state.

4.3. Performance Analysis of LSTM Networks for Different Sampling Frequency

Due to the high integration of the wheel loader and the high signal density, the sampling frequency is severely restricted by the storage capacity of the host. The appropriate sampling frequency should be as low as possible while ensuring prediction accuracy. The low sampling frequency can reduce the amount of data, thereby reducing the cost of data storage and the consumption of computation resources. Therefore, to reduce the cost, it is necessary to study the relationship between the signal sampling frequency and the prediction accuracy. The sampling frequency is reduced to 100, 50, 20, and 10 Hz, respectively.

Table 2. The RMSE of throttle value for different operating materials and models.

Phase	Multiple LSTM of Using LCG	Multiple LSTM of Using SCG	Single LSTM of Using LCG	Single LSTM of Using SCG	BPNN of Using LCG	BPNN of Using SCG
V1	0.64	0.54	0.71	0.57	0.75	0.73
V2	1.07	0.89	1.67	1.09	1.36	1.14
V3	0.83	0.78	0.86	0.92	1.18	0.91
V4	1.00	0.91	1.02	1.00	1.28	1.13
V5	0.95	0.87	1.52	1.01	1.26	1.10
V6	0.96	0.90	1.10	1.06	1.16	1.01

Table 3. The RMSE of state for different operating materials and models.

Phase	Multiple LSTM of Using LCG	Multiple LSTM of Using SCG	Single LSTM of Using LCG	Single LSTM of Using SCG	BPNN of Using LCG	BPNN of Using SCG
V1	2.36	1.53	2.57	1.68	4.53	1.95
V2	3.44	1.34	4.56	1.75	4.49	1.85
V3	2.36	1.44	2.66	1.67	3.89	2.64
V4	3.03	1.98	3.36	2.09	4.17	2.32
V5	3.08	1.18	4.30	1.43	4.01	1.89
V6	2.28	1.81	2.44	1.96	3.08	2.48

Figures 14 and 15 show the relationship between prediction performance and signal sampling frequency. Table 4 shows the RMSE of throttle value and state under different sampling frequencies. It can be seen that the prediction effect improves with the increase of the signal sampling frequency. This result may be explained by the fact that the higher sampling frequency can provide sufficient feature information in time. However, the too-high sampling frequency may bring more noise, making it difficult for the neural network model to learn the correct mapping from input to output. At the same time, when the sampling frequency is higher than 50 Hz, the increase in frequency does not significantly improve the prediction performance. In practice, although the increase in sampling frequency will improve the prediction accuracy, it will also lead to an increase in storage costs and a decrease in the real-time calculation rate. Therefore, a trade-off is necessary for the selection of sampling frequency. For example, if a fully automated system is required, a higher sampling frequency is necessary to reduce the prediction error. However, for the assisted driving, a lower sampling frequency should be considered to reduce the storage and computing costs.

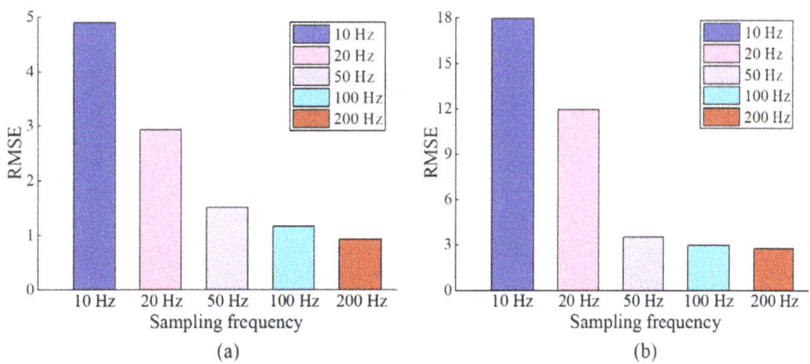

Figure 14. Relationship between sampling frequency and prediction performance using large coarse gravel: (**a**) prediction of throttle value (**b**) prediction of state.

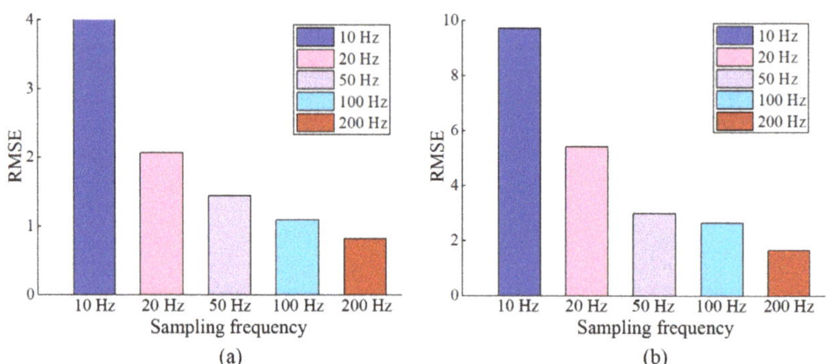

Figure 15. Relationship between sampling frequency and prediction performance using small coarse gravel: (**a**) throttle prediction (**b**) state prediction.

Table 4. The RMSE of throttle value and state under different sampling frequency.

Sampling Frequency	RMSE of Throttle Value Using LCG	RMSE of State Using LCG	RMSE of Throttle Value Using SCG	RMSE of State Using SCG
10 HZ	4.90	17.95	4.00	9.72
20 HZ	2.93	11.93	2.07	5.43
50 HZ	1.51	3.53	1.44	3.00
100 HZ	1.17	3.00	1.09	2.64
200 HZ	0.93	2.75	0.82	1.65

5. Conclusions

This paper proposed a deep-learning-based method to predict throttle value and state for wheel loaders. The prediction model can help achieve autonomous operation and reduce the need for remote intervention during remote operation. Additionally, the proposed model can be applied to model predictive control and energy management to achieve a good performance in terms of efficiency and fuel consumption.

The prediction model consists of three main parts, namely, data collection and pre-processing, LSTM and BPNN. Six LSTM networks are used to extract the temporal features of six stages of the V-cycle for wheel loaders. Based on the extracted temporal features, two BPNNs are employed to predict the throttle value and state of wheel loaders, respectively. The data obtained from two different working materials and pavements are used to train and test the proposed prediction model. The results show that the proposed prediction model can achieve a good prediction effect under different working conditions and out-perform BPNNs. Moreover, compared with end-to-end deep learning, which only uses a single LSTM network for prediction, the prediction model of multiple LSTM networks shows better prediction performance. However, the prediction model of multiple neural networks requires more hand-designed features. The relationship between signal sampling frequency and prediction accuracy is also studied. In the range of 10 Hz to 200 Hz, as the frequency increases, the prediction performance improves. However, when the signal sampling frequency exceeds 50 Hz, the improvement effect of prediction accuracy is not obvious as the frequency increases. Therefore, in engineering practice, it is necessary to weigh the prediction accuracy and cost. Although this paper takes the wheel loader as the research object, the proposed prediction model can be adapted to other construction machinery. In future, the prediction network will be deployed to a physical wheel loader to improve the efficiency and real-time fuel efficiency using reinforcement learning.

Author Contributions: Conceptualization, J.H. and X.C.; methodology, J.H. and Y.S.; software, J.H. and X.C.; validation, J.H. and Y.S.; formal analysis, D.K.; investigation, X.C.; resources, J.W.; data curation, Y.S. and X.C.; writing—original draft preparation, J.H.; writing—review and editing, J.W. and D.K.; visualization, X.C. and Y.S.; supervision, J.W.; project administration, J.W.; funding acquisition, J.W. and D.K. All authors have read and agreed to the published version of the manuscript.

Funding: This research was funded by the National Natural Science Foundation of China grant number 51875239.

Institutional Review Board Statement: Not applicable.

Informed Consent Statement: Not applicable.

Conflicts of Interest: The authors declare no conflict of interest.The funders had no role in the design of the study; in the collection, analyses, or interpretation of data; in the writing of the manuscript, or in the decision to publish the results.

References

1. Davila Delgado, J.M.; Oyedele, L.; Ajayi, A.; Akanbi, L.; Akinade, O.; Bilal, M.; Owolabi, H. Robotics and automated systems in construction: Understanding industry-specific challenges for adoption. *J. Build. Eng.* **2019**, *26*, 100868. [CrossRef]
2. Frank, B.; Kleinert, J.; Filla, R. Optimal control of wheel loader actuators in gravel applications. *Autom. Construct.* **2018**, *91*, 1–14. [CrossRef]
3. Hemami, A.; Hassani, F. An overview of autonomous loading of bulk material. *Int. Symp. Autom. Rob. Constr.* **2009**, 405–411. [CrossRef]
4. Dadhich, S.; Bodin, U.; Sandin, F.; Andersson, U. From Tele-Remote Operation to Semi-Automated Wheel-Loader. *Int. J. Electr. Electron. Eng. Telecommun.* **2018**, 178–182. [CrossRef]
5. Hemami, A. Fundamental Analysis of Automatic Excavation. *J. Aerosp. Eng.* **1995**, *8*, 175–179. [CrossRef]
6. Bobbie, F.; Lennart, S.; Reno, F.; Anders, F. On Increasing Fuel Efficiency by Operator Assistant Systems in a Wheel Loader. In Proceedings of the International Conference on Advanced Vehicle Technologies and Integration (VTI 2012), Changchun, China, 16–19 July 2012; pp. 155–161.
7. Roberts, J.M.; Duff, E.S.; Corke, P.I. Reactive navigation and opportunistic localization for autonomous underground mining vehicles. *Inf. Sci.* **2002**, *145*, 127–146. [CrossRef]
8. Dadhich, S.; Bodin, U.; Andersson, U. Key challenges in automation of earth-moving machines. *Autom. Construct.* **2016**, *68*, 212–222. [CrossRef]
9. Marshall, J.A.; Murphy, P.F.; Daneshmend, L.K. Toward Autonomous Excavation of Fragmented Rock: Full-Scale Experiments. *IEEE Trans. Autom. Sci. Eng.* **2008**, *5*, 562–566. [CrossRef]
10. Sotiropoulos, F.E.; Asada, H.H. A Model-Free Extremum-Seeking Approach to Autonomous Excavator Control Based on Output Power Maximization. *IEEE Robot. Autom. Lett.* **2019**, *4*, 1005–1012. [CrossRef]
11. Dobson, A.A.; Marshall, J.A.; Larsson, J. Admittance Control for Robotic Loading: Design and Experiments with a 1-Tonne Loader and a 14-Tonne Load-Haul-Dump Machine. *J. Field Robot.* **2017**, *34*, 123–150. [CrossRef]
12. Dadhich, S.; Sandin, F.; Bodin, U.; Andersson, U.; Martinsson, T. Field test of neural-network based automatic bucket-filling algorithm for wheel-loaders. *Autom. Construct.* **2019**, *97*, 1–12. [CrossRef]
13. Sun, W.; Iwataki, S.; Komatsu, R.; Fujii, H.; Yamashita, A.; Asama, H. Simultaneous tele-visualization of construction machine and environment using body mounted cameras. In Proceedings of the 2016 IEEE International Conference on Robotics and Biomimetics (ROBIO), Qingdao, China, 3–7 December 2016; pp. 382–387. [CrossRef]
14. Yamada, H.; Muto, T. Development of a Hydraulic Tele-Operated Construction Robot using Virtual Reality. *Int. J. Fluid Power* **2003**, *4*, 35–42. [CrossRef]
15. Fernando, C.L.; Saraiji, M.Y.; Seishu, Y.; Kuriu, N.; Minamizawa, K.; Tachi, S. Effectiveness of Spatial Coherent Remote Drive Experience with a Telexistence Backhoe for Construction Sites. In Proceedings of the ICAT-EGVE, Kyoto, Japan, 28–30 October 2015; pp. 69–75.
16. Feng, H.; Yin, C.; Ma, W.; Yu, H.; Cao, D. Parameters identification and trajectory control for a hydraulic system. *ISA Trans.* **2019**, *92*, 228–240. [CrossRef]
17. Yoo, S.; Park, C.G.; You, S.H.; Lim, B. A Dynamics-Based Optimal Trajectory Generation for Controlling an Automated Excavator. *Proc. Inst. Mech. Eng. Part C-J. Eng. Mech. Eng. Sci.* **2010**, *224*, 2109–2119. [CrossRef]
18. Zou, Z.; Chen, J.; Pang, X. Task space-based dynamic trajectory planning for digging process of a hydraulic excavator with the integration of soil–bucket interaction. *Proc. Inst. Mech. Eng. Part K-J. Multi-Body Dyn.* **2018**, *233*, 598–616. [CrossRef]
19. Bauza, M.; Hogan, F.R.; Rodriguez, A. A Data-Efficient Approach to Precise and Controlled Pushing. In Proceedings of the Conference on Robot Learning, Zürich, Switzerland, 29–31 October 2018; pp. 336–345.
20. Byravan, A.; Fox, D. SE3-nets: Learning rigid body motion using deep neural networks. In Proceedings of the 2017 IEEE International Conference on Robotics and Automation (ICRA), Singapore, 29 May–3 June 2017; pp. 173–180. [CrossRef]

21. Sotiropoulos, F.E.; Asada, H.H. Autonomous Excavation of Rocks Using a Gaussian Process Model and Unscented Kalman Filter. *IEEE Robot. Autom. Lett.* **2020**, *5*, 2491–2497. [CrossRef]
22. Sandzimier, R.J.; Asada, H.H. A Data-Driven Approach to Prediction and Optimal Bucket-Filling Control for Autonomous Excavators. *IEEE Robot. Autom. Lett.* **2020**, *5*, 2682–2689. [CrossRef]
23. Koo, B.; La, S.; Cho, N.W.; Yu, Y. Using support vector machines to classify building elements for checking the semantic integrity of building information models. *Autom. Construct.* **2019**, *98*, 183–194. [CrossRef]
24. Hasan, M.A.M.; Ahmad, S.; Molla, M.K.I. Protein subcellular localization prediction using multiple kernel learning based support vector machine. *Mol. Biosyst.* **2017**, *13*, 785–795. [CrossRef]
25. Shi, Y.; Xia, Y.; Zhang, Y.; Yao, Z. Intelligent identification for working-cycle stages of excavator based on main pump pressure. *Autom. Construct.* **2020**, *109*. [CrossRef]
26. Li, Y.; Xu, M.; Wei, Y.; Huang, W. A new rolling bearing fault diagnosis method based on multiscale permutation entropy and improved support vector machine based binary tree. *Measurement* **2016**, *77*, 80–94. [CrossRef]
27. Arabi, S.; Haghighat, A.; Sharma, A. A deep-learning-based computer vision solution for construction vehicle detection. *Comput.-Aided Civil. Infrastruct. Eng.* **2020**, *35*, 753–767. [CrossRef]
28. Park, J.; Lee, B.; Kang, S.; Kim, P.Y.; Kim, H.J. Online Learning Control of Hydraulic Excavators Based on Echo-State Networks. *IEEE Trans. Autom. Sci. Eng.* **2017**, *14*, 249–259. [CrossRef]
29. Kim, J.; Chi, S. Action recognition of earthmoving excavators based on sequential pattern analysis of visual features and operation cycles. *Autom. Construct.* **2019**, *104*, 255–264. [CrossRef]
30. Yao, Z.; Huang, Q.; Ji, Z.; Li, X.; Bi, Q. Deep learning-based prediction of piled-up status and payload distribution of bulk material. *Autom. Construct.* **2021**, *121*. [CrossRef]
31. Luo, H.; Wang, M.; Wong, P.K.Y.; Tang, J.; Cheng, J.C.P. Construction machine pose prediction considering historical motions and activity attributes using gated recurrent unit (GRU). *Autom. Construct.* **2021**, *121*. [CrossRef]
32. Shi, J.; Sun, D.; Hu, M.; Liu, S.; Kan, Y.; Chen, R.; Ma, K. Prediction of brake pedal aperture for automatic wheel loader based on deep learning. *Autom. Construct.* **2020**, *119*. [CrossRef]
33. Xing, Y.; Lv, C.; Mo, X.; Hu, Z.; Huang, C.; Hang, P. Toward Safe and Smart Mobility: Energy-Aware Deep Learning for Driving Behavior Analysis and Prediction of Connected Vehicles. *IEEE Trans. Intell. Transp. Syst.* **2021**, *22*, 4267–4280. [CrossRef]
34. Dai, S.; Li, L.; Li, Z. Modeling Vehicle Interactions via Modified LSTM Models for Trajectory Prediction. *IEEE Access* **2019**, *7*, 38287–38296. [CrossRef]
35. Fang, W.; Ding, L.; Zhong, B.; Love, P.E.D.; Luo, H. Automated detection of workers and heavy equipment on construction sites: A convolutional neural network approach. *Adv. Eng. Inform.* **2018**, *37*, 139–149. [CrossRef]
36. Kim, J.; Chi, S.; Seo, J. Interaction analysis for vision-based activity identification of earthmoving excavators and dump trucks. *Autom. Construct.* **2018**, *87*, 297–308. [CrossRef]
37. Zhang, S.; Yao, L.; Sun, A.; Tay, Y. Deep Learning Based Recommender System: A Survey and New Perspectives. *ACM Comput. Surv.* **2019**, *52*, 1–38. [CrossRef]
38. Hochreiter, S.; Schmidhuber, J. Long short-term memory. *Neural Comput.* **1997**, *9*, 1735–1780. [CrossRef]
39. Kingma, D.P.; Ba, J.L. Adam: A Method for Stochastic Optimization. *arXiv* **2018**, arXiv:1412.6980.
40. Nair, V.; Hinton, G.E. Rectified linear units improve Restricted Boltzmann machines. In Proceedings of the ICML, Haifa, Israel, 21–24 June 2010; pp. 807–814.

MDPI

Article

Comparing Different Levels of Technical Systems for a Modular Safety Approval—Why the State of the Art Does Not Dispense with System Tests Yet

Björn Klamann * and Hermann Winner

Institute of Automotive Engineering, Department of Mechanical Engineering, Technical University of Darmstadt, 64287 Darmstadt, Germany; hermann.winner@tu-darmstadt.de
* Correspondence: bjoern.klamann@tu-darmstadt.de

Abstract: While systems in the automotive industry have become increasingly complex, the related processes require comprehensive testing to be carried out at lower levels of a system. Nevertheless, the final safety validation is still required to be carried out at the system level by automotive standards like ISO 26262. Using its guidelines for the development of automated vehicles and applying them for field operation tests has been proven to be economically unfeasible. The concept of a modular safety approval provides the opportunity to reduce the testing effort after updates and for a broader set of vehicle variants. In this paper, we present insufficiencies that occur on lower levels of hierarchy compared to the system level. Using a completely new approach, we show that errors arise due to faulty decomposition processes wherein, e.g., functions, test scenarios, risks, or requirements of a system are decomposed to the module level. Thus, we identify three main categories of errors: insufficiently functional architectures, performing the wrong tests, and performing the right tests wrongly. We provide more detailed errors and present examples from the research project UNICAR*agil*. Finally, these findings are taken to define rules for the development and testing of modules to dispense with system tests.

Keywords: safety validation; automated driving systems; decomposition; modular safety approval; modular testing; fault tree analysis

Citation: Klamann, B.; Winner, H. Comparing Different Levels of Technical Systems for a Modular Safety Approval—Why the State of the Art Does Not Dispense with System Tests Yet. *Energies* **2021**, *14*, 7516. https://doi.org/10.3390/en14227516

Academic Editors: Arno Eichberger, Zsolt Szalay, Martin Fellendorf and Henry Liu

Received: 16 October 2021
Accepted: 4 November 2021
Published: 11 November 2021

Publisher's Note: MDPI stays neutral with regard to jurisdictional claims in published maps and institutional affiliations.

1. Introduction

The safety validation in the automotive industry still focuses on the vehicle or system level. Their systems have become increasingly complex so that today's processes require comprehensive tests to be carried out even at a low hierarchical level, e.g., on component level. This is justified by advances in Software- and Hardware-in-the-Loop testing [1]. Despite resulting improvements in the reliability of systems, the final safety validation is still only permissible at the system level [2]. With growing complexity, especially due to the implementation of automated driving functions, it has been shown that field operation tests on system level can no longer be managed economically [3]. For automated driving functions, scenario-based testing has proved to be a promising approach (see also e.g., [4,5]). However, with the validation at the system level, even minor changes and system variations require testing all scenarios again. Furthermore, Amersbach and Winner [6] point out that a feasible application of scenario-based testing is still challenging due to the required number of scenarios. With the new concept of a modular safety approval, we provide the opportunity to dispense with system tests. The safety approval of modules, as subsystems of the superordinate system "vehicle", has the advantage that modifications only require repeated approval of the modified modules, not the whole system. In that case, different sets of modules may also be combined to various use cases without the need for a safety approval for every different combination [7]. To realize such a modular approach, it is crucial to differentiate between the development and testing of systems and of said modules.

Therefore, we analyze the development and testing process in the automotive industry by using a novel developed concept based on well-established risk analysis methods combined with new findings in system theory. Consequently, we indicate possible uncertainties during the development and testing of modules in comparison to the whole system. While the state of the art only tries to reveal errors from the decomposition process in system tests, we present a method that helps to avoid specific errors already during development. Thereby, newly provided awareness of possible uncertainties and rules to avoid them reduces errors in the development process and may also lead to the development of further methods that reduce uncertainty of a modular safety approval. Additionally, the results of this paper can be taken as a starting point to identify more detailed possible errors for a specific development process as well as for a specific system. Further rules can be defined to avoid these errors. Moreover, for some of these errors, module tests can be defined to reveal if they have been put into effect. In summary, the presented approach not only leads, at least, to fewer faults in the integrated system, thus, to fewer revision efforts on system level, but can also guide the way to a modular safety approval.

While various development processes are published in the literature, usually depending on the purpose of the given process, we are going to use the process according to ISO 26262 [2]. The standard focuses on functional safety which is also a mandatory part of the safety approval for an automated vehicle. The additional publicly available specification SOTIF further considers potential insufficiencies of the intended system functionality but does not include substantial modifications of the process regarding the derivation of subsystems or components [5]. As ISO 26262 contains a well-established and detailed process description, it is better suited for our following analysis.

The main process in ISO 26262 starts with the definition of the item under consideration. An item is defined as a "system or combination of systems", while a system is defined as a "set of components or subsystems that relates at least a sensor, a controller and an actuator with one another" [2]. Similarly, Leveson [8] (p. 187), as well as Schäuffele et al. [9] (p. 125), describe three levels of hierarchy: the system, subsystem and component. Both authors point out that systems are always part of a greater system and, thus, subsystem at the same time. Thus, the definition of the hierarchical level depends on the view the hierarchy is used for. On system level, the cumulated functionality that is fulfilled by different subsystems, can be observed in its behavior expressed through its interaction with its operating environment [10] (p. 6). ISO 26262 uses the vehicle level as the highest considered hierarchical view and the system level as being part of the vehicle. After the analysis of ISO 26262 in this section, we use the term system instead of vehicle, and subsystem (or module) instead of the system to assure the applicability to other domains than the automotive domain.

To understand how the development process of the state of the art lacks in safety validation on lower hierarchical levels, awareness of how these levels are already addressed in the validation is key. ISO 26262 demands the final safety validation to be executed on the vehicle level. Additionally, the standard recommends different test methods linked to particular ASIL (Automotive Safety Integrity Level) classifications on a vehicle, system, and hardware/software level. Hardware/software levels can be components or sets of components being part of the system. With this, potential weaknesses regarding a modular safety approval can be identified. The recommended test methods for the safety validation in ISO 26262 for different hierarchical levels are shown in Table 1 [2].

Table 1. Recommended test methods for different hierarchical levels according to ISO 26262. Crosses indicate recommendation of the test method for the hierarchical level.

Test Method	Hardware/Software Level	System Level	Vehicle Level
Requirements-based test	x	x	x
Fault-Injection test	x	x	x
Back-to-Back test	x	x	–
Performance test	x	x	x
Test of external interfaces	x	x	x
Test of internal interfaces	x	x	x
Interface consistency check	x	x	x
Error guessing test	x	x	x
Resource usage test	x	x	x
Stress test	x	x	x
Test for interface resistance and robustness under certain environmental conditions	–	x	x
Test of interaction/communication	–	x	x
Test derived from field experience	–	x	x
Long-term test	–	–	x
User test under real-life conditions	–	–	x

It can be seen that the vehicle level is only missing the back-to-back test. Back-to-back tests are mainly used for the validation of the simulation model. Mostly, a simulation model of the whole vehicle is non-existent or not in the same detail as it is for the investigated system. Thus, the back-to-back test may not be purposeful on the vehicle level. The user test under real-life conditions, missing on hardware/software and system level, can only be performed on the whole vehicle by its definition. However, the real-life conditions and the user can be modeled for simulation, too. With this, at least a user test with uncertainties of validity can be performed on lower levels. The long-term test may conflict with the schedule of the development process, but can also be derived to lower levels with the consideration of uncertainties due to the decomposition.

Two tests that are missing on the hardware/software level in comparison to the system level consider the communication to other hardware/software components. These are only regarded to be possible with finally developed components on system level. Tests from field experience on the hardware/software level are also not recommended by the standard. This may be reasonable because field experiences are only possible in a late stage of the development process. The made experiences can then already be performed on vehicle level. However, breaking down these experiences or derived test cases to lower levels can reduce uncertainty in following development processes.

In conclusion, some test methods may provide the potential to be performed already on the hardware/software level. A deeper analysis of the exact test methods is not the focus of this paper but will be part of our following work. Still, the decomposition of necessary information for tests on lower hierarchical levels is shown to be mandatory to achieve a modular safety approval.

Except for the ISO 26262, the mentioned literature describes subsystems or components as parts of a system, as we will do in the following sections. Figure 1 explains the relation between the different hierarchical levels following ISO 26262 [2] and Steimle et al. [11] by using the class diagram of the Unified Modeling Language (UML) [12]. Conforming to Steimle et al., we also include a hierarchy level below with components that are a hardware part and a software unit to show a broader picture of the hierarchies. In this work, we see the vehicle level equivalent to the system level to achieve an approach that can also be applied to other domains. For a subsystem with modular properties, the term module is used. We define a module in the context of this paper as follows:

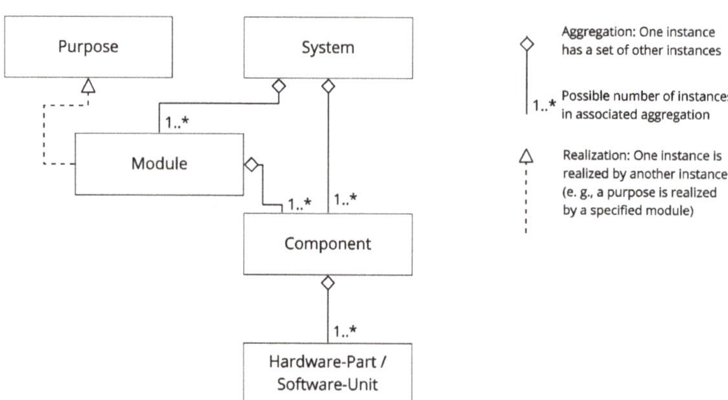

Figure 1. Relation between different hierarchical levels represented as a UML class diagram.

A **module** is a set of components building an encapsulated unit with common properties regarding the associated purpose of the modular architecture. With this purpose, a module is relatively independent of other modules but has well-defined interfaces to them.

This definition conforms to Walden et al. [10] (p. 209) and Göpfert [13] (p. 27) but adds remarks about the associated purpose. Thereby, it highlights the fact that different modular architectures exist due to different purposes. A system may thus be decomposed into different sets of modules depending on the purpose of the decomposition. In the project UNICAR*agil*, a new modular automated vehicle with a service-oriented service architecture is developed, built and safety approved [7]. The project uses different modular architectures for function, software, hardware and the safety validation. As introduced, the purpose of modularization in this initiative is a modular safety approval. It is remarkable that the consequently developed vehicle system uses different modular architectures with different purposes, which are still interdependent. All architectures are connected to the final implementation of the system and are thus at least connected indirectly. In conclusion, the design of each modular architecture is always a trade-off between the objectives of the different purposes.

2. Materials and Methods

In this chapter, we describe the methods we use to derive rules to avoid errors due to the use of modules, but not the system for the safety validation. For this, revealing possible uncertainty during decomposition processes leading to uncertainty in modular testing is needed. Decomposition is the refinement of a greater entity to smaller entities. In systems engineering, smaller entities are assumed to fulfill the greater entity's properties [9] (p. 124). The main view of decomposition lies in the functional decomposition and the decomposition of hardware and software parts. In this work, we also analyze the refinement or decomposition of information, associated with function, hardware and software, such as requirements or parameter spaces [8] (p. 191). In general, we do not analyze other steps than the ones that derive information from system to module level. Additionally, more than one decomposition step is likely, depending on the level of hierarchy. However, with the assumption that the same methodical steps produce the same categories of uncertainty, we expect only one decomposition step to be sufficient in the analysis.

ISO 26262 is not clear on the point how the overall system or system properties should be decomposed. Although the item definition should be described on vehicle level, the following hazard analysis requires knowledge about the system architecture which can be concluded from the given examples in ISO 26262 [2] (part 3). Furthermore, it is said that the item definition should contain all already existent ideas of the item and should continuously be refined during development [2] (part 10). Schäuffele et al. point out that the decomposition process (explicitly referring to hierarchy building and modularization) is often an intuitive process [9] (p. 126). Thus, when analyzing the way to module development and testing, following the process described in the standard or other literature does not result in a robust structured procedure. Instead, we compare the information mandatory at the system level to the ones derived at the module level. Furthermore, we analyze possible uncertainty in modular testing which may partly be a solution of insufficiencies in the decomposition process.

While the automotive industry focuses on the test procedure for the safety validation [14] (p. 75), our approach leads to an entire consideration of the development as well as the test procedure. For a modular safety approval, we assume the control of the development process to be mandatory, in particular when it comes to complex functions of an automated vehicle. However, we only consider possible errors in the decomposition process during development until the start of the technical implementation. The specific errors depend on the specific implementation of the development process. These processes vary widely between domains and companies and are analyzed in other publications (cf., e.g., [15,16]).

The term error is defined in accordance to Avizienis et al. [17] as a deviation from a correct state and caused by a fault. A fault as an "abnormal condition that can cause an element or an item to fail" [18] in decomposition processes can, e.g., be the unawareness of an engineer about possible vehicle states or a misleading used word in the definition of a requirement. An error may lead to a failure, which we define as the termination of an intended behavior in accordance to Stolte et al. [18], but does not necessarily have to. This is also true for insufficiencies in a decomposition process that may cause a failure but do not necessarily have to. In software engineering, errors are defined as human actions that cause incorrect results [19]. This definition contradicts the sequence of fault and error of Avizienis [17] and the ISO 26262 [2] (part 10). Still, for this work, both definitions of error are convenient. Thus, errors in this work are defined as insufficiencies in decomposition processes (caused by human actions) being deviations from a correct state that may lead to a failure.

The identification of errors is done by using a fault tree analysis starting with an undesired top event, which is then broken down to causes that may lead to it [20] (IV-1). These causes are seen as another undesired event and broken down into further causes. This is repeated until a general level is reached where the causes can still be applied to arbitrary systems. Causes in a fault tree analysis are faults of the top level event. However, the definition of the terms depends on the perspective taken. With the perspective and scope of this analysis, the identified causes are errors due to the decomposition process or due to the identified differences that appear on module but not on system level. Thus, the findings of this paper are not assumed to be sufficient for a modular safety approval. Further approaches for module and system development as well as testing need to be added. For the supplementation of the fault tree, we are using a similar approach to the System Theory and Process Analysis (STPA) from Leveson et al. [21] For each of the two identified superordinate errors, a simplified control structure is modeled. The errors identified by the fault tree analysis are assigned to the modeled entities or connections. Entities or connections without assigned errors are analyzed in more detail for possible errors due to the decomposition processes. Newly revealed errors are then added to the fault tree to increase its completeness.

3. Results

On system level according to the item definition in ISO 26262, a functional description, e.g., based on possible use cases of the system, is given [2] (part 3). This functional description can already be decomposed into a set of functions, building the functional architecture. On the other side, requirements can be derived from the functional description.

In ISO 26262, these requirements on vehicle level are called safety goals. The third important information category is hazards, usually described by scenarios that may lead to these hazards. In the domain of automated vehicles, they are called critical scenarios as described by Junietz [22]. In a previous publication [23], we already showed how hazards, that may i.a. be derived by the consideration of violated safety goals, can be broken down to the module level. Still, this process can be as erroneous as any other decomposition process that needs to be considered in the safety validation.

On module level, the functional description is derived directly from the functional description of the system or from the requirements on module level that are derived from requirements on system level.

For a modular safety approval, we identified two main requirements that need to be fulfilled. The first requirement is that the module's function together with the functions of other modules fulfills the requirements on system level. Additionally, the specified requirements of the module need to be fulfilled by the implementation of the module and is defined as the second requirement.

Taking the abovementioned in mind, we use the described fault tree analysis approach with the top event that a *module is not properly safety approved individually*, in order to derive possible hazards. On the second level, we assume that at least one of the two main requirements is not fulfilled. For the first error A, it is assumed that *the specified function of the module is not fulfilling the requirements on system level together with the functions of other modules*. This is illustrated by a simplified functional architecture of a system with an input and an output in Figure 2. An error of the specified function means that the implemented function of the module works as specified, but the specification of the function is faulty.

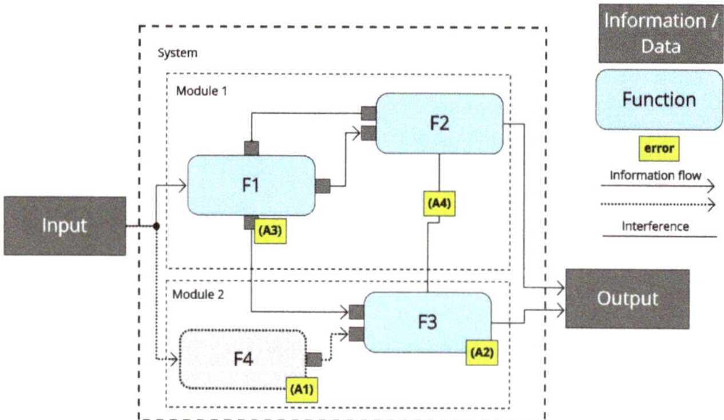

Figure 2. Exemplary abstracted functional architecture with functions F1–F4 including system and module boundaries visualized as UML activity diagram. Arrows are showing information flow, the line without arrows represents an interference. Dotted lines represent the added function F4 and its information flow after identification of error A1. Green boxes show the location of errors A1–A4 from the fault tree of error A.

For the case that the second requirement is not fulfilled, one error can simply be that the specified function is not fulfilled by the module. However, this broad generalization is not the aim of this work (e.g., analyzing or even changing the development process of the modules after their functional description in the process). However, testing the module function still serves as a major approval method. Thus, we take the testing procedure of modules into account as a potential uncertainty, because it uses information from the system level to try to falsify the absence of unreasonable risk. Due to this, the second error B leading to the top event of the fault tree is defined such that *the specified function of the module is not sufficiently tested*.

3.1. Errors Due to Insufficiently Specified Functions

The fault tree for causes derived by error A is shown in Figure 3. All multiple connections are OR-gates so that we do not use the associated symbol to reduce necessary space and increase readability. Figure 2 supports the findings of the following branches in the fault tree derived from error A.

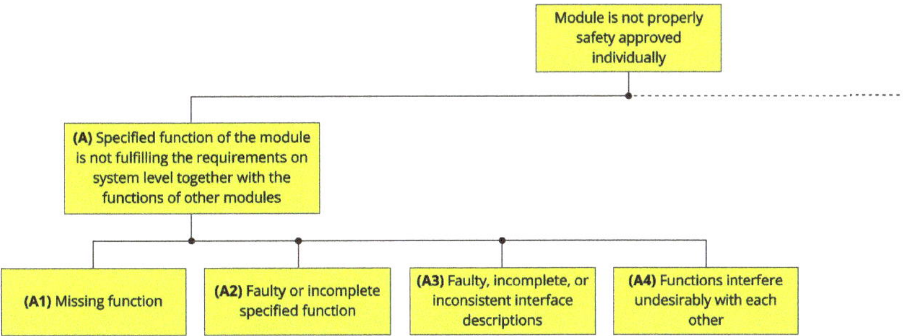

Figure 3. Fault tree of error A.

In the architecture, errors are assigned to their location, visualized by small boxes, and named by a letter and a number in accordance with the fault tree. The system contains module 1 and module 2, which are containing the functions F1, F2 and F3 that are receiving and sending information, illustrated by arrows.

The function F4 in Figure 2 is visualized by a gray box to represent a *missing function* as error A1. This function is not necessarily mandatory to fulfill the overall requirements, the absence of which would be obvious but can be a supporting function that only leads to unreasonable risk in specific scenarios. In UNICAR*agil*, we experienced this with the initial absence of the pose offset correction as described by Homolla et al. [24] which would have led to an increasing control error in cases of inconsistent localization data. The simplified functional architecture of UNICAR*agil* showing this exemplary error is illustrated in Figure 4. In the architecture, the trajectory calculation uses localization data x(t) from a camera-based source as described by [25] (pp. 1541–1542). The trajectory control uses the localization data x(t) from the vehicle dynamic state estimator module which is based on the sensor data of the global navigation satellite system receiver (NovAtel Inc., Calgary, AB, Canada) (GNSS), odometry and an inertial measurement unit (IMU). The added function offset correction is then eliminating the offset between these two localization data to x(t) (corrected) while maintaining a stable provision of localization data. Reasons for this and further advantages of this architecture are described in [24].

Error A2 describes a *faulty or incompletely specified function* that therefore inhibits the correct function of the system. However, a function may not only be faulty or incomplete but the communication between functions can be an issue due to *faulty, incomplete or inconsistent interface descriptions*, described by error A3. One exemplary fault is the inconsistent use of the angle definition that we have also seen for the yaw angle in our project as also shown in Figure 4. Even though a missing function usually results in missing interfaces for some other functions, only some minor information on an existing interface may either be missing to fulfill the function correctly (missing information about the interface is also an issue for module testing, but will be part of the test process analysis later on).

Interfaces can not only provide the desired information, energy, or material, some may also *interfere undesirably with the function* of another function leading to error A4. Figure 2 shows interference between functions F2 and F3. A possible example of this is the influence of a function F2 to a function F3 that is using the same bus system. When F2 is faulty (e.g., by sending a big amount of faulty data), it may not be safety relevant, depending on its functionality. However, the sent data may jam the bus so that the safety-relevant functionality of F3 is not able to send its data anymore.

These abovementioned general causes for error A serve as a starting point for a more detailed analysis. Such analysis is strongly dependent on the exact development process, which varies widely between products or companies.

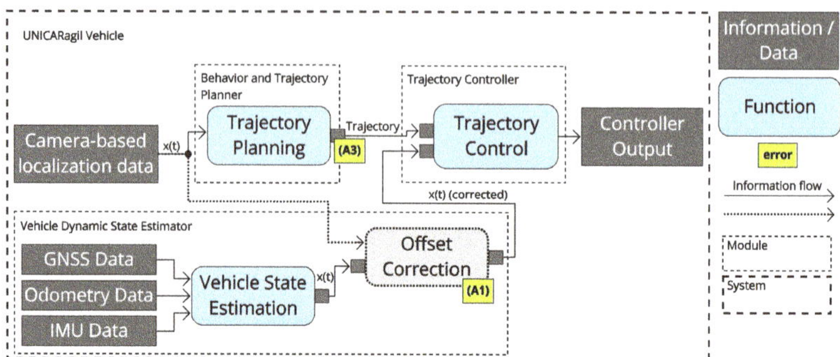

Figure 4. Simplified functional architecture of the module's *behavior and trajectory planner, trajectory controller* and *vehicle state estimator* together with the added function *offset correction* reproduced with the permission from Homolla et al. [24]. Information/data flow, functions and modules of the UNICAR*agil* vehicle are not completely illustrated. Green boxes show the location of the experienced errors.

3.2. Errors Occurring in Module Tests

The following section is going to focus on the testing process, but will also reveal possible uncertainty in the development process that causes uncertainty in testing. The error that a *module is not sufficiently tested* (B) can only be due to *right tests that are performed wrongly* (B1) or because *the wrong tests are performed* (B2).

Again, we use a visualization of the possible errors to increase the completeness of the possible errors. For this, Figure 5 shows a generalized test bench, including the module under test and the specified test environment. The module is assumed to be the same as it is used in the overall system. For some test cases or in earlier development phases, it might be reasonable to use simplified versions of the module under test. However, the same challenges apply on system level and are therefore not considered any further. Additionally, the process of test case development is illustrated. As before, green boxes show the location of the associated errors which are developed by the fault tree. In addition, blue boxes are associated with components or connections that do not have an associated error after the initial fault tree development. In conclusion, the fault tree is extended by revealing possible errors at these remaining components and connections.

3.2.1. Right Tests Are Performed Wrongly

One possible derived error B1-1 for wrongly performed tests is an *invalid stimulation of the module* under test. Figure 5 shows that the stimulation of an object under test is commonly divided into stimulation from other modules of the considered system and the remaining environment. Therefore, we also differ between the errors and derive two errors from B1-1. Firstly, error B1-1-1, which occurs when the *specified and desired stimulation by other modules of the overall system is invalid*. Secondly, error B1-1-2, which is caused in the case of the *remaining environment that influences the function of the module as invalid* (e.g., the aforementioned thermal energy from other sources). In simulation tests, non-decomposed systems often only require a model of their environment. If they are tested in their original environment (e.g., by unsupervised field operation tests (see also e.g., [26])), they might not need any models at all. In comparison, decomposed modules possess crucial interfaces to other modules and beyond that, direct or indirect interfaces to the remaining environment. Today, common practices struggle with the validity of such models due to technical or knowledge limitations. Overall, dealing with modular decomposed systems and simulations of connected modules as well as environments (in software or hardware) can increase or decrease the complexity of the validity challenge. Both errors B1-1-1 and B1-1-2 may therefore be caused by module boundaries that do not provide a valid stimulation, e.g., due to technical limitations of simulation practices based on the state of the art.

Validity does not mean that a model is representing an exact clone of the simulated entity. The required validity properties of the models in the test bench rather depend on each specific test case [27]. As illustrated by Figure 6, in accordance with Steimle et al. [11], the required validity of the test case is derived by its test targets. If, for example, their purpose is to test the consistency of interfaces or communication between modules, models of participating modules and their functions usually do not require high validity. If on the other side, test targets demand evaluation of the module functionality, highly valid models of the communication might not be necessary for all test cases.

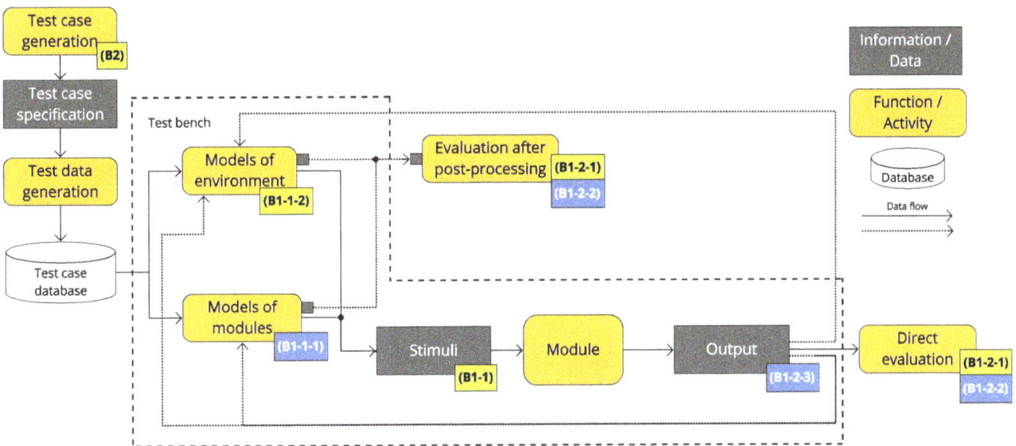

Figure 5. Generalized test bench of a module visualized as UML activity diagram including the steps towards test case identification and generation as well as the evaluation. The dashed line represents the test bench boundary. Dotted lines show feedback data for evaluation. Green boxes show the location of errors discovered by the fault tree analysis. Blue boxes show the location of errors found by the analysis of the test bench and the test procedure.

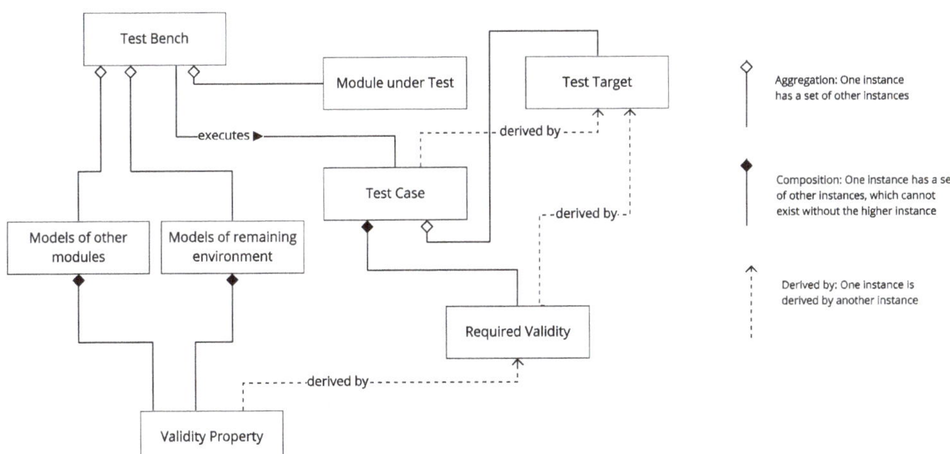

Figure 6. Dependencies between test case and validity, visualized as UML class diagram.

However, it might be crucial to evaluate the influence of time on the implemented behavior of time-critical functions. In conclusion, the connection between test targets and derived test cases needs to be considered even as early as the definition of the module boundaries.

Stimulation can be invalid, but so can the evaluation of its results leading to error B1-2 as *invalid evaluation*. Evaluation of module tests can be performed directly at the output of the tested modules. Additionally, indirect assessment using models of other modules or the environment and evaluating their outputs represents an alternative. This second way requires models of other modules or the remaining environment, which might again cause the errors B1-1-1 and B1-1-2.

For both ways of evaluation, specific metrics and corresponding pass-/fail criteria (also referred to as evaluation criteria by Steimle et al. [11] or verification criteria by [28] (p. 165)) are needed. Hereby, it should be noted that a convenient interface for evaluation depends on the available metrics and their ability to predict the behavior at the system level. This is usually not done directly by the metric, but by evaluating compliance with requirements of investigated modules. In conclusion, module boundaries may be defined so that a valid evaluation is not possible. For example, the evaluation of a decomposed environmental sensor, as depicted by Rosenberger et al. [29], and its intended functionality is almost impossible by raw sensor data. In that case, decomposing the sensor so that raw sensor data is taken as the output would result in a module boundary that does not allow a valid evaluation. However, in our project, we chose a modular architecture, wherein the controller and its controlled process are not part of the same module. This decision is made because the actuators (dynamic modules), as a part of the controlled process, are intended to work on several platforms, potentially stimulated by different controllers, while the controller (trajectory controller) is parameterized to these different platforms [7,30]. Although this procedure is still under research, it allows to approve the safety of said controller individually when valid models of the dynamic modules and the further environment can be built in simulation. Using the actual external module, instead of a representation of it in tests of the module under consideration, cannot be seen as an option for a modular safety approval. The change may mean that a change of the external module requires a new safety approval if it is not clear how the changes influence the tests of the module under consideration. We see this as a boundary condition and categorize it as an invalid stimulation or evaluation. Therefore, it is not introduced as another error.

Figure 5 reveals that the output, respectively, its interface, is not assigned to an error after the fault tree analysis. Regarding the decomposition process, another error B1-2-3 is identified at this point, when the *output is insufficient for evaluation*, e.g., when information is missing. In contrast to an insufficient metric (B1-2-1), this case might have a metric, but it is technically not possible to provide the necessary information for this metric. For example, the test target, whether a planner is able to correctly detect traffic regulations, can be reached by using a metric that is simply comparing the actual traffic rules to the ones that the planner detects (e.g., a stop sign demanding to stop the vehicle in front of a defined line). However, a planning algorithm is usually a black box, mainly providing the trajectory as an output. The information about the detected traffic rules can therefore only be estimated indirectly by the given trajectory. This results in a more complex or even impossible required metric for evaluation.

3.2.2. Wrong Tests Are Performed

With error B2-1, we introduce another error type for situations when *test cases are not sufficiently derived* for module testing. In regard to the decomposition process, Amersbach and Winner [31] propose a way to reduce the parameter space of test scenarios by decomposing the system (an automated vehicle) and only considering parameters from its system level that are relevant to each separate decomposed functional layer. Their individual relevance can be specified by experts or, e.g., determined by sensitivity analyses. In order to specify relevance by experts, the performing experts need to obtain enough knowledge about the module's interfaces, functionalities and implementation. Knowledge about the implementation is important for cases where a parameter on system level p_{S1} cannot directly be transferred to the interface of the considered module with a parameter set p_M so that $p_{S1} \neq p_M$ applies. When p_{S1} influences other parameters p_M that influence the module, $p_M(p_{S1})$ applies so that p_{S1} needs to be considered in module tests. Missing information at the interfaces is already covered by error A3 in Section 3.2.1. Amersbach and Winner [31] provide the example that the difference between a hedge and a wall is not relevant for the functionality of the planner if it only uses the information that both are static objects that cannot be driven through. However, a planner may also use the class of the static objects to calculate a probability that dynamic objects (e.g., pedestrians) can get through these objects. For example, a hedge may have low density or some open parts so that objects can get through while a wall of 2 m height may have a lower probability for another object getting through. Even though the class of an object is no interface specified by the planner, it may be that the planning algorithm (especially one that is based on machine learning) uses other information to classify the object without the direct input of the class. Furthermore, for testing the actual implementation of the planner, timing differences between the detection of a wall and the detection of a hedge may occur. At least the time range that a planner needs to handle should be known for the specification of test cases. In general, this can be seen as hidden knowledge in the given information. This hidden knowledge can be revealed by knowledge about the implementation of the module or by specific testing (e.g., the aforementioned sensitivity analysis).

In our project, we revealed a parameter, which was specified as relevant by experts but does not significantly influence the behavior of the trajectory controller. Simultaneously, another parameter shows high influence which was not specified as relevant beforehand. In a test set with said module, circles with varying radii were driven in simulation. This was done due to the assumption that the controller can be challenged by higher lateral acceleration. However, the more relevant parameter that caused a higher control deviation was the yaw acceleration. Yaw acceleration is caused at the entrance from the previous straight line into the circle where the highest deviation from the planned trajectory can be observed.

Consequently, not only a functional description and knowledge of module implementation is required (as it is for a system as well), but also knowledge about the processing of parameters within the entire modular architecture from system level to module level. This is commonly achieved through interface specifications which provide detailed information about parameters and their possible values. However, for a modular safety approval, even more detailed descriptions become necessary. In order to clarify which information or behavior on system level is causing certain inputs to the individual modules, all transformations in each flow of information need to be defined. Conversely, it requires specification of all system level behavior which is caused by information outputs of modules. This is essentially what metrics (for direct outputs of modules) as well as simulation models of the environment or other modules are supposed to do. We already cover possible errors to this matter by error types B1-1, B1-2 and the following.

With the proposed test methods of ISO 26262, as listed in Table 1, interfaces, requirements, use cases and possible hazards should be addressed by testing particularly. These tests can be derived from system level information or other information, which has already been derived to module level. Interfaces are generally defined on module level. However, for a sufficient test case generation, not only the technical description of each interface is important, but also specified expectations for its values, their time behavior and possible sequences. This can be derived from information on other modules or the environment. Requirements on system level can be decomposed to the module level with the aid of the functional architecture. Use cases on system level can be decomposed in similar ways so that modular functionalities for each specific use case are formally described. On that basis, requirements for the module level can be derived additionally. Moreover, requirements are derived from the concrete technical design, especially when safety functions are needed due to the chosen design. These requirements do not have a direct connection to system requirements, initially. Still, they do have a connection to the system level through functions or requirements on module level. The same applies to possible hazards that can be identified directly on module level or derived from system to module level. Principally, module tests are derived such that the module-specific causes for the identified hazards are likely to occur. The final test cases which are designed for the system level can finally be broken down to module level as well, similar to use cases or, as called in the automated driving domain, scenarios. While all derived information may be insufficiently derived leading to error types B2-1-1-1, B2-1-2 and B2-1-3, certain parameters may also be neglected due to an invalid stimulation or evaluation. All errors of this chapter are illustrated in the fault tree in Figure 7.

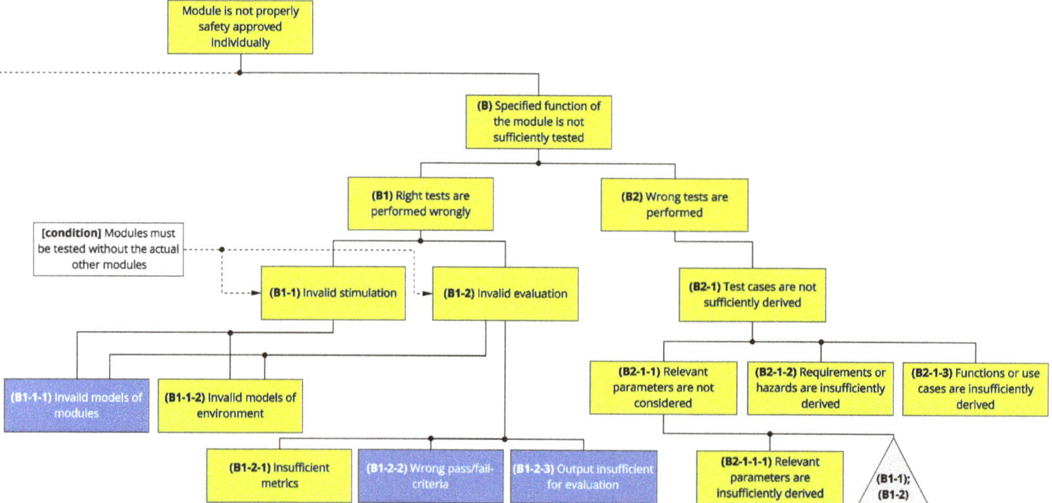

Figure 7. Fault tree of error B. Green boxes represent errors found by fault tree analysis. Blue boxes represent errors found by additional analysis of the test bench or test process. The grey triangle is referring to errors that were already introduced in the fault tree, but are also a cause for error B2-1-1. Dashed lines refer conditions for which the errors apply.

However, a lot of information that is relevant on system level is not relevant for every single particular module, but only for other specific modules. In UNICAR*agil*, we test the trajectory controller using trajectories that are previously generated based on maneuvers considering the vehicle's driving capabilities. Analyzing all possible scenarios on system level would only reveal that these are combinations of differently parameterized maneuvers. For example, a scenario wherein the regulations are not clear due to changes to the infrastructure (e.g., multiple traffic signs or lane markings) may be challenging for the environment perception and the planner but not for the controller. Still, an abrupt change of the trajectory needs to be considered in this scenario, which then also needs to be tested on controller level. Driving through road spray is another scenario on system level which could be challenging for module systems of the environment perception. Consequently, for the trajectory planner, this physically interfering factor could result simplified in a reduced range of view. This further leads to fast reaction requirements of the controller to achieve low braking distances or evasion capabilities. In another perspective, spray is a result of wet conditions and therefore correlates with a reduced friction coefficient in comparison to dry conditions. In conclusion, scenarios on system level can be used to derive test cases on module level directly, but need to be categorized to the functionalities of the considered module. Thus, a previously defined functional architecture is mandatory. This results in similar uncertainties as in the test case definition that is based on already decomposed information on module level, considering scenarios on system level may extend test cases generated from information on module level. However, it is likely to cause a less efficient process in comparison to the test case generation by module requirements, functions, or risks.

3.3. Rules to Avoid Errors Due to Decomposition

In order to avoid aforementioned errors, we provide certain rules that can be applied during development and testing. Therefore, each discovered error on the lowest level of the fault tree is used to define a rule, which then helps to avoid this specific error. These rules might serve as a general checklist and are listed in Table 2. Besides aiming to prevent the error, they also include previously described findings in regard to the validity of simulation models and the required information for test cases.

Table 2. List of rules (R) for the avoidance of errors due to decomposition.

No.	Rules
R1	The functional architecture shall contain all functions to fulfill the system's requirements. Thus, all functions shall be traceable to requirements of the system and vice versa.
R2	The modular architectures shall describe the interfaces between modules in detail to ensure that the functions together fulfill the system's requirements. The interface description shall at least contain a functional description, the permitted values, their time behavior and possible sequences including a failure model of each.
R3	All architectural views shall consider dependencies in all possible conditions of the modules or the overall system.
R4	All architectural views shall be analyzed for undesirable interference between its modules.
R5	The modular architectures shall provide interfaces that allow valid representations of other modules and the environment in regards to all desired test cases and their test targets. Thus, test cases and test targets shall be defined before the final definition of the modular architectures.
R6	The modular architectures shall provide interfaces on which all required metrics and associated pass-/fail criteria can be applied for evaluation. Representations of other modules or the environment might be considered for this. Thus, metrics and pass-/fail criteria shall be defined before the final definition of the modular architectures.
R7	All information on module level (e.g., as mentioned in this paper, functionalities, requirements, use cases, parameters, hazards and test cases) shall be traceable to the system level by at least one view of the modular architecture and vice versa.

Covering errors A1 and A2, the first rule R1 demands that module functions in a system fulfill all system requirements. While this alone seems to be trivial, it leads to the overarching recommendation to design functions and requirements traceable from system to module level and vice versa. This is particularly important when single changes of one module create an impact on other modules. All connected and possibly necessary changes then need to be traceable.

The second rule R2 demands a detailed description of the interfaces of each module in all different architectural views to cover error A3. Such a description shall not only contain a technical and a functional perspective as usual but instead cover all possible circumstances of the interface in its intended environment. Consequently, at least a functional description, permitted values, the time behavior and possible sequences, as well as including a failure model of each are mandatory. Failure models, including all probable or even allowed failures, are important for other affiliated modules in order to avoid failures on system level.

Here also, the next rule ties in. Rule R3 demands that all possible conditions of each module are already included in the architectural descriptions. Derived from these specifications, relevant conditions for other modules can be considered in all ensuing development and testing processes.

As a further guideline, rule R4 avoids error A4 by demanding an analysis of possible interferences between modules, which might have been overlooked during the specification of modular architectures. The descriptions demanded by the rules R2 and R3 support this analysis and are supplemented by its results.

For a valid stimulation rule R5, which covers errors B1-1-1 and B1-1-2, demands that the modular architecture provides interfaces that allow a valid representation of other modules and the environment. This leads to the additional need for an early definition of test targets and test cases, which clarify the necessary representations.

In a similar way, rule R6 demands interfaces that allow for a valid evaluation with metrics and associated pass-/fail criteria to cover errors B1-2-1, B1-2-2 and B1-2-3. Thus, they also need to be defined together with the associated test targets and test cases before the final definition of the modular architectures.

Finally, rule R7 covers B2-1-1-1, B2-1-2 and B2-1-3, demanding information on each module level to be traceable to the system level by at least one modular architecture and vice versa. Analogous to traceable requirements in R1, transparent information references are specifically important in order to accomplish sufficient development of test cases.

Yet, not all rules are measurable and thus cannot be verified directly. These rules should be broken down for an actual technical implementation to become verifiable, while others offer as indicators for developers and testers on what to pay attention to. Furthermore, every discovered violation of one of these rules can be taken for improving the applied development and test process. Primarily, the demanded traceability is well measurable but challenging in assessing whether the information at module level is sufficient for a modular safety approval.

4. Discussion

The presented approach of revealing errors due to the decomposition process does provide the opportunity to avoid and reduce these. Adjusting the development process or its methods in order to identify error consequences for modules or the entire system could be done by testing specifically for our discovered error types. While we only consider one general decomposition step, in a complete development process more specific faults may occur. This process as a whole is not part of our analysis, because processes in different companies and different domains vary widely. Still, we assume that the derived errors represent a wide coverage independently from specific faults that cause the errors. This is in particular supported by our new approach of using FTA with the help of generic architectures representing a wide range of systems. Even though this novel approach is not assumed to reveal all possible errors, it improves awareness for likely shortcomings and can therefore increase completeness in development and testing. The rules derived from our found errors serve as help for developers and testers for a modular safety approval, but due to their generalization cannot be verified until they are broken down into more specific rules or requirements for a specific system. Finally, focusing the analyses on the decomposition process is just one part of the safety approval, while faults and errors as they also occur for the system level need to be considered as well. However, the decomposition of the system is the mandatory difference between the two levels. In regard to the described necessary steps in development and testing, the analysis effort is expected to increase and needs to be weighed against the conventional approaches on system level.

5. Conclusions

In this paper, we have shown that the new concept of a modular safety approval requires a structured and specified procedure in development and testing that considers the differences between system and modules. A fault tree analysis together with the analysis of an associated functional architecture and a generalized test bench is presented as a new approach to identify possible errors in modular development projects. Firstly, the system level is compared to the module level in regard to available information and test methods. Resulting conclusions are used to derive errors that may occur during the decomposition of this information for the development and testing of systems, respectively, of its modules with applied decomposition.

Subsequently derived rules for the decomposition process can be broken down into more specific rules or requirements for a specific system with its modules in order to support developers and to become verifiable themselves. Therefore, we provide our findings as a starting point to identify further errors and rules, which assist to reach a successful modular safety approval.

The initial question of why the state of the art does not dispense on system tests yet can be traced back from the described processes according to the state of the art. Industry norms and standards, at least in the automotive industry, explicitly require to perform the safety validation on system level (or for ISO 26262 equivalently seen at the vehicle level). One reason for this is the lack of supervision processes for the development steps of modules. The focus of information aggregation, process control and testing instead still lies on system level. Since the system level provides less uncertainty than lower levels, fewer analyses of uncertainties are required for the safety validation. Still, we expect interest in modular development to increase despite the additional analysis effort. The potential reduction of the enormous testing efforts by a modular safety approval, in which a module can be used for a variety of vehicles and does not require regression testing after modifications of other modules, is key for an economically feasible introduction of automated vehicles. Looking ahead, implementing modular approaches in norms and industry standards might prove to be advantageous for other vehicle functions as well. Finally, modular practices might be particularly beneficial for the automotive supplier industry, enabling them to develop and sell modules without requiring validation on a system level.

Author Contributions: Conceptualization, B.K. and H.W.; methodology, B.K. and H.W.; formal analysis, B.K. and H.W.; investigation, B.K. and H.W.; resources, B.K. and H.W.; writing—original draft preparation, B.K.; writing—review and editing, H.W. and B.K.; visualization, B.K.; supervision, H.W.; project administration, H.W.; funding acquisition, H.W. All authors have read and agreed to the published version of the manuscript.

Funding: Federal Ministry of Education and Research: FKZ 16EMO0286; Federal Ministry for Economic Affairs and Energy: FKZ 19A19002S; Deutsche Forschungsgemeinschaft (DFG—German Research Foundation) and the Open Access Publishing Fund of Technical University of Darmstadt.

Institutional Review Board Statement: Not applicable.

Informed Consent Statement: Not applicable.

Data Availability Statement: Not applicable.

Acknowledgments: The authors would like to thank Johannes Krause, Felix Glatzki, and Moritz Lippert for proofreading.

Conflicts of Interest: The authors declare no conflict of interest.

References

1. Raghupatruni, I.; Burton, S.; Boumans, M.; Huber, T.; Reiter, A. Credibility of software-in-the-loop environments for integrated vehicle function validation. In *20. Internationales Stuttgarter Symposium*; Bargende, M., Reuss, H.-C., Wagner, A., Eds.; Springer Fachmedien Wiesbaden: Wiesbaden, Germany, 2020; pp. 299–313, ISBN 978-3-658-30995-4.
2. International Organization for Standardization. *Road Vehicles—Functional Safety (ISO 26262:2018)*; International Organization for Standardization: London, UK, 2018.
3. Wachenfeld, W.; Winner, H. The Release of Autonomous Vehicles. In *Autonomous Driving: Technical, Legal and Social Aspects*; Maurer, M., Lenz, B., Winner, H., Gerdes, J.C., Eds.; Springer: Berlin/Heidelberg, Germany, 2016; pp. 425–449, ISBN 978-3-662-48847-8.
4. Wood, M.; Robbel, P.; Maass, M.; Tebbens, R.D.; Meijs, M.; Harb, M.; Reach, J.; Robinson, K. Safety First for Automated Driving. 2019. Available online: https://www.daimler.com/dokumente/innovation/sonstiges/safety-first-for-automated-driving.pdf (accessed on 16 September 2021).
5. International Organization for Standardization. *Road Vehicles—Safety of the Intended Functionality (ISO DIS 21448:2021)*; International Organization for Standardization: London, UK, 2021.
6. Amersbach, C.; Winner, H. Defining Required and Feasible Test Coverage for Scenario-Based Validation of Highly Automated Vehicles. In Proceedings of the 22nd IEEE Intelligent Transportation Systems Conference (ITSC), Auckland, New Zealand, 27–30 October 2019; IEEE: Piscataway, NJ, USA, 2019; pp. 425–430, ISBN 978-1-5386-7024-8.

7.   Woopen, T.; Lampe, B.; Böddeker, T.; Eckstein, L.; Kampmann, A.; Alrifaee, B.; Kowalewski, S.; Moormann, D.; Stolte, T.; Jatzkowski, I.; et al. UNICARagil—Disruptive Modular Architectures for Agile, Automated Vehicle Concepts. In Proceedings of the 27th Aachen Colloquium, Aachen, Germany, 10 October 2018.

8.   Leveson, N. *Engineering a Safer World: Systems Thinking Applied to Safety*; MIT Press: Cambridge, MA, USA, 2012; ISBN 9780262016629.

9.   Schäuffele, J.; Zurawka, T.; Carey, R. *Automotive Software Engineering: Principles, Processes, Methods, and Tools*, 2nd ed.; SAE International: Warrendale, PA, USA, 2016; ISBN 9780768083347.

10.   Walden, D.D.; Roedler, G.J.; Forsberg, K.; Hamelin, R.D.; Shortell, T.M. (Eds.) *Systems Engineering Handbook: A Guide for System Life Cycle Processes and Activities*, 4th ed.; INCOSE-TP-2003-002-04; Wiley: Hoboken, NJ, USA, 2015; ISBN 978-1-118-99940-0.

11.   Steimle, M.; Menzel, T.; Maurer, M. Towards a Consistent Terminology for Scenario-Based Development and Test Approaches for Automated Vehicles: A Proposal for a Structuring Framework, a Basic Vocabulary, and Its Application. 2021. Available online: https://arxiv.org/pdf/2104.09097 (accessed on 5 October 2021).

12.   Object Management Group. OMG Unified Modeling Language Specification (OMG UML): Version 2.5.1. Available online: https://www.omg.org/spec/UML/ (accessed on 4 October 2021).

13.   Göpfert, J. *Modulare Produktentwicklung: Zur Gemeinsamen Gestaltung von Technik und Organisation*; Deutscher Universitätsverlag: Wiesbaden, Germany, 1998; ISBN 978-3-8244-6827-0.

14.   Ross, H.-L. *Functional Safety for Road Vehicles*; Springer International Publishing: Cham, Switzerland, 2016; ISBN 978-3-319-33360-1.

15.   Sanchez, R. Building real modularity competence in automotive design, development, production, and after-service. *IJATM* **2013**, *13*, 204. [CrossRef]

16.   Sanchez, R.; Shibata, T. *Modularity Design Rules for Architecture Development: Theory, Implementation, and Evidence from Development of the Renault-Nissan Alliance "Common Module Family" Architecture*; WorkingPaper; Data Science and Service Research Discussion Paper No. 80; IDEAS: Tokyo, Japan, 2018.

17.   Avizienis, A.; Laprie, J.-C.; Randell, B.; Landwehr, C. Basic concepts and taxonomy of dependable and secure computing. *IEEE Trans. Dependable Secur. Comput.* **2004**, *1*, 11–33. [CrossRef]

18.   Stolte, T.; Ackermann, S.; Graubohm, R.; Jatzkowski, I.; Klamann, B.; Winner, H.; Maurer, M. A Taxonomy to Unify Fault Tolerance Regimes for Automotive Systems: Defining Fail-Operational, Fail-Degraded, and Fail-Safe. 2021. Available online: http://arxiv.org/pdf/2106.11042v3 (accessed on 20 September 2021).

19.   Institute of Electrical and Electronics Engineers; IEEE Computer Society; IEEE-SA Standards Board. *IEEE Standard Classification for Software Anomalies*; Institute of Electrical and Electronics Engineers: New York, NY, USA, 2010; ISBN 978-0-7381-6114-3.

20.   Vesely, W.E. *Fault Tree Handbook*; Nuclear Regulatory Commission: Washington, DC, USA, 1981.

21.   Leveson, N.G.; Thomas, J.P. STPA Handbook. Available online: http://psas.scripts.mit.edu/home/get_file.php?name=STPA_handbook.pdf (accessed on 16 September 2021).

22.   Junietz, P.M. Microscopic and Macroscopic Risk Metrics for the Safety Validation of Automated Driving. Ph.D. Thesis, TU Darmstadt, Darmstadt, Germany, 2019.

23.   Klamann, B.; Lippert, M.; Amersbach, C.; Winner, H. Defining Pass-/Fail-Criteria for Particular Tests of Automated Driving Functions. In Proceedings of the 22nd IEEE Intelligent Transportation Systems Conference (ITSC), Auckland, New Zealand, 27–30 October 2019; IEEE: Piscataway, NJ, USA, 2019; pp. 169–174, ISBN 978-1-5386-7024-8.

24.   Homolla, T.; Gottschalg, G.; Winner, H. Verfahren zur Korrektur von Inkonsistenten Lokalisierungsdaten in modularen technischen Systemen. In *Uni-DAS 13. Workshop Fahrerassistenz und Automatisiertes Fahren. FAS 2020*; RWTH Aachen University: Aachen, Germany, 2021.

25.   Buchholz, M.; Gies, F.; Danzer, A.; Henning, M.; Hermann, C.; Herzog, M.; Horn, M.; Schön, M.; Rexin, N.; Dietmayer, K.; et al. Automation of the UNICARagil vehicles. In Proceedings of the 29th Aachen Colloquium Sustainable Mobility, Aachen, Germany, 5–7 October 2020.

26.   Junietz, P.; Wachenfeld, W.; Klonecki, K.; Winner, H. Evaluation of Different Approaches to Address Safety Validation of Automated Driving. In Proceedings of the 21st International Conference on Intelligent Transportation Systems (ITSC), Maui, HI, USA, 4–7 November 2018; pp. 491–496.

27.   Sargent, R.G. Simulation model verification and validation. In Proceedings of the 1991 Winter Simulation Conference, Phoenix, AZ, USA, 8–11 December 1991; Nelson, B.L., Kelton, W.D., Clark, G.M., Eds.; IEEE: Piscataway, NJ, USA, 1991.

28.   Hood, C.; Fichtinger, S.; Pautz, U.; Wiedemann, S. *Requirements Management: The Interface Between Requirements Development and All Other Systems Engineering Processes*; Springer: Berlin/Heidelberg, Germany, 2008; ISBN 978-3-540-68476-3.

29.   Rosenberger, P.; Holder, M.; Zofka, M.R.; Fleck, T.; D'hondt, T.; Wassermann, B.; Prstek, J. Functional Decomposition of Lidar Sensor Systems for Model Development. In *Validation and Verification of Automated Systems: Results of the ENABLE-S3 Project*; Leitner, A., Watzenig, D., Ibanez-Guzman, J., Eds.; Springer International Publishing: Cham, Switzerland, 2020; pp. 135–149, ISBN 978-3-030-14628-3.

30.   Martens, T.; Pouansi Majiade, L.B.; Li, M.; Henkel, N.; Eckstein, L.; Wielgos, S.; Schlupek, M. UNICARagil Dynamics Module. In Proceedings of the 29th Aachen Colloquium Sustainable Mobility, Aachen, Germany, 5–7 October 2020.

31.   Amersbach, C.; Winner, H. Functional Decomposition: An Approach to Reduce the Approval Effort for Highly Automated Driving. In Proceedings of the 8. Tagung Fahrerassistenz, München, Germany, 22–23 November 2017.

Article

Increasing the Safety of Adaptive Cruise Control Using Physics-Guided Reinforcement Learning

Sorin Liviu Jurj *, Dominik Grundt, Tino Werner, Philipp Borchers, Karina Rothemann and Eike Möhlmann

OFFIS e.V. Institute for Information Technology, Escherweg 2, 26121 Oldenburg, Germany; dominik.grundt@offis.de (D.G.); tino.werner@offis.de (T.W.); philipp.borchers@offis.de (P.B.); karina.rothemann@offis.de (K.R.); eike.moehlmann@offis.de (E.M.)
* Correspondence: sorin.jurj@offis.de; Tel.: +49-441-9722-493

Abstract: This paper presents a novel approach for improving the safety of vehicles equipped with Adaptive Cruise Control (ACC) by making use of Machine Learning (ML) and physical knowledge. More exactly, we train a Soft Actor-Critic (SAC) Reinforcement Learning (RL) algorithm that makes use of physical knowledge such as the jam-avoiding distance in order to automatically adjust the ideal longitudinal distance between the ego- and leading-vehicle, resulting in a safer solution. In our use case, the experimental results indicate that the physics-guided (PG) RL approach is better at avoiding collisions at any selected deceleration level and any fleet size when compared to a pure RL approach, proving that a physics-informed ML approach is more reliable when developing safe and efficient Artificial Intelligence (AI) components in autonomous vehicles (AVs).

Keywords: adaptive cruise control; informed machine learning; physics-guided reinforcement learning; safety; autonomous vehicles

---

Citation: Jurj, S.L.; Grundt, D.; Werner, T.; Borchers, P.; Rothemann, K.; Möhlmann, E. Increasing the Safety of Adaptive Cruise Control Using Physics-Guided Reinforcement Learning. *Energies* **2021**, *14*, 7572. https://doi.org/10.3390/en14227572

Academic Editors: Arno Eichberger, Zsolt Szalay, Martin Fellendorf and Henry Liu

Received: 3 September 2021
Accepted: 13 October 2021
Published: 12 November 2021

Publisher's Note: MDPI stays neutral with regard to jurisdictional claims in published maps and institutional affiliations.

1. Introduction

According to a recent study, [1], around 94% of road accidents are happening due to human errors. For this reason, considerable efforts are made by the scientific research institutions and the automotive industries in order to reach autonomous cars that are safer than human drivers [2]. These efforts are driven also by the fact that AVs are becoming influential on the social and economic development of our society [3]. Nevertheless, because usually, the AI models used in AVs are dependent on huge amounts of data and labeling efforts, which are mostly expensive and hard to obtain, this can result in so-called "black box" AI models which are limited not only due to the size of the dataset they were trained on but also due to imperfect labeling. This is a very crucial problem regarding safety because the resulting AI models which are agnostic to real physical relations and principles found in the real world, being unable to generalize well to unseen scenarios [4]. This is especially the case for accidents as the frequency of critical situations is very low, and, thus, the number of such situations in datasets collected from real-world recordings tends to be low as well.

Thus, there is a need for a new kind of AI models that are more efficient regarding safety, interpretability, and explainability, with a promising viable solution in this direction being represented by the use of so-called Informed ML [5] approaches where AI models can be improved by using additional prior knowledge into their learning process. Recently, this approach is proving to be successful in many fields and applications such as lake temperature modeling [4], MRI reconstruction [6], real-time irrigation management [7], structural health monitoring [8], fusion plasmas [9], fluid dynamics [10] and machining tool wear prediction [11]. However, regarding autonomous driving, this approach was not fully explored, with recent research projects such as KI Wissen [12] funded by the German Federal Ministry for Economic Affairs and Energy being one of of the first, if not, the first one that tries to bring knowledge integration into Automotive AI in order to increase their safety.

With regard to autonomous driving, one of many safety-critical components is considered to be the ACC, mainly due to its ability to increase safety and driving comfort by automatically adjusting the speed of the ego-vehicle according to the position and speed of a leading vehicle while following it. ACCs are also known for having several advantages over human driving such as reducing the energy consumption [13] in a vehicle or improving the traffic dynamics [14], to name only a few. Despite being available in many modern vehicles, ACCs are still heavily dependent on the available sensors equipped on them. These sensors differ for each manufacturer and model, such as radar and LIDAR, which can either have a malfunction or their sensor data readings are affected by noisy and low accuracy data [15] which can lead to instability, severe conditions regarding speed, discomfort, and even risks of collisions [16]. More than that, because ACCs are typically approached as a model-based controller design based on an Intelligent Driver Model (IDM), despite performing decently on highways, they lack the ability to adapt to environments or driving preferences, and thus, an RL-based ACC approach is seen as more favorable towards fully autonomous cars which can be fully trusted by humans. Some of the main reasons for this are the advantages of an RL-based ACC approach such as that it does not require a dataset and that training can be realized irrespective of the environment [17].

Considering these aspects, in this paper, we show, to the best of our knowledge, for the first time in literature, a PG RL approach, which is able to increase the safety of vehicles equipped with ACC by a large margin for any deceleration level and at any fleet size when compared to a pure RL approach, also in the case when the input data is perturbed. Despite the fact that platooning scenarios, even the ones using RL, have already been considered in the literature, many works focus on the yet unrealistic scenario of communicating vehicles so that each vehicle in the queue immediately receives non-perturbed information about the intended actions of all other vehicles, as seen in the work presented by the authors in [18], or which perform joint optimization as seen in the work by the authors in [19]). Similarly, the work in [20] restricts the communication between the individual vehicles but they consider platooning scenarios that differ from ours by using other control schemes (e.g., averages of four controllers) as well as by the goal of focussing on the lead vehicle of a platoon. The novelty of our approach presented in this paper is the combination of RL with deep state abstractions, reward shaping w.r.t. a safety requirement (i.e., jam-avoiding distance), perturbed inputs as well as individual behavior in an AVs platoon regarding car-following scenarios. By using the proposed PG RL approach for ACC, we demonstrate that it is possible to improve an AI model's performance (less collisions and more equidistant travel) only by using physical knowledge as part of a pre-processed input, without the need of extra information.

The paper is organized as follows. In Section 2, we present the related work regarding different implementations of ACCs using physics or using RL. Section 3 details the proposed PG RL solution for increasing the safety of ACCs. Section 4 presents the simulation details of the car-following scenario implementation. In Section 5, we present the experimental setup and results. Finally, in Section 6, we present the conclusions and future work of this paper.

2. Related Work

Recently, the advancement of AVs technology has resulted in unique concepts and methods that allow the successful deployment of vehicles capable to drive in different levels of autonomy. However, different authors used different approaches to target safe self-driving control speed and learning navigation. In addition, there are several works that propose solutions regarding safer ACCs either using only physical knowledge or by using ML methods such as RL [21,22].

In the field of transportation engineering, the work in [23] serves as an introduction and analysis of the theoretically successful AI frameworks and techniques for AVs control in the age of mixed automation. They conclude that multi-agent RL algorithms are being preferred for long-term success in multi-AVs. The authors in [24] introduce a cooperative ACC method that makes use of an ACC controller created using the concept of RL in order to manage traffic efficiency and safety, showing impressive results in their experiments with a low-level controller. The work in [25] successfully implemented a method for Society of Automotive Engineers (SAE) low-cost modular AV by designing a vehicle unique in the industry, and which proves to be able to transport persons successfully. The approach in this work leads to the realistic application of behavioral replication and imitation learning algorithms in a stable context. The authors in [14] proposed a physics-based jam-avoiding ACC solution based on an IDM and proved that by using physical knowledge, the traffic congestion can be drastically improved by employing even a small number of vehicles equipped with ACCs. The authors in [13] propose an end-to-end vision-based ACC solution based on deep RL using the double deep Q-networks method, and which is able to generate a better gap regulated as well as a smoother speed trajectory when compared to a traditional radar-based ACC or human-in-the-loop simulation. Also, the authors in [17] proposed an RL-based ACC solution that is capable of mimicking human-like behavior and is able to accommodate uncertainties, requiring minimal domain knowledge when compared to traditional non-RL-based ACCs in congested traffic scenarios in a crowded highway as well as countryside roads. The work in [26] evaluates the safety impact of ACCs in traffic oscillations on freeways also by using a modified version of IDM in order to simulate the car-following movements using Matlab2014b software, concluding that an ACC system can significantly improve safety only when parameter settings such as larger time gaps, smaller time delays, and larger maximum deceleration rates are maintained. Physical and world knowledge was used also in other deep learning models such as regarding the off-road loss in [27] and models that respect dynamic constraints [28], both of these approaches being combined in the work presented in [29]. In addition, the authors in [30] add a kinematic layer to the model which produces kinematically conform trajectory points that serve as additional training points for prediction. World knowledge, in terms of social rules, has been integrated into deep learning models in [31] where residuals are added to knowledge-driven trajectories in order to realistically reflect pedestrian behavior, and in [32] where social interaction is invoked in order to make collision-free trajectory predictions for pedestrians. A similar work is presented also in [33], where interaction-aware trajectory predictions for vehicles are computed. Concerning the violation of traffic rules, the work in [34] uses a penalty term for adversarial agents, with the work in [35] also adding a collision reward term as well as a penalty for unrealistic scenarios. Regarding safety distance, this has been considered by the authors in [36] who added a safety distance violation penalty and a collision penalty, among others, to a hierarchical RL model, by the authors in [37], who consider a fixed safety distance in overtaking maneuvers, and also by the authors in [38], where a distance reward is invoked in car-following maneuvers.

The works mentioned in this chapter highlight the importance of safety in ACCs in the literature, indicating that by using either physics or ML-based solutions such as RL, considerably better results can be obtained. However, to the best of our knowledge, there is no work in literature that combines both physics and deep RL in order to increase the safety of ACCs. For this reason, in this paper, we combine the two approaches of physics knowledge as well that of RL into a stand-alone PG RL solution, providing a basis for future researchers to build upon.

3. Physics-Guided Reinforcement Learning for Adaptive Cruise Control

In this section, we describe the proposed approach that combines the physical knowledge in the form of jam-avoiding distance together with the SAC RL algorithm [39] in order to increase the safety of ACCs. First, we briefly introduce the SAC algorithm, followed by the physical model used, and finally, we also show their merging approach and how the integration of prior knowledge is realized in this work.

3.1. Soft Actor-Critic Algorithm

In this paper, we make use of the RL framework for training our ACC model, more exactly, of the SAC RL algorithm [39]. RL refers to a collection of learning techniques that train an agent through experience. Here, the experience is collected as a simulation in the forms of states, actions, and rewards in order to find the policy that maximizes the expected cumulative reward it obtains. One of the main advantages of RL is that it does not require a specific dataset for training, the data used for its training being generated as experience in the simulation. However, many of the existent RL algorithms found in the literature have limitations during on-policy learning such as sample inefficiency as well as during off-policy learning such as hyperparameter sensitivity and increased time required for tuning them in order to achieve convergence.

SAC [39] is an off-policy state-of-the-art RL algorithm that does not have the limitations mentioned above. This is the reason we choose to use SAC in our work. Furthermore, we deal with continuous action spaces where SAC is not efficient in maximizing the reward but still in maximizing the entropy of the policy. This is important as a higher entropy encourages a higher exploration of the state space by the agent and improves the convergence [39]. In order to achieve such improvements using a random strategy over other RL algorithms that use deterministic strategy, SAC according to [39] makes use of soft Q-learning, relying on two different function approximators such as a soft Q-value function as well as a stochastic policy which are optimized alternately. The soft Q-function $Q_\theta(s_t, a_t)$ with s_t describing the state at time t and a_t the action at time t, is parametrized by θ. The tractable policy $\pi_\phi(s_t|a_t)$, containing the state-action pair, is parametrized with ϕ.

3.2. Prior Knowledge

From traffic experiences as well as from governmental traffic rules, it is known that traffic participants have to ensure a sufficient safety distance to each other, to avoid possible collisions. Besides this prior world knowledge, there is also conjunctive physical knowledge on how the distance between an agent and a leading vehicle can be controlled. An example regarding this aspect is given by the authors in [14] who extend an existing IDM-Model in order to realize an ACC lane following controller with model parameters in Table 1. Based on that desired parameters such as velocity, acceleration constraints, and minimum distance for jam-avoiding, the authors present the desired acceleration for a jam-free lane as seen in Equation (1) [14].

$$acceleration = a_m \left[1 - \left(\frac{v}{v_0} \right)^4 - \left(\frac{s^*(v, \Delta v)}{s} \right)^2 \right] \tag{1}$$

Here, s is the distance to the leading vehicle and $s^*(v, \Delta v)$ describes the minimum jam-avoiding distance depending on the current agent velocity v and the velocity difference Δv to the leading vehicle.

Table 1. Static model parameters used in the proposed approach for increasing the safety of ACC [14].

Static Model Parameter	Symbol	Value
Desired velocity	v_0	120 km/h
Save time headway	T	1.5 s
Maximum acceleration	a_m	1.0 m/s^2
Desired deceleration	b	2.0 m/s^2
Jam distance	s_0	2 m

Together with a static minimum distance s_0 for low velocities, the save time headway T and desired deceleration and maximum acceleration b and a_m respectively, $s^*(v, \Delta v)$ results to the Equation (2) [14]:

$$s^*(v, \Delta v) = s_0 + max(0, vT + \frac{v\Delta v}{2\sqrt{a_m b}}). \tag{2}$$

Considering the goal of this paper, we make use of Equation (2) when integrating prior knowledge into the SAC RL algorithm.

3.3. Integration of Prior Knowledge

The main goal of the integration is to help the autonomous agent learn the correct control actions that result in reasonable trajectories. Regarding this aspect, one can distinguish essentially between supervised and non-supervised algorithms for such problems. Supervised approaches like constrained control algorithms [40] optimize a particular objective function constrained to some hard constraints which formalize safety requirements. By solving the respective constrained optimization problem, the ego-trajectory is guaranteed to satisfy the safety constraints. A major drawback of these supervised strategies is that they require target/reference trajectory points and velocities, to name only a few, which are usually difficult to obtain [41]. In contrast, RL cannot cope with hard constraints but poses them as soft constraints (where their severity depends on the regularization parameter λ) onto the objective function. The main advantage is, however, that RL does not require any data but generates the data during training where it learns which trajectories and therefore, which control actions are reasonable in which situation by receiving reward feedback, so actions that severely violate the constraints lead to very low rewards. In our work, we design the regularization term of the reward function, i.e., the soft constraint, so that it represents the safety constraint in order to keep the optimal jam-avoiding distance, encouraging the agent to respect this safety constraint.

Following, we present our merging approach between the SAC architecture and the physical knowledge. As can be seen in Figure 1, a typical regular RL approach for ACC (black arrows) considers information about relative velocity and front vehicle distance based on radar systems. In addition, the current velocity is taken into account by the actor networks. According to this raw data by the sensors, the normal RL approach is deciding about the next acceleration steps. However, regarding the overall ACC goal of driving in perfect target separation as often as possible, the proposed PG RL approach is taking an important relation, namely the jam-avoiding distance (red arrows), between the raw data into account. By considering the jam-avoiding distance with the sensor data and the model parameters seen in Table 1, the actor-network is better prepared than the normal RL approach on finding an optimal policy for the ACC.

A more detailed explanation for the choice of this physical knowledge is explained later in the States subsection of this paper. In addition, a comparison and evaluation between both approaches are detailed in Section 5.

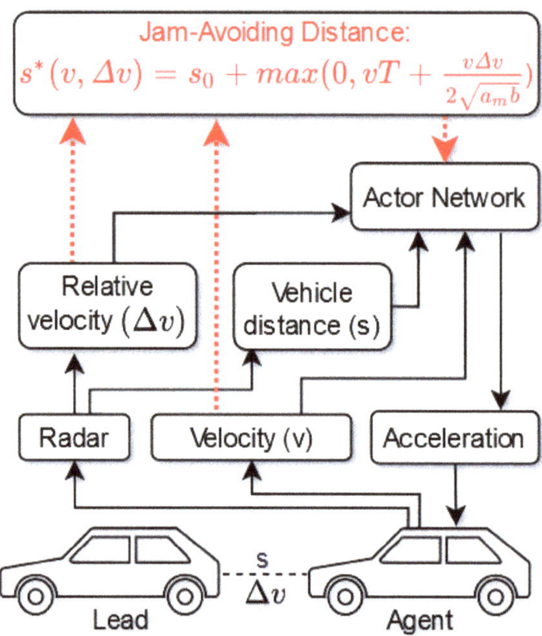

Figure 1. Proposed PG RL approach for increasing the safety of ACC by integrating prior knowledge in the form of the Jam-Avoiding Distance.

4. Simulation

Regarding the ACC system, we implemented a car-following scenario. For the scenario, we consider an urban road and normal weather conditions without influencing any fraction coefficients. In the main simulation, we assume perfect perception without perturbations s.t. all required data and information are available at any time; however, afterwards, we also performed the same procedures but introducing perturbations. The basic setup consists of two vehicles, a leading vehicle and one following vehicle, which contains the acting agent calculating the acceleration of the agent vehicle. Initially, the distance between the vehicles is 20 m. Based on the adapted physical IDM [14], the static model parameters applied are the ones presented earlier in Table 1. Following the RL approach and the physical model, for each simulation step, the acceleration to be executed is determined by the actor-network by extrapolating the current state to the next partial state based on the current position and velocity. Here, the resulting velocity and position, which are the relation values for the used physical model and the environment, are determined by the Eulers method seen in Equation (3):

$$f(t + h) = f(t) + h\frac{df}{dt}(t) \tag{3}$$

with the step size $h = 0.1$.

The resulting velocity of an agent is thus determined in each simulation step with Equation (4):

$$v_{t+1} = v_t + h \cdot a_{t+1} \tag{4}$$

where v_t is the velocity at time t and a_{t+1} is the acceleration determined by the artificial neural network at time $t + 1$.

The same procedure is also used to determine the new position of an agent with Equation (5):

$$x_{t+1} = x_t + h \cdot v_{t+1} \tag{5}$$

where x_t is the position at time t.

4.1. Leading Agent Acceleration

The only parameter that is not directly handled by the physical model and the virtual environment is the acceleration of the leading vehicle. In order to enable a simulation also for the leading vehicle, we need an acceleration replacement, such as one of the following heuristics presented in Table A1:

- Random acceleration at each time step (*randomAcc*),
- Constant acceleration with random full stops (setting lead velocity with $v = 0$) (*randomStops9* accelerates by 90% of its capacity and *randomStops10* accelerates full throttle)
- Predetermined acceleration for each time step (*predAcc*).

Based on the simulation performance with the test results presented in Section 4.6, the predetermined acceleration heuristic was chosen in the following manner: first, the vehicle will accelerate at 0.8 of its maximum acceleration until reaching half of its maximum speed. In this part, the agent will have to learn to accelerate but will not be able to accelerate at maximum capacity, being forced to also learn some control. Secondly, the vehicle will decelerate constantly until it stops. This will force the agent to learn to brake. Finally, it will repeat the first two steps, but accelerating at 0.9 of its maximum capacity, thus forcing the agent to accelerate at a greater capacity and then brake from a higher velocity as well.

4.2. States

The overall MDP is given by the tuple $(\mathcal{S}, \mathcal{A}, T, r)$ with the state space $\mathcal{S}$, the action space $\mathcal{A}$, a deterministic transition model $T : \mathcal{S} \times \mathcal{A} \rightarrow \mathcal{S}$ and rewards r. A discount factor is not considered in this work. The goal is to learn a deterministic parametric policy $\pi_\phi : \mathcal{S} \rightarrow \mathcal{A}$.

Regarding the state space, the simulation is fundamentally driven by three different parameters. One parameter to consider is the separation between agents. The second parameter is the speed difference between the agent and the lead vehicle (approaching velocity). Lastly, the speed of the acting agent is observed. Here, because the Q-function is modeled as an expressive neural network in the SAC algorithm [39], for faster processing, the value domains of the parameters were normalized to the interval $[0, 1]$.

Based on the integration of the physical model from [14] and the consequently relevant target separation s^*, two further indicators are introduced for the simulation. First, the target separation itself is observed as a parameter. This was also normalized to the interval $[0, 1]$. Secondly, a Boolean was introduced, which indicates whether the current separation is smaller (0) or larger (1) than the target separation. The reason for introducing this value was to provide the agent with an additional indicator for improving the determination of the acceleration that needs to be executed. In Section 5, we will evaluate the impact of adding this physical knowledge as inputs to the agent in the learning process.

Regarding the action space, we translated the asymmetric interval ranging from the maximum negative deceleration to the maximum acceleration into the symmetric interval $\mathcal{A} = [-1, 1]$.

4.3. Penalization

Based on the present scenario with an ACC system, in case of a collision with the agent in front an agent is penalized with a negative reward. In this work, different magnitudes from 0 to 10^6 for the execution of the penalization were tested. We discovered that if the penalization for the collision of an agent is too large, in order to avoid collisions, the agent may learn not to move at all. On the other hand, if the penalization is too small, the agent may ignore this misbehavior. In order to handle the adjustment of the correct penalization magnitude, a test attempt was made to introduce another penalization for not moving. Finally, after experimenting with different magnitudes we discovered that a relatively small collision penalization of 3000 has been working the best, this penalization value being applied when the agent collides (meaning that the resulting reward will be reduced by 3000). We observed that a good but riskier policy achieves a better learning result due to the chosen reward function than the search for a possibly fundamentally new strategy due to a high collision penalization. Thus, in the case of a good but risky policy, the selected reward function is taking a collision risk more into account. More detailed information about the test results can be seen in Section 4.6.

4.4. Reward

In the course of several simulations, several different reward functions were considered. First, a target distance reward that evaluates the absolute difference between the current separation of the vehicle and the target separation was tested (named as *absoluteDiff* in Table A1). This metric was not useful due to the bias introduced into being closer to the lead vehicle rather than farther behind. A second reward function tested was related to velocity (named as *velocity* in Table A1): the faster the vehicles follow each other without collision, the better the strategy was. We observed that, when considering possible speed limits, this reward function can only slightly lead to an improvement of an already good but not optimal strategy in the search space. The last reward function examined does not contain an evaluation of a strategy but only the penalization if a collision occurs or if the agent does not move (named as *None* in Table A1). It is important to mention that a liveness reward that encourages the agent to move must however be designed with caution since the maximal liveness reward should be very small regarding absolute value in comparison to penalties for violating the speed limit.

Across our simulations, the following target distance reward function (named as *symmetric* in Table A1) performed the best. More exactly, the reward of a performed action was then determined by Equation (6):

$$r = -\frac{|s - ts|}{ts} - \frac{|s - ts|}{2s} \tag{6}$$

with $ts = s_0 + max(0, vT)$ being the target separation at the given speed. This reward has only one optimal point at $s = ts$. The reward is also symmetrical to its variables, so for example if $s = 2 \cdot ts$ or $ts = 2 \cdot s$ are considered, the reward value is the same in both cases, as can be also observed in Figure 2 with $ts = 10$ (for different values of ts the function has the same properties). More detailed information about the test results can be seen in Section 4.6.

4.5. Termination Conditions

As termination conditions for a simulation run, we consider the goal of the system to be that of traveling as fast as possible while producing no collisions. This requires a suitable termination criterion for the simulation to be found. In our tests, we observed that the sole inclusion of collision in the termination criterion leads to the fact that the simulation does not end when the agents have found an optimal policy. We also found that, if a fixed period of time is included in the termination criterion, an acting agent can find a good policy by merely driving slowly within this fixed period. Finally, we observed that a termination after a certain number of simulation steps was more reasonable. Therefore, the chosen termination criterion for the simulation is a combination of a collision consideration and a certain number of simulation steps.

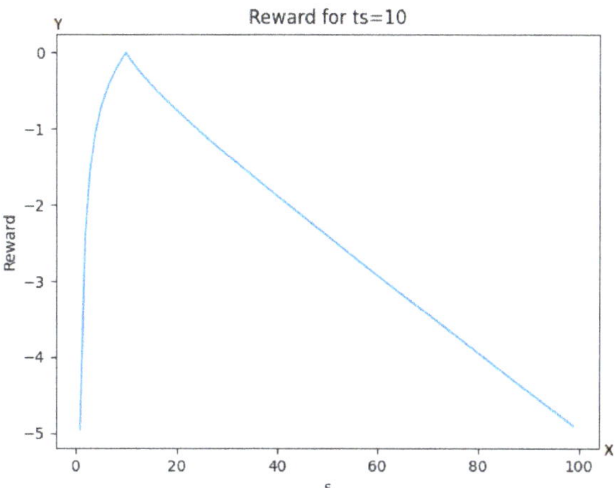

Figure 2. Graph of the reward function for different values of s at ts = 10 m. The *y*-axis is representing the reward value, while the *x*-axis is representing the number of simulation steps.

4.6. Parameter Search Test

Considering the high number of choices there are to make in the realms of reward, penalization and lead vehicle behavior, we performed trainings for each possible combination of a set of parameters. More exactly, for the lead vehicle behavior, the possibilities considered are the ones described earlier in Section 4.1 (*randomAcc, randomStops9, randomStops10,* and *predAcc*). Regarding the reward, the possibilities considered are the ones described earlier in Section 4.4 (*symmetric, velocity, absoluteDiff* and *None*). Regarding the penalization, the values considered were 0, 100, 3000 and 100,000. This gives us a total of 64 different combinations of parameters. In order to find the best set of parameters, for each of the 64 different combinations, we train the model for 1,000,000 iterations and then perform an evaluation in order to find the Headway (HW) and Time Headway (THW) criticality metrics [42] of a single agent following a lead vehicle accelerating and suddenly stopping at two different points (this is done 10 times in order to average the results) and also an evaluation involving 12 agents and a lead vehicle (behaving similarly to the previous test) in order to evaluate the final positions of the vehicles (this is also performed 10 times and averaged). At any of the 20 tests, if there is a collision, the test ends and a collision is counted before continuing to the following test. From the results of these tests, we not only want to see a low collision count but also a reasonable THW (preferably between 1 and 3 s). Here, higher THW indicates that the agent is very slow, while low THW is risky. As we can see in Table A1, row 33, the model 3000/*predAcc*/*symmetric* (meaning penalization is 3000, the lead agent performs predetermined accelerations and reward is the *symmetric* one) is the only model with 0 collisions while having a good THW (2.25).

We can also see that penalizations of 0 and 100 are too low and because of that the models tend to have more collisions, while a penalization of 100,000 produces little to no collisions but barely accelerates (as seen in the case of high THW values presented in Table A1). We can observe that the random accelerations for the lead vehicle produce extremely cautious agents (as seen in the high THW values presented in Table A1). The same thing can be said about random stops, even though the impact is not as drastic. Finally we can see that *velocity* and *absoluteDiff* rewards lead to a lot of collision, while no reward as expected produces really slow agents. Considering these results, we will use a penalization of 3000, a predetermined acceleration lead agent behavior, and the custom reward *symmetric*.

4.7. Perturbed Inputs

The main focus of our work is on the perfect scenario where the velocity, the distance to the leading vehicle, and its velocity are accurately known at every step. However, we also run all the experiments in a secondary simulation in which a perturbation is introduced in the form of a random multiplier of uniform value between 0.9 and 1.1 which is applied to each of the three mentioned variables.

4.8. Training Setup

Regarding training setup, we made use of the Ubuntu distribution of Linux, version 20.04, together with Tensorflow 2.4.1; here, we also made use of Reverb 0.2.0 framework [43] as the experience replay system for RL. Regarding training, we trained each model for 1 million iterations of the simulation, using the same architectures (except for the fact that one has more input and in turn more connections) for both neural networks composed of a single hidden layer with 500 neurons. The reason for choosing these architectures was their low-dimensional feature space.

5. Evaluation

The objective of this section is to show the advantages of adding physical knowledge to the RL model found in ACC and to prove that vehicles equipped with a PG RL-based ACC are safer. With that in mind, we will compare results in different tasks between a traditional RL model and our proposed PG RL model, in which we introduce prior knowledge, as explained in the previous section. The tasks will consist of a lead agent with a predetermined acceleration being followed by one or more agent vehicles controlled by one of our models, at a predetermined initial separation distance. For this, we will evaluate how likely each of the models is to collide, and how well the agents controlled by the models spread out.

5.1. Task 1

For the first task, it is important to mention that the first agent has no obstacles at all, nor a front vehicle to follow, this being a task to be learned by the other agents. Here, the acceleration of the lead vehicle will be $0.5\,\mathrm{m/s^2}$ for 1100 steps, however, between steps 400 and 500 it will be at $-0.6\,\mathrm{m/s^2}$. The lead vehicle will be followed by 11 agent-controlled vehicles initially separated by 20 m and with the initial velocities and accelerations being 0. In order to observe the difference between the two models we are evaluating, for simplicity, after 1100 steps we will capture the positions of the agent vehicles. Here, the chosen reward function is the constant collision penalization plus the reward r, leading to Equation (7):

$$3000 - \frac{|s - ts|}{2ts} - \frac{|s - ts|}{2s}. \tag{7}$$

The results of this task are presented in Figure 3.

Here, the y-axis represents the position relative to the first agent and the x-axis represents the order of vehicles from the last one to the first one. For instance, if there is a point in the graph at x = 1 and y = 1000, that means that vehicle 1 ended 1000 m behind the first agent.

As can be observed in Figure 3, in the traditional RL model (blue color line), the final positions form a convex curve. In contrast, the proposed PG RL model (red color line) finds the agents spread more evenly than the traditional RL model. In order to put a magnitude to this appreciation of curvature/linearity in Figure 3, we calculated the distance of every point in the graphs to the corresponding points in a straight line connecting the first and the last point, this measure is also known as the Gini index. Then, we added the absolute values of these differences for each of the models and, as can be observed, our appreciation is correct, the sum of the distances being 1128 for the proposed PG RL model as compared to 1584 for the traditional RL model. While this doesn't necessarily prove that one model is better than the other, it shows that the addition of physical knowledge in the model does have an effect on the behavior of the agents.

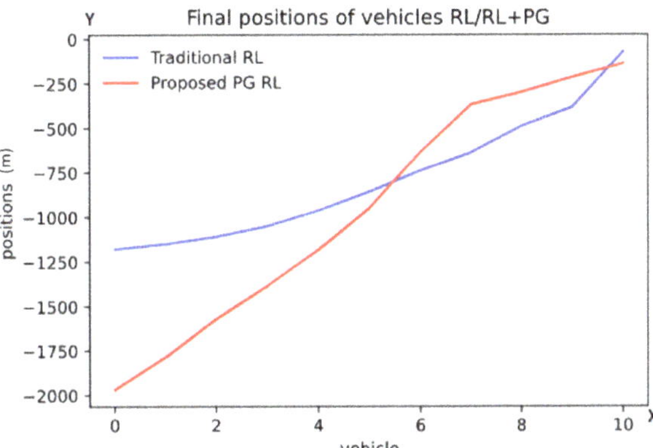

Figure 3. Graph of the finals positions after the first task using the traditional RL (blue color) and the proposed PG RL (red color) models. The y-axis is representing the position behind the lead vehicle in meters while one point on the x-axis is referring to exactly one vehicle. These are the average final positions at the end of the scenarios, with the numbers referring to the vehicles (from back to front).

Next, we decided to study what would happen if there was some imprecision with the readings from the radar. To do this we introduce to each input a random uniform multiplier from 0.9 to 1.1. For example, if the real value of the reading were 10.0, the observed value would be a random value between 9.0 and 11.0 uniformly distributed. This randomness is applied to each input or simulated reading individually.

As we see in Figure 4 in comparison to Figure 3, we can observe that the perturbation of the inputs doesn't change the innate behavior of the result for this task, but just smooths out each curve.

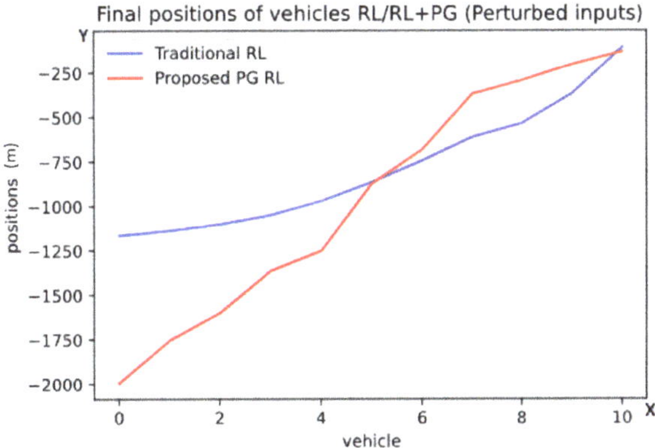

Figure 4. Graph of the finals positions after the first task with randomized inputs using the traditional RL (blue color) and the proposed PG RL (red color) models. The y-axis is representing the position behind the lead vehicle in meters while one point on the x-axis is referring to exactly one vehicle. These are the average final positions at the end of the scenarios, with the numbers referring to the vehicles (from back to front).

In Figure 5, we observe that the behavior of both models trained with perturbed inputs differs from its original counterparts. The average THW following a lead agent is of 21 s for the traditional model and 17 s for the PG RL model. This indicates a very slow behavior of the perturbed trained models, which is expected considering their experienced uncertainty. Lastly, we will reduce the number of agents to one in order to measure some of the criticality metrics introduced in [42] that apply to ACC such as THW, TTHW, and

Deceleration to Safety Time (DST). The HW criticality metric (referenced as s in our work) is the distance between a vehicle and its leading vehicle.

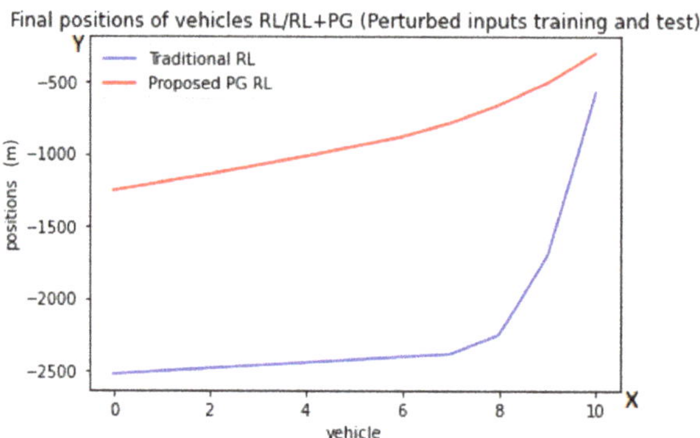

Figure 5. Graph of the finals positions after the first task with randomized inputs using the traditional RL (blue color) and the proposed PG RL (red color) models trained with perturbations. The *y*-axis is representing the position behind the lead vehicle in meters while one point on the *x*-axis is referring to exactly one vehicle. These are the average final positions at the end of the scenarios, with the numbers referring to the vehicles (from back to front).

As we see in Figure 6, the HW in our proposed PG RL model (red color line) is at most steps higher than in the traditional RL model (blue color line), with its lowest points being also higher.

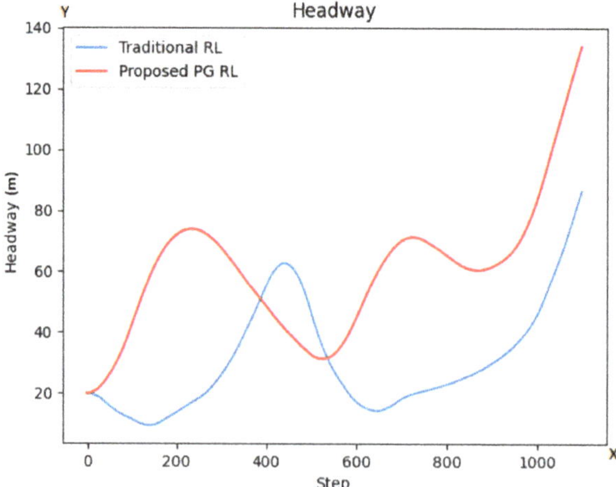

Figure 6. HW values at each step for both models (traditional RL in blue color line; our proposed PG RL model in red color line). The *y*-axis is representing the HW values while the *x*-axis represents the number of simulation steps.

The THW criticality metric is the time a vehicle would take at a given step to reach its leading vehicle if its own velocity was constantly the same as the velocity at the given step and the leading vehicle remained still at its current position.

As we see in Figure 7, the THW in our PG RL model (red color line) is also at most steps higher than in the traditional RL model, with even its lowest points being higher. These results combined with the HW results suggest a more safe driving by our PG RL model than the traditional RL one.

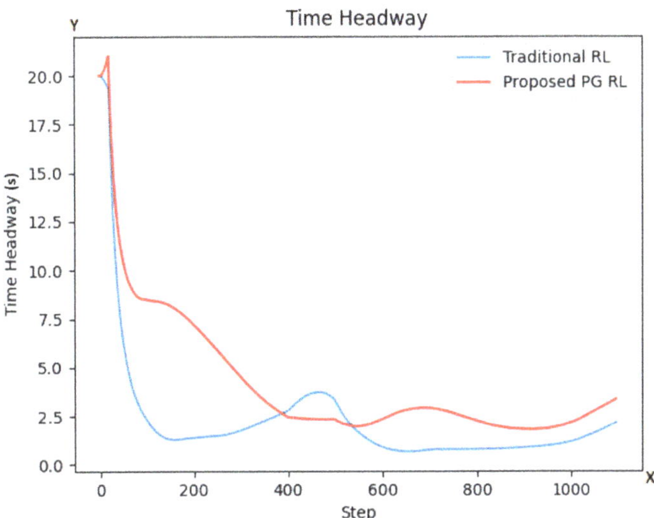

Figure 7. THW values at each step for both models (traditional RL in blue color line; our proposed PG RL model in red color line). The *y*-axis is representing the THW values while the *x*-axis represents the number of simulation steps.

Finally, the DST metric calculates the deceleration required by the agent vehicle in order to maintain a safety time of *ts* seconds under the assumption of constant lead vehicle velocity. At a given step, the DST criticality metric is calculated as seen in Equation (8):

$$DST(v1, v2, s, ts) = \frac{3(v_1 - v_2)^2}{2(s - v_2.ts)} \tag{8}$$

where v_1 is the agent's velocity, v_2 is the lead vehicle's velocity, and s is the distance between the vehicles, everything measured at a given step.

In Figure 8, we observe a strange behavior for both agents. The DST function spikes around steps 250, 550, and 1050. The values it reaches suggests impossible values for accelerations and decelerations, for instance requesting going from 400 m/s to 0 (at step 250) or from 0 to 300 m/s (at step 1050) in 0.1 s. The reason for these spikes is that the function for DST is linearly proportional to $\frac{1}{s - v_2.ts}$, which would suggest that the deceleration should be greater the closer the distance s is to $v_2.ts$, and more than that, that it should be infinite (with indeterminate sign) if $s = v_2.ts$, which, at the very least doesn't coincide with the supposed objective of this function.

5.2. Task 2

In the next tasks, the scenario will be the same as in Section 5.1, except that, here, we will introduce increasingly more dramatic brakes for the predetermined lead agent. More exactly, in the first of these tasks the deceleration rate will be $-0.7\,\text{m/s}^2$, in the second one, $-0.71\,\text{m/s}^2$, then $-0.75\,\text{m/s}^2$ and finally $-1.0\,\text{m/s}^2$.

These considerations are very important for our evaluation because, for each of these tasks, we are able to observe which is the first agent vehicle that collides against the vehicle right in front of it, thus giving us a sense of how safe the platoon of vehicles is for both RL and PG RL models considered in this work. Thus, if the nth vehicle is the first vehicle that collides, we can say that the platoon is safe for $n - 1$ vehicles in the given scenario.

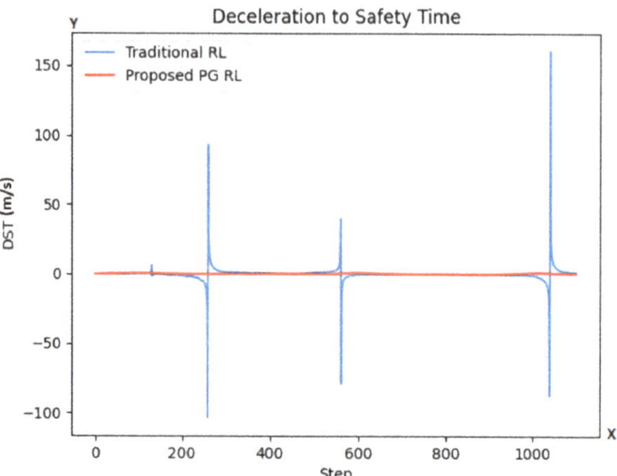

Figure 8. DST values at each step for both models (traditional RL in blue color line; our proposed PG RL model in red color line). The *y*-axis is representing the DST while the *x*-axis represents the number of simulation steps.

As can be seen in Table 2, the proposed PG RL model is proving to be safer in every single scenario when compared to the traditional RL model.

Table 2. Collision safety comparison between the traditional RL and the proposed PG RL models.

Lead Deceleration	Collision Vehicle PG RL	Collision Vehicle RL
1.0	10th	1st
0.75	No collisions	1st
0.71	No collisions	6th
0.7	No collisions	No collisions

Here, the first column shows the deceleration of the lead agent in each scenario, with the second and third columns showing at each scenario which car was the first to c for each of the models respectively. The presented values in Table 2 come from testing the same scenario 20 times and obtaining the worst result.

We run the same experiment introducing perturbations to the inputs as in Section 5.1; however, since the agents were trained with perfect inputs, both of the models performed considerably worse. We did 20 attempts for each of the models and each of the lead deceleration values in Table 2, but no matter the number of vehicles, the first agent always collide against the lead vehicle in at least one of the 20 attempts.

Performing this experiment with the perturbed inputs, the trained models yield safer results due to the cautious nature of these models, however, this shouldn't be taken as a virtue of these models because, upon qualitative analysis, we observe that they barely accelerate due to the uncertain nature of their training when compared to the regular trained ones.

6. Conclusions

Despite AI paving the way towards fully automated driving, its development is mostly driven by data without taking into consideration prior knowledge. This paper presents a novel approach in increasing the safety of ACCs by merging these two approaches, more exactly, by making use of physical knowledge in the form of jam-avoiding distance as part of a more processed input for a SAC RL algorithm. The advantage over constrained-based optimal control algorithms is that RL approaches do not require any data while the advantage over common rule-based driving is the greater flexibility of an RL-based agent thanks to the state abstractions learned by the underlying deep neural network. In our evaluation, we show that a PG RL agent is able to learn how to behave in its scenario better than a traditional RL approach, showing less collisions and more equidistant travel, providing a basis for future work to build upon. Another important result is the encouragingly good performance of our RL-based agents in the platooning scenario as well as in the scenario with perturbed input data. We want to emphasize that the agents do not have the opportunity to communicate but once one of the vehicles brakes, the vehicles behind it learn to brake as well by only observing the (perturbed) distance to the respective vehicle in front and their (perturbed) velocity. In addition, the proposed PG RL approach achieves considerable better results also when evaluated with criticality metrics such as TW and THW, proving that safety in AVs can be increased by making use of prior knowledge into AI components. As future work for improving the performance of an ACC, we plan to identify and integrate additional knowledge into the PG RL model by increasing the complexity of scenarios. We want to realize this by using additional traffic participants such as pedestrians crossing the road in front of the lead vehicle. Promising future directions are to consider adjacent domains such as Car2Car and Car2X communications which are able to provide information about better traffic predictions, as well as to integrate additional and diverse knowledge by other approaches such as extending the reward function.

Author Contributions: Conceptualization, S.L.J. and D.G.; methodology, S.L.J., D.G., T.W. and P.B.; software,S.L.J., D.G., P.B. and K.R.; validation, S.L.J., D.G., T.W. and E.M.; formal analysis, S.L.J., D.G., T.W. and P.B.; investigation, S.L.J., D.G., T.W., P.B., K.R. and E.M.; resources, S.L.J., D.G., T.W., P.B., K.R. and E.M.; data curation, S.L.J. and D.G.; writing—original draft preparation, S.L.J., D.G. and T.W.; writing—review and editing, S.L.J., D.G., T.W., P.B., K.R. and E.M.; visualization, S.L.J. and K.R.; supervision, E.M.; project administration, D.G., T.W. and E.M.; funding acquisition, E.M. All authors have read and agreed to the published version of the manuscript.

Funding: This work received funding from the German Federal Ministry of Economic Affairs and Energy (BMWi) through the KI-Wissen project under grant agreement No. 19A20020M.

Institutional Review Board Statement: Not applicable.

Informed Consent Statement: Not applicable.

Data Availability Statement: Not applicable.

Conflicts of Interest: The authors declare no conflict of interest.

Abbreviations and Nomenclature

The following abbreviations and symbols are used in this manuscript:

ACC	Adaptive Cruise Control
ML	Machine Learning
SAC	Soft Actor-Critic Algorithm
RL	Reinforcement Learning
PG	Physics-guided
AI	Artificial Intelligence
AV	Autonomous vehicle
MRI	Magnetic Resonance Imaging
LIDAR	Light detection and ranging
IDM	Intelligent Driver Model
MDP	Markov decision process
HW	Headway
THW	Time Headway
DST	Deceleration to Safety Time
$Q_\theta(s_t, a_t)$	Soft Q-function
s_t	State at time point t
a_t	Action at time point t

θ	Soft Q-function parameter
$\pi_\phi(s_t\|a_t)$	Policy with state-action pair
ϕ	Policy parameter
v_0	Desired velocity
T	Save time headway
a_m	Maximum acceleration
b	Desired deceleration
s_0	Jam distance
s	Distance to the leading vehicle
s^*	Minimum jam-avoiding distance
v	Current agent velocity
Δv	Velocity difference to the leading vehicle
t	Time point t
v_t	Velocity at time point t
a_t	Acceleration at time point t
x_t	Position at time point t
h	Step size
$\mathcal{S}$	State Space in MDP
$\mathcal{A}$	Action Space in MDP
T	Deterministic transition model in MDP
r	Reward
$\pi_\phi : \mathcal{S} \to \mathcal{A}$	Deterministic parametric policy
ts	Target seperation
$DST(v1, v2, s, ts)$	Deceleration to Safety Time
$v1$	Vehicle 1
$v2$	Vehicle 2

Appendix A

Table A1. Test results for the different combinations of parameters regarding the RL simulation.

	Model	Collisions	HW (m)	THW (s)	Separation (m)
0	0/predAcc/symmetric	2	21.986616	4.746649	17.932934
1	0/predAcc/symmetric	20	21.523313	4.598207	17.974341
2	0/predAcc/velocity	20	12.238809	6.684551	17.714547
3	0/predAcc/absoluteDiff	20	13.512952	7.932466	18.123747
4	0/predAcc/None	10	552.604185	22.768953	21.750825
5	0/randomAcc/symmetric	0	120.883542	5.407901	147.354975
6	0/randomAcc/velocity	20	12.467875	7.977922	17.749884
7	0/randomAcc/absoluteDiff	0	185.853298	8.008320	171.210850
8	0/randomAcc/None	20	30.555453	4.365431	49.872451
9	0/randomStops9/symmetric	10	111.509197	5.244257	17.899641
10	0/randomStops9/velocity	20	17.324510	3.036292	13.902887
11	0/randomStops9/absoluteDiff	20	13.116112	7.374774	17.958673
12	0/randomStops9/None	0	818.533661	32.071159	229.345633
13	0/randomStops10/symmetric	20	11.578542	2.877303	19.522080
14	0/randomStops10/velocity	0	96.373722	4.650809	103.114341
15	0/randomStops10/absoluteDiff	20	13.114346	7.399617	17.963344
16	0/randomStops10/None	0	818.434954	32.064495	229.345483
17	100/predAcc/symmetric	20	64.280247	7.214039	73.357278
18	100/predAcc/velocity	20	12.276690	4.083880	13.097846
19	100/predAcc/absoluteDiff	20	13.453637	7.653036	18.272531
20	100/predAcc/None	10	795.550077	30.799090	22.119458
21	100/randomAcc/symmetric	9	38.420217	2.480810	54.983419

Table A1. *Cont.*

	Model	Collisions	HW (m)	THW (s)	Separation (m)
22	100/randomAcc/velocity	4	201.610520	7.274791	91.056809
23	100/randomAcc/absoluteDiff	10	85.191944	4.217098	62.292352
24	100/randomAcc/None	0	229.362628	10.193637	213.209962
25	100/randomStops9/symmetric	0	88.737306	3.578192	89.672347
26	100/randomStops9/velocity	20	9.988424	2.657730	13.041972
27	100/randomStops9/absoluteDiff	20	12.637536	7.668094	17.778202
28	100/randomStops9/None	0	794.717912	31.126503	229.345636
29	100/randomStops10/symmetric	2	119.310468	5.001601	95.460083
30	100/randomStops10/velocity	20	26.619944	4.218386	53.601016
31	100/randomStops10/absoluteDiff	20	40.782619	4.782894	21.429448
32	100/randomStops10/None	0	815.799968	31.938171	229.311862
33	3,000/predAcc/symmetric	0	30.806897	2.256980	104.479287
34	3,000/predAcc/velocity	18	16.539250	2.489905	10.804999
35	3,000/predAcc/absoluteDiff	20	13.571941	4.840553	18.439911
36	3,000/predAcc/None	0	800.393554	31.542861	229.345636
37	3,000/randomAcc/symmetric	0	557.436259	21.370929	189.098820
38	3,000/randomAcc/velocity	1	314.244283	11.907776	174.433635
39	3,000/randomAcc/absoluteDiff	0	519.028676	19.635769	198.585035
40	3,000/randomAcc/None	0	487.673884	18.984621	227.800876
41	3,000/randomStops9/symmetric	0	121.610496	5.503739	145.529442
42	3,000/randomStops9/velocity	0	71.099924	3.601645	96.712122
43	3,000/randomStops9/absoluteDiff	20	12.240173	2.740492	16.209725
44	3,000/randomStops9/None	0	736.721510	29.327481	229.345636
45	3,000/randomStops10/symmetric	0	95.886739	4.443669	103.119817
46	3,000/randomStops10/velocity	14	9.988159	4.999919	33.352000
47	3,000/randomStops10/absoluteDiff	20	13.616155	6.805060	18.093497
48	3,000/randomStops10/None	0	160.721774	7.162411	154.905059
49	100,000/predAcc/symmetric	16	76.071190	3.418030	16.016787
50	100,000/predAcc/velocity	1	134.614581	6.869882	149.200815
51	100,000/predAcc/absoluteDiff	0	818.685845	32.077105	229.344549
52	100,000/predAcc/None	0	761.794150	29.485078	229.345636
53	100,000/randomAcc/symmetric	0	160.713442	8.460585	225.571123
54	100,000/randomAcc/velocity	0	817.021321	31.972462	229.345213
55	100,000/randomAcc/absoluteDiff	0	576.845622	23.913704	229.345636
56	100,000/randomAcc/None	0	509.216830	21.192002	229.345242
57	100,000/randomStops9/symmetric	0	90.975791	4.659583	135.021949
58	100,000/randomStops9/velocity	0	76.346305	4.586848	130.914674
59	100,000/randomStops9/absoluteDiff	0	292.012632	14.117718	229.345634
60	100,000/randomStops9/None	0	816.244478	31.961323	229.345221
61	100,000/randomStops10/symmetric	0	174.359445	7.620220	144.156391
62	100,000/randomStops10/velocity	0	430.033252	18.551971	229.345570
63	100,000/randomStops10/absoluteDiff	0	379.953210	16.932130	229.340859
64	100,000/randomStops10/None	10	688.564158	26.086424	54.548802

References

1. Singh, S. Critical Reasons for Crashes Investigated in the National Motor Vehicle Crash Causation Survey 2015. Available online: https://crashstats.nhtsa.dot.gov/Api/Public/ViewPublication/812115 (accessed on 24 October 2021)
2. Ni, J.; Chen, Y.; Chen, Y.; Zhu, J.; Ali, D.; Cao, W. A Survey on Theories and Applications for Self-Driving Cars Based on Deep Learning Methods. *Appl. Sci.* **2020**, *10*, 2749. [CrossRef]
3. Clements, L.M.; Kockelman, K.M. Economic Effects of Automated Vehicles. *Transp. Res. Rec.* **2017**, *2606*, 106–114. [CrossRef]
4. Karpatne, A.; Watkins, W.; Read, J.S.; Kumar, V. Physics-guided Neural Networks (PGNN): An Application in Lake Temperature Modeling. *arXiv* **2017**, arXiv:1710.11431.
5. von Rueden, L.; Mayer, S.; Beckh, K.; Georgiev, B.; Giesselbach, S.; Heese, R.; Kirsch, B.; Walczak, M.; Pfrommer, J.; Pick, A.; et al. Informed Machine Learning—A Taxonomy and Survey of Integrating Prior Knowledge into Learning Systems. *IEEE Trans. Knowl. Data Eng.* **2021**, 1. [CrossRef]
6. Yaman, B.; Hosseini, S.A.H.; Moeller, S.; Ellermann, J.; Uğurbil, K.; Akçakaya, M. Self-supervised learning of physics-guided reconstruction neural networks without fully sampled reference data. *Magn. Reson. Med.* **2020**, *84*, 3172–3191. [CrossRef]

7. Gumiere, S.J.; Camporese, M.; Botto, A.; Lafond, J.A.; Paniconi, C.; Gallichand, J.; Rousseau, A.N. Machine Learning vs. Physics-Based Modeling for Real-Time Irrigation Management. *Front. Water* **2020**, *2*, 8. [CrossRef]

8. Zhang, Z.; Sun, C. Structural damage identification via physics-guided machine learning: A methodology integrating pattern recognition with finite element model updating. *Struct. Health Monit.* **2020**, *20*, 1675–1688. [CrossRef]

9. Piccione, A.; Berkery, J.; Sabbagh, S.; Andreopoulos, Y. Physics-guided machine learning approaches to predict the ideal stability properties of fusion plasmas. *Nucl. Fusion* **2020**, *60*, 046033. [CrossRef]

10. Muralidhar, N.; Bu, J.; Cao, Z.; He, L.; Ramakrishnan, N.; Tafti, D.; Karpatne, A. Physics-Guided Deep Learning for Drag Force Prediction in Dense Fluid-Particulate Systems. *Big Data* **2020**, *8*, 431–449. [CrossRef] [PubMed]

11. Wang, J.; Li, Y.; Zhao, R.; Gao, R.X. Physics guided neural network for machining tool wear prediction. *J. Manuf. Syst.* **2020**, *57*, 298–310. [CrossRef]

12. AI Knowledge Consortium. AI Knowledge Project, 2021. Available online: https://www.kiwissen.de/ (accessed on 24 October 2021).

13. Wei, Z.; Jiang, Y.; Liao, X.; Qi, X.; Wang, Z.; Wu, G.; Hao, P.; Barth, M. End-to-End Vision-Based Adaptive Cruise Control (ACC) Using Deep Reinforcement Learning. *arXiv* **2020**, arXiv:2001.09181.

14. Kesting, A.; Treiber, M.; Schönhof, M.; Kranke, F.; Helbing, D. Jam-Avoiding Adaptive Cruise Control (ACC) and its Impact on Traffic Dynamics. In *Traffic and Granular Flow'05*; Schadschneider, A., Pöschel, T., Kühne, R., Schreckenberg, M., Wolf, D.E., Eds.; Springer: Berlin/Heidelberg, Germany, 2007; pp. 633–643.

15. Kral, W.; Dalpez, S. Modular Sensor Cleaning System for Autonomous Driving. *ATZ Worldw.* **2018**, *120*, 56–59. [CrossRef]

16. Knoop, V.L.; Wang, M.; Wilmink, I.; Hoedemaeker, D.M.; Maaskant, M.; der Meer, E.J.V. Platoon of SAE Level-2 Automated Vehicles on Public Roads: Setup, Traffic Interactions, and Stability. *Transp. Res. Rec.* **2019**, *2673*, 311–322. [CrossRef]

17. Pathak, S.; Bag, S.; Nadkarni, V. A Generalised Method for Adaptive Longitudinal Control Using Reinforcement Learning. In *International Conference on Intelligent Autonomous Systems*; Springer: Cham, Switzerland, 2019; pp. 464–479.

18. Farag, A.; AbdelAziz, O.M.; Hussein, A.; Shehata, O.M. Reinforcement Learning Based Approach for Multi-Vehicle Platooning Problem with Nonlinear Dynamic Behavior 2020. Available online: https://www.researchgate.net/publication/349313418_Reinforcement_Learning_Based_Approach_for_Multi-Vehicle_Platooning_Problem_with_Nonlinear_Dynamic_Behavior (accessed on 24 October 2021)

19. Chen, C.; Jiang, J.; Lv, N.; Li, S. An intelligent path planning scheme of autonomous vehicles platoon using deep reinforcement learning on network edge. *IEEE Access* **2020**, *8*, 99059–99069. [CrossRef]

20. Forbes, J.R.N. *Reinforcement Learning for Autonomous Vehicles*; University of California: Berkeley, CA, USA, 2002.

21. Sallab, A.E.; Abdou, M.; Perot, E.; Yogamani, S. Deep Reinforcement Learning framework for Autonomous Driving. *arXiv* **2017**, arXiv:1704.02532.

22. Kiran, B.; Sobh, I.; Talpaert, V.; Mannion, P.; Sallab, A.; Yogamani, S.; Perez, P. Deep Reinforcement Learning for Autonomous Driving: A Survey. *IEEE Trans. Intell. Transp. Syst.* **2021**, 1–18. [CrossRef]

23. Di, X.; Shi, R. A survey on autonomous vehicle control in the era of mixed-autonomy: From physics-based to AI-guided driving policy learning. *Transp. Res. Part Emerg. Technol.* **2021**, *125*, 103008. [CrossRef]

24. Desjardins, C.; Chaib-draa, B. Cooperative Adaptive Cruise Control: A Reinforcement Learning Approach. *IEEE Trans. Intell. Transp. Syst.* **2011**, *12*, 1248–1260. [CrossRef]

25. Curiel-Ramirez, L.; Ramirez-Mendoza, R.A.; Bautista, R.; Bustamante-Bello, R.; Gonzalez-Hernandez, H.; Reyes-Avendaño, J.; Gallardo-Medina, E. End-to-End Automated Guided Modular Vehicle. *Appl. Sci.* **2020**, *10*, 4400. [CrossRef]

26. Li, Y.; Li, Z.; Wang, H.; Wang, W.; Xing, L. Evaluating the safety impact of adaptive cruise control in traffic oscillations on freeways. *Accid. Anal. Prev.* **2017**, *104*, 137–145. [CrossRef]

27. Niedoba, M.; Cui, H.; Luo, K.; Hegde, D.; Chou, F.C.; Djuric, N. Improving movement prediction of traffic actors using off-road loss and bias mitigation. In *Workshop on 'Machine Learning for Autonomous Driving' at Conference on Neural Information Processing Systems*; 2019. Available online: https://djurikom.github.io/pdfs/niedoba2019ml4ad.pdf (accessed on 24 October 2021).

28. Phan-Minh, T.; Grigore, E.C.; Boulton, F.A.; Beijbom, O.; Wolff, E.M. Covernet: Multimodal behavior prediction using trajectory sets. In Proceedings of the IEEE/CVF Conference on Computer Vision and Pattern Recognition, Seattle, WA, USA, 13–19 June 2020; pp. 14074–14083.

29. Boulton, F.A.; Grigore, E.C.; Wolff, E.M. Motion Prediction using Trajectory Sets and Self-Driving Domain Knowledge. *arXiv* **2020**, arXiv:2006.04767.

30. Cui, H.; Nguyen, T.; Chou, F.C.; Lin, T.H.; Schneider, J.; Bradley, D.; Djuric, N. Deep kinematic models for physically realistic prediction of vehicle trajectories. *arXiv* **2019**, arXiv:1908.0021.

31. Bahari, M.; Nejjar, I.; Alahi, A. Injecting Knowledge in Data-driven Vehicle Trajectory Predictors. *arXiv* **2021**, arXiv:2103.04854.

32. Mohamed, A.; Qian, K.; Elhoseiny, M.; Claudel, C. Social-stgcnn: A social spatio-temporal graph convolutional neural network for human trajectory prediction. In Proceedings of the IEEE/CVF Conference on Computer Vision and Pattern Recognition, Seattle, WA, USA, 13–19 June 2020; pp. 14424–14432.

33. Ju, C.; Wang, Z.; Long, C.; Zhang, X.; Chang, D.E. Interaction-aware kalman neural networks for trajectory prediction. In Proceedings of the 2020 IEEE Intelligent Vehicles Symposium (IV), Las Vegas, NV, USA, 19 October–13 November 2020; IEEE: Piscataway, NJ, USA, 2019; pp. 1793–1800.

34. Chen, B.; Li, L. Adversarial Evaluation of Autonomous Vehicles in Lane-Change Scenarios. *arXiv* **2020**, arXiv:2004.06531.

35. Ding, W.; Xu, M.; Zhao, D. Learning to Collide: An Adaptive Safety-Critical Scenarios Generating Method. *arXiv* **2020**, arXiv:2003.01197.

36. Qiao, Z.; Tyree, Z.; Mudalige, P.; Schneider, J.; Dolan, J.M. Hierarchical reinforcement learning method for autonomous vehicle behavior planning. *arXiv* **2019**, arXiv:1911.03799.
37. Li, X.; Qiu, X.; Wang, J.; Shen, Y. A Deep Reinforcement Learning Based Approach for Autonomous Overtaking. In Proceedings of the 2020 IEEE International Conference on Communications Workshops (ICC Workshops), Dublin, Ireland, 7–11 June 2020; IEEE: Piscataway, NJ, USA, 2020; pp. 1–5.
38. Wu, Y.; Tan, H.; Peng, J.; Ran, B. A Deep Reinforcement Learning Based Car Following Model for Electric Vehicle. *Smart City Appl.* **2019**, 2, 1–8. [CrossRef]
39. Haarnoja, T.; Zhou, A.; Hartikainen, K.; Tucker, G.; Ha, S.; Tan, J.; Kumar, V.; Zhu, H.; Gupta, A.; Abbeel, P.; et al. Soft Actor-Critic Algorithms and Applications. *arXiv* **2019**, arXiv:1812.05905.
40. Hermand, E.; Nguyen, T.W.; Hosseinzadeh, M.; Garone, E. Constrained control of UAVs in geofencing applications. In Proceedings of the 2018 26th Mediterranean Conference on Control and Automation (MED), Zadar, Croatia, 19–22 June 2018; IEEE: Piscataway, NJ, USA, 2018; pp. 217–222.
41. Wang, P.; Gao, S.; Li, L.; Sun, B.; Cheng, S. Obstacle avoidance path planning design for autonomous driving vehicles based on an improved artificial potential field algorithm. *Energies* **2019**, *12*, 2342. [CrossRef]
42. Westhofen, L.; Neurohr, C.; Koopmann, T.; Butz, M.; Schütt, B.; Utesch, F.; Kramer, B.; Gutenkunst, C.; Böde, E. Criticality Metrics for Automated Driving: A Review and Suitability Analysis of the State of the Art. *arXiv* **2021**, arXiv:2108.02403.
43. Cassirer, A.; Barth-Maron, G.; Brevdo, E.; Ramos, S.; Boyd, T.; Sottiaux, T.; Kroiss, M. Reverb: A Framework For Experience Replay. *arXiv* **2021**, arXiv:2102.04736.

Article

Automated Conflict Management Framework Development for Autonomous Aerial and Ground Vehicles

David Sziroczák * and Daniel Rohács

Department of Aeronautics and Naval Architecture, Budapest University of Technology and Economics, 1111 Budapest, Hungary; rohacs.daniel@kjk.bme.hu
* Correspondence: sziroczak.david@kjk.bme.hu

Abstract: The number of aerial- and ground-based unmanned vehicles and operations is expected to significantly expand in the near future. While aviation traditionally has an excellent safety record in managing conflicts, the current approaches will not be able to provide safe and efficient operations in the future. This paper presents the development of a novel framework integrating autonomous aerial and ground vehicles to facilitate short- and mid-term tactical conflict management. The methodology presents the development of a modular web service framework to develop new conflict management algorithms. This new framework is aimed at managing urban and peri-urban traffic of unmanned ground vehicles and assisting the introduction of urban air mobility into the same framework. A set of high-level system requirements is defined. The incremental development of two versions of the system prototype is presented. The discussions highlight the lessons learnt while implementing and testing the conflict management system and the introduced version of the stop-and-go resolution algorithm and defines the identified future development directions. Operation of the system was successfully demonstrated using real hardware. The developed framework implements short- and mid-term conflict management methodologies in a safe, resource efficient and scalable manner and can be used for the further development and the evaluation of various methods integrating aerial- and ground-based autonomous vehicles.

Keywords: autonomous conflict management; UTM; UAV; UGV; U-Space; framework development

Citation: Sziroczák, D.; Rohács, D. Automated Conflict Management Framework Development for Autonomous Aerial and Ground Vehicles. *Energies* **2021**, *14*, 8344. https://doi.org/10.3390/en14248344

Academic Editor: Arno Eichberger

Received: 9 November 2021
Accepted: 8 December 2021
Published: 10 December 2021

1. Introduction

The use of autonomous vehicles, both road going and aerial, is expanding rapidly these days. Just in the case of unmanned aerial vehicles, between 2016 and 2021 Goldman Sachs [1] predicted a USD 100 billion market opportunity, with a further 16–24% annual growth predicted between 2026–2028 [2,3]. The market opportunities for autonomous road vehicles are no less significant. To serve a market of this size, a significant amount of traffic needs to be managed. In controlled airspaces, by 2050 EUROCONTROL predicts that the total flight hours of UAS (unmanned aerial systems) in Europe will account for 20% of total traffic, an estimated 7 million flight hours [4]. Higher demand will be seen at the very low level (VLL) airspace, below 120 m (400 ft) AGL (above ground level), often referred to as the U-Space in the EU. In U-Space, about 250 million commercial flight hours are predicted in urban environments, 20 million in rural settings and 80 million for hobby use. This demand is a magnitude higher than what the current manned air traffic control systems handle.

In the case of road vehicles, as there is no central control, the system relies on individual drivers to manage conflict situations. In the EU, the ACEA (European Automobile Manufacturers Association, Brussels, Belgium) estimates 268 million road vehicles in 2018, with an average growth rate of 2% compared to the previous year [5]. It is further predicted by the Victoria Transport Policy Institute (Victoria, BC, Canada) that by 2045 half of new vehicles will be autonomous, and by 2060 half of the active vehicle fleet will be autonomous [6]. Based on the defined autonomous driving levels [7] autonomous road vehicles are designed to achieve safe and efficient driving individually, and thus control and conflict management of these vehicles could also benefit road safety. Based on the high volume of traffic to be managed, it is predicted that a system similar to today's air traffic control will not be sufficient for these vehicles either. Current ATC (air traffic control) procedures are human labour intensive and offer poor automation possibilities.

Aircraft traffic management for unmanned systems, or UTM in short, is an active research field today; both national and international organizations actively develop and test various concepts for the management of this novel class of airspace users. ICAO has already published the third edition of their UTM harmonization guidelines [8]. The document is aimed at providing a common framework along which nations can develop harmonised solutions for UTM activities. In addition to working towards international compatibility, the initiative also helps to reduce the costs associated with the development of UTM solutions. Essentially, all nations have to find solutions for this novel challenge, and it would be beneficial if the development activities were shared in some form as well. In the USA both NASA (National Aeronautics and Space Administration, Washington, DC, USA) and the FAA (Federal Aviation Administration, Washington, DC, USA) are actively developing the nation's Next Generation Air Transportation System, planned to be in operation by 2025. While the system is aimed at improving various aspects of airspace use, integrating UAS is a significant part of the challenge. Both organizations have been actively engaged in the UAS Integration Pilot Program (IPP) (ended in 2020, now continued in the BEYOND programme) and the UTM Pilot Program (UPP) [9] since 2019, which are aimed at integrating the current estimated 350,000 UAS and future systems into the national airspace. ASTM (American Society for Testing and Materials, West Conshohocken, PA, USA) has developed standards for the remote identification of UAS, published as ASTM F3411 [10]. In the EU, the development of the U-Space system (part of the SESAR programme) is aimed at implementing UTM solutions [11]. At the national level, but also as part of the U-Space initiative, UTM systems have already been introduced, for example, in Finland (GOF, Gulf of Finland [12]), Switzerland (SUSI, Swiss U-Space Implementation, Zug, Switzerland, [13]) and the UK (CAA Innovation Hub, London, UK, [14]). There is also activity at the private level; companies such as Altitude Angel, Skyguide, AirMap and sees.ai have already developed UTM solutions, and some of them are already providing these solutions today. There are also various consortia, such as DOMUS, USIS, DIODE, EuroDRONE, SAFIR, VUTURA, GeoSAFE, PODIUM, SAFEDRONE and CORUS developing various aspects of UTM and related activities. In Hungary, where the research presented was performed, HungaroControl has founded the UTM Innovation HUB and CybAIR Cluster, which are both contributing towards UTM systems development.

In the autonomous ground vehicle (often referred to as self-driving car) industry primary emphasis is placed on collision avoidance systems in individual vehicles. As was previously mentioned, autonomous operation is being developed as a stepped approach, and many of these concepts also appear as driver assistance systems in conventional ground vehicles. These systems include collision warning systems (front and rear) adaptive cruise control, lane assist, road sign detection and similar systems. Approaches such as smart city or more generally smart mobility are aimed at increasing the safety, sustainability and efficiency of current traffic systems, but they do not necessarily rely on the control of vehicles. Solutions include the analysis and evaluation of traffic, variable speed limits, dynamic traffic signal timing or dynamic lanes. In the case of cooperating and connected vehicles, such systems could provide smart routing and speed control for optimum traffic efficiency, essentially reducing time spent in congestions and improving travel experience for drivers and passengers.

This paper is aimed at developing a system prototype that can be used to implement the short- and mid-term conflict management of a local U-Space system containing both autonomous aerial and ground vehicles, along with other potential users. The inclusion of both aerial and ground autonomous systems in a common management framework can be considered as a novel idea. The pairing of self-driving cars and UAS has been experimented with, but only for specific applications such as food delivery [15], search and rescue [16] and infrastructure inspection [17]. There are also research works available that focus on the technological specifics of such integration [18,19]. Published studies of conflict management approaches that combine autonomous ground and aerial vehicles are unknown to the author.

As the whole concept of unmanned traffic management, both for UGV and UAV, is new, there are no set standards or solutions available today, only development initiatives. The only regulation available in this field is that UGV must abide by the applicable highway code of a given country in order to be compatible with current (manned) traffic. In the case of UAV, there are regulations concerning the individual use of vehicles (such as the EASA or FAA regulations), but there are no standards regarding the urban environment, or any sort of other traffic regulation for that matter, as this type of dense traffic simply does not exist today. As a result, systems being developed today do not need to be prepared to comply with any particular standard; rather, they can form the base of developing standards and regulations for the future.

While the concept of interaction between aerial and ground can be questionable, situations do arise where they can come into conflict. The first category is related to take-off and landing type manoeuvres, in which case UAS voluntarily moves close to the ground to where it can be affected by ground traffic. A typical use of this would be collection and delivery type scenarios. Additionally, this category includes vehicle–aircraft interactions, such as when a drone takes off or lands on the top of a delivery van, which can be either static or moving. The second category includes involuntary near-ground activities, such as emergency landing or descent due to failure, weather effects or while evading other airspace users or obstacles. The third category relates to environmental effects in man-made environments. Concepts such as urban canyons [20], essentially artificial channels that have a significant local impact on wind and gust characteristics (also on GPS and communications), are already known to researchers studying built environments. Additionally, large vehicles, such as buses, HGVs or trains can generate significant air flow in a constrained environment, which is often referred to as the piston effect. These effects can be significant enough that UAS would have to consider the presence, movement and potential path of ground vehicles even when operating at altitudes where direct physical contact is of no concern.

Developing a conflict management system involving both unmanned aerial and ground vehicles is essential for the advancement of autonomous vehicle use, especially in urban or peri-urban environments. In the field of aviation, the use of vehicles in an urban environment is referred to as urban air mobility, or UAM for short. Systems being developed for autonomous ground vehicles utilise sensors that primarily detect in the ground plane, as that is where the road traffic is expected. As a result, they are not equipped to detect and deal with aerospace traffic. Based on the roadmaps and published research available today, this is not likely to change in the near future. In the case of aerospace vehicles, as it was mentioned, current ATC procedures are not applicable, mainly due to the performance constraints and poor automation opportunities of current procedures. They key capability such a combined traffic system must provide are the following:

- Adequate level of safety: system must eliminate (significantly reduce) number and severity of accidents.
- Treatment of a wide range of users: ground and air vehicles must be included, along with potential other users (pedestrians, obstacles, etc.).
- High performance: management of conflicts must not significantly affect the mission times of the users or result in unacceptably long detours.
- Acceptable cost of operation: where cost includes system development, installation, maintenance, labour and cost to system users.

Automation in conflict management is the only possibility to provide the required capabilities, mainly due to the high number of users and conflicts envisioned. As such, a common and automated treatment for both classes of vehicles is essential. This paper presents a prototype solution developed to provide adequate capability for the management of conflicts between these users.

The first part of this paper deals with the methodology used to define the requirements and key components of the proposed conflict management system. It begins with an introduction to conflict management concepts and their applicability to the defined purpose. Then, high-level requirements are defined for the proposed conflict management system. In the results section, the system architecture of the implemented conflict management system is presented, based on two iterations and live testing of the system prototype. The discussion chapter presents the results and insights gained from the testing of the prototype system and proposes further development directions.

2. Materials and Methods

2.1. Conflict Management Methodologies

The proposed conflict management system implements a short- and mid-term tactical conflict management methodology. Conflict management approaches can be generally divided into 2 categories: strategic and tactical management. The basis of the categorization relates to the time factor of conflict detection and the available means to manage the event. The discussion of conflict management in this chapter is primarily based on concepts found in the field of aviation.

Strategic management refers to measures taken before the actual operation has begun. The activities performed during strategic management are aimed at minimizing the potential effects arising from the lack of information or faulty planning of routes. In aviation, typical planning issues include the availability of airspace (including routing through restricted and forbidden airspace), lack of air traffic control capacity (overburdened control officers) and flight plans received from other users. In all these cases, the conflict situation can be predicted in advance of the flight operation. The usual tools for strategic conflict management are the rerouting of flight paths or rescheduling of the operations, or when required, even cancelling the flight altogether. As an addendum, considering from a more abstract perspective, the long-term infrastructure planning and traffic policy making can also be considered as strategic level conflict management, just at the timescale of many years. As an example, NTU (Nanyang Technological University) Singapore proposed airspace design methods for UAV traffic in an urban environment, including vertical separation, "traffic light effect", digital unidirectional lanes and other features [21]. In the future, implementing these designs could contribute towards minimizing the situations where tactical level conflict arises. The longer-term strategy also has relevance regarding the infrastructure available to detect users and conflicts during operations (the tactical level). The system proposed in this research does not implement strategic conflict management, and as such it will not be discussed further.

Tactical conflict management is responsible for preventing collisions in the following situations:

- Between aircraft and other aircraft;
- Between aircraft and ground;
- Between aircraft and airspace (or geofence);
- Between aircraft and obstacles.

As these conflict situations arise during operation, most of the time they are not predictable at the planning and path approval stages. As a result, the time available to detect and resolve these conflict situations is significantly shorter than in the case of strategic management. The management of conflict situations usually requires active commands, which will result in deviation from the planned path for at least one user involved in the conflict. In the world of manned aviation these commands indicate a change in direction, altitude or speed. The command can come from 3 sources: ground-based systems (voice command from air traffic control officer through radio), on-board systems (TCAS—Traffic Collision Avoidance System—electronic indicators on flight display) or conflict analysis and decision from the pilot itself.

In the case of tactical management, the most critical factor is time. For human pilots (or drivers), processing information and coming to decisions is directly related to the time taken to do so. In the case of urban environments, the typical distances involved in conflict situations are generally small, which leaves little time for human decision making.

The tactical conflict management can be divided into 2 stages: conflict detection and conflict resolution. Detection methods can be grouped into 3 categories: deterministic, probabilistic and worst-case methods. While each of the detection methods have advantages and disadvantages compared to each other, their applicability is only valid in the short term, regardless of the method. An overview and classification of published classic conflict detection methods is presented in Table 1.

Table 1. Conflict management methods overview.

Type	Conflict Management Methods/Areas Addressed	Author and Reference
Deterministic	Optimal avoidance manoeuvres, conflict zones	Krozel et al. [22]
	Expert system for avoidance manoeuvres	Iijima et al. [23]
	Rule-based conflict management	Coenen et al. [24]
	Aircraft ground collision	GPWS [25]
	Manoeuvres	Bilimoria et al. [26]
	Optimal tactical and strategic manoeuvres	Krozel and Peters [27]
	Generalised conflict zones	Havel and Husarcik [28]
	Aerial collision avoidance	TCAS [29]
	Conflict alert evaluation	Ford [30]
Worst case	Optimal manoeuvres	Tomlin et al. [31]
	Uncertainty of planned flightpaths	Shepard et al. [32]
	Worst case turns or velocity changes	Shewchun and Feron [33]
	TCAS system limitations	Ratcliffe [34]
Probabilistic	Rapid conflict prediction	Paielli and Erzberger [35]
	Probability-based detection	Carpenter and Kuchar [36]
	Markov chain-based probability estimation	Bakker and Blom [37]
	Trajectory confidence model	Williams [38]
Hybrid	Genetic algorithm-based avoidance	Durand et al. [39]
	Stochastic hybrid model including wind effects	Glover and Lygeros [40]
	Monte Carlo simulation based	Visintini et al. [41]
	Non-cooperating target classification	Palme et al. [42]

In terms of conflict resolution, the earlier a conflict is detected and acted on, the more efficient and safer the conflict resolution will be. The conflict management approach studied in this paper focuses on short- and mid-term tactical management and as such, other methods will not be discussed further. For the sake of completeness, Table 2 presents the compiled overview of generalised approaches to conflict management, as defined by the researchers in this study. The time considered spans the complete time horizon of an impact event, including the possible very long term preceding activities and the post-impact treatment, which are not classically included in conflict management. The definitions and boundaries of the various levels can vary between researchers; the ones presented here represent the definitions used in this study and by other research activities by the researchers. Here, impact refers to the potential consequence of not managing the conflict, e.g., it can be actual physical impact between vehicles or environment or damage due to abrupt manoeuvres.

Table 2. Generalised conflict management timeline.

Time Scale Relative to Impact	Conflict Management Level	Aim of Activities	Available Tools
Up to about 10 years before	Strategic level system and technology planning	Develop safe and efficient procedures, integrate new technologies	Policy making, Research and development
Up to about 5–10 years before	Strategic level infrastructure planning	Deploy and develop infrastructure both ground based and onboard systems	City planning, infrastructure planning, vehicle design codes
Up to about 1–5 years before	Strategic level traffic planning	Determine and influence modes and volume of traffic	Policies, incentives, regulations, market development
Hours before	Strategic conflict management	Route planning	Rerouting, rescheduling, cancelling operations
5–10 s before	Tactical conflict management (mid-term)	Detection and alternative path planning	Active trajectory change: direction or speed adjustment
1–5 s before	Tactical conflict management (short-term)	Attempt to avoid impact	Evasive manoeuvres
Fraction of second before	Active safety systems	Preventive steps to minimise impact	Passenger safety systems, seat belt tensioners, airbags, etc.
During impact	Passive safety systems	Minimise impact	Crashworthy structures, optimal occupant positioning, etc.
Post impact	Post-impact treatment	Recover from effects of impact	Emergency services, warning for other traffic

When dealing with short- and mid-term conflict management in the case of autonomous vehicles, challenges arise in the resolution step. None of the current methods relying on human awareness or accepting and effecting commands are appropriate in this case, reducing the options available to digital commands either from a ground-based or on-board computational system.

2.2. High-Level Conflict Management System Requirements

The following chapter discusses the core principles and the high-level requirements that were defined to develop the conflict management system.

2.2.1. Performance Requirements

The proposed conflict management system is aimed at serving a wide variety of users, both aerial and ground based, implementing short- and mid-term tactical conflict management solutions. As such, the system needs to be able to detect conflicts and issue appropriate commands within the available time frame, which is in the order of 10 s or less. Due to the high magnitude of unmanned traffic predicted in the future, all steps of the conflict management process need to work autonomously, without human interaction.

2.2.2. Technology and Solution Independence

In addition to the high number of operations and users, it is also predicted that a diverse variety of users will need to be served. While there are initiatives that are (at least in theory) standardised in the automotive world (OBD—on board diagnostics—connectors for example) it is very unlikely that autonomous capabilities and communication solutions developed by individual manufacturers will be standardised in the near term. The case is similar in the UAS industry. There are initiatives, mostly in the open-source world, to standardise communication and control protocols, for example the MAVLink (micro air vehicle link) protocol. Individual manufacturers, however, still mostly rely on proprietary closed solutions, where even different models from the same manufacturer might be using different communication and control systems. This is the case for DJI for example, the most widespread civilian-use COTS (commercial off-the-shelf) drone solution. It is unknown whether any one particular technology will be adopted in the future as a standard solution, and at the moment there seem to be no initiatives either. Regarding communication solutions, there are also no dominant approaches; however, V2X (vehicle-to-anything) communication methods are rapidly changing active research field today. Potential solutions include mobile network-based (4G/LTE, 5G), LORA, WLAN, Bluetooth and other technologies in use today. While 5G solutions have been demonstrated, linked to this project [43], and it is likely that mobile network-based solutions will become more and more widespread in the future, it cannot be said with certainty that it will become the definite solution in the future. As such, the conflict management system needs to be technology and solution independent.

2.2.3. Wide Variety of System Users and Components

The system needs to accommodate a wide range of users. In addition to the technological aspects already discussed, the specific type of users also represents a wide range. In case of UAVs, the smaller multi-copter designs are of primary concern, but VTOL (Vertical Take-off and Landing—either small or aerial taxi sized) and potentially fixed-wing aircraft and others (balloons, airships, etc.) could be present in the airspace. In terms of ground-based users, in addition to self-driving cars, manned vehicles, cyclists and pedestrians can also be present in a traffic situation. The environment and obstacles can be regarded as a third source of "users" in a conflict management system, as impact with these elements must also be avoided. As such, from the planning stage the system is aimed at integrating data from GIS (Geographic Information System) systems, including Geofence definitions.

2.2.4. No Controller Development

A very important criterium for the system is that the individual users must be able to operate using their own control algorithms. That practically means all hardware must work in a plug and play manner. It is infeasible to rely on the conflict management system for low-level control for 2 reasons. For one, it would involve extensive development for each vehicle as a dynamic system, not even considering payload, centre of gravity and other configuration possibilities. This is an activity that the provider of a conflict management system cannot afford to do. Another aspect is responsibility. The conflict management system must provide a command to follow to avoid the conflict, but it should not decide how to follow the command. Individual control tuning, dynamics, limitations and assistance systems (collision prevention, obstacle detection, lane assist, etc.) need to function individually and in addition to the conflict management system. Some, such as the assistance systems or geofencing, essentially act as additional safety barriers for the management of conflicts. The same logic applies to mission planning software. For UAV these tools are referred to as ground control stations (GCS or GC). All mission (autonomous flight)-capable UAVs have some form of GCS for the operator to setup the desired mission. The concept is very similar for UGV, but the mission planning software will likely appear as a map application integrated into the dashboard or similar. For UAV the open-source world (PX4 and ArduPilot are the most common control software) has well-established solutions (QGroundControl, MissionPlanner or APM), and other manufacturers use their own solutions (DJI Pilot, for example). There are also dedicated commercial solutions aimed mainly at commercial and enterprise level users, such as UGCS or Auterion Mission Control. These GCS software must be used by the operator, as the responsibility of planning and executing their operations cannot be taken up by the conflict management system provider.

2.2.5. Modular Software Solution

The software solution needs to have modular architecture. This enables the integration of the various types of users by developing type-specific communication interfaces for a given solution (MAVLink, DJI SDK, etc.). This also futureproofs the system as upcoming new types can be included by adding new modules to connect them. The conflict management algorithms also benefit from the modular architecture, as different types of detection and resolution methods can be developed as separate modules and their performance tested. The development process follows LEAN software development principles, aiming to minimise the time of each plan–build–test–evaluate cycle when new modules and functionality are implemented. To achieve this, it is important to start the development with the most widespread and available technologies, solutions and software stacks.

2.2.6. XITL Tools Integration

XITL [44] (X = anything in the loop) integration is a generalisation of using testing tools such as HITL (hardware), SITL (software), VITL (vehicle) and so on. This enables the development, testing and even operation of systems using a wide variety of sources, real world or simulated. Advanced concepts such as digital twins [45] or scenarios in the loop [43] have already been demonstrated for UGV as part of the Hungarian Autonomous Systems National Laboratory, in cooperation with the ZalaZone Automotive Proving Ground facility [46,47]. Simulation tools are especially beneficial in the case of conflict management, primarily because the consequences of mismanaging the conflict are minimal as opposed to physical impact between real-world vehicles. There are also additional benefits in terms of development time, cost, flexibility, safety, security and also convenience. While UGV operations can also be affected, especially in the case of UAVs, simulation tools also remove constraints arising from weather. Another major benefit is that significantly more data can be collected or generated from simulations, with more accuracy; measurement does not disrupt the phenomenon to be measured, and there is no measurement of uncertainty and noise. As an added benefit it is very easy to generate "fake" data programmatically, which can be used to test situations that would be difficult to orchestrate using real hardware or even high-fidelity simulators. The proposed conflict management tool needs to be designed so that it can integrate with XITL tools seamlessly.

2.2.7. Summary of Requirements

In summary, the key requirements towards proposed system are:

- Detecting and resolving tactical conflicts in an approximately 10 s timeframe;
- Fully autonomous operation;
- Technology and solution independent;
- Wide range of system users integrated;
- Plug and play integration;
- Modular software solution;
- XITL tools integration.

Figure 1 shows the architecture of the proposed framework concept. The key development areas are coloured in blue; these are the components that need to be built from scratch to achieve the desired functionality. The communication and user layers are items that are intentionally left to use only COTS solutions. The detection and resolution methods are coloured differently. The framework can work by integrating these algorithms, even if they are already developed or available only as a black box tool. The tools used can be changed in a modular way in the system. The final piece of the system is the GUI (graphical user interface), which enables a human supervisor to monitor the system and interact with it if necessary. It is not coloured blue, because there is no strict need to develop an independent GUI; the proposed system could integrate with other tools with existing GUIs, as long as appropriate interfaces are developed for data sharing.

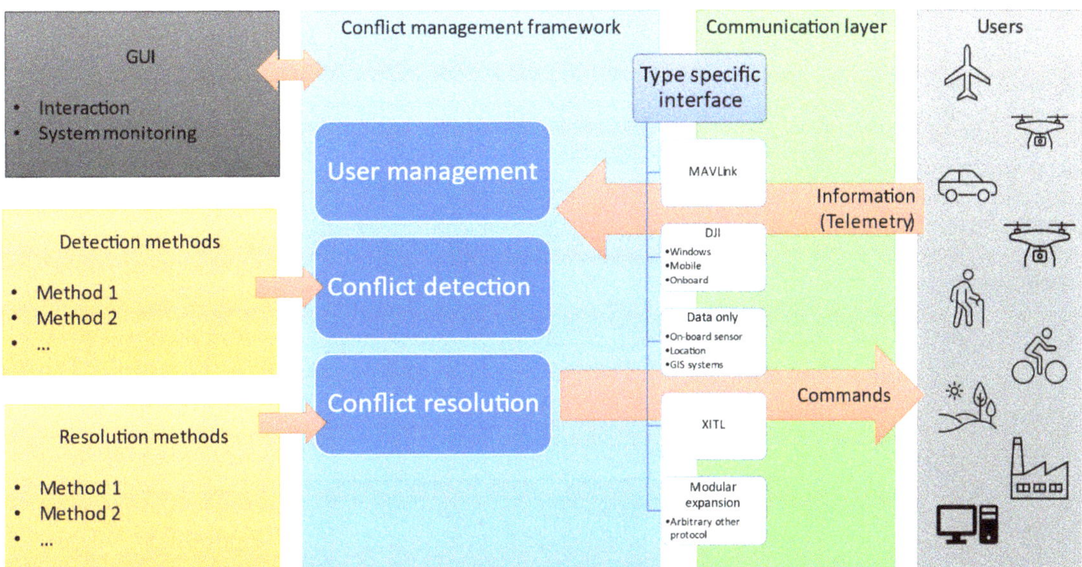

Figure 1. Conflict management system high level architecture.

3. Results

This chapter discusses the software implementation of the proposed conflict management framework, and the results achieved testing the framework with simulation and real hardware. Up to this point in the research project two versions of the framework were implemented, both of which will be presented along with the reasons for the development of the second iteration.

The programming language chosen for the system is Python (v3.8), as it is a language well suited for prototyping, and a wide variety of modules are available both for IT and engineering purposes. In the following, "system" always refers to the implemented conflict management system, unless otherwise noted.

3.1. First Version of Software Implementation

3.1.1. Overview

The first version of the conflict management software was developed as a self-contained computer application, where the conflict management framework is started at program execution, and the system users' missions were also started from the application. Essentially, this would enable the system to provide conflict management for pre-defined scenarios for testing purposes. Note that this is not equivalent to a scripted scenario, as the users are only given commands to begin their missions, after which the system only manages the conflicts, not the individual paths.

Figure 2 shows a top-level overview of the first framework version's workflow. First, in the system initialization step, the system goes through five tasks, after which it initiates an infinite loop where conflict management is facilitated. The system components are presented and discussed following the order presented in the flow diagram.

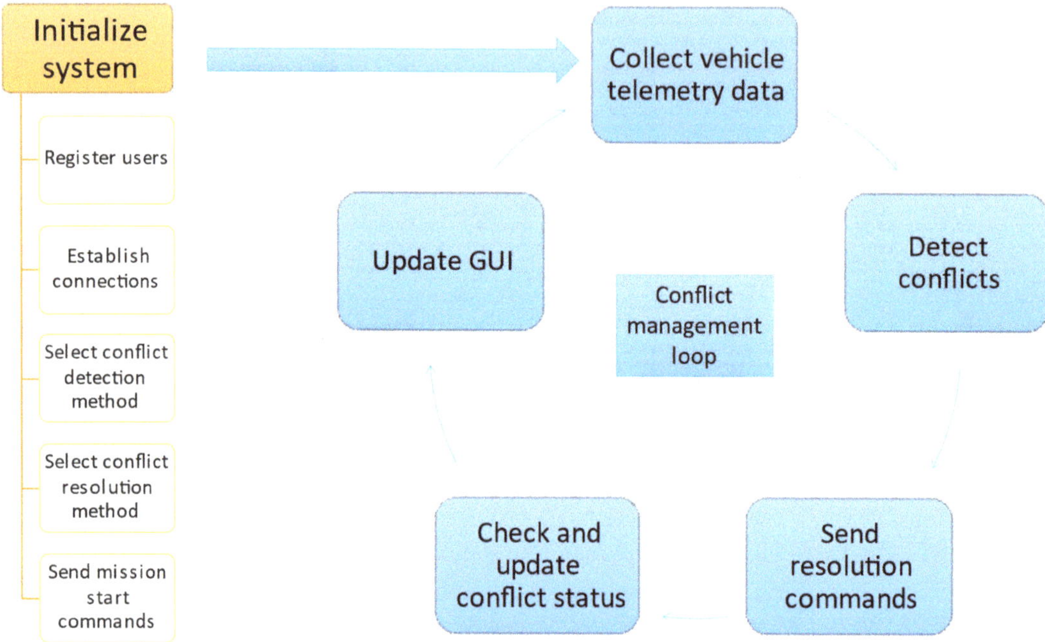

Figure 2. Top level flow diagram of first framework version.

3.1.2. System Users and Connections

The users of the system are the individual vehicles, people, etc., for which the system provides the conflict management service. Following the defined requirements and key principles, the users are represented as an abstract class, from which the individual user types are inherited, thus ensuring modularity. The first version implements the following user types:

- Algorithmically generated ("fake") location data;
- MAVLink connection:
  - ○ SITL simulation
  - ○ Real hardware.

The algorithmically generated data essentially calculates a time series of position data, and provides it for the conflict management system, without considering the actual vehicle type or its dynamics. Additionally, the vehicle receives no control commands from the system. This class of users represents a vehicle, which is not a cooperative member of the conflict management framework; however, its position is known. For example, positions could be received through protocols and channels not implemented in the system yet, or its position could be inferred with using computer vision or similar methods and the position provided by other users of the system. In this project, the class was initially implemented to enable the research group's self-driving smart car (Figure 3) to integrate into the system.

Figure 3. BME Automated Drive self-driving Smart [48].

The Smart currently operates as a self-contained unit; it is able to follow pre-specified paths in an autonomous way. However, with its current implementation, it cannot receive commands from an off-board source, nor does it provide telemetry data in any standardised format. As such, to integrate it into the system, the Smart provides its current position along the pre-defined path as simple coordinates data, which conceptually works the same way as the algorithmically generated data from the system's point of view. The algorithmically generated class is also very useful to test specific scenarios, as actual vehicle control and dynamics do not constrain the possible timings and paths.

The MAVLink connection class implements two-way communication between MAVLink-capable vehicles and the system. There are three options available in the system:

- PX4 SITL simulator;
- X500 UAV with PX4 flight control;
- "PX4 in a box" unit.

PX4 [49] is an open-source control software that is ported to a wide variety of hardware. It is capable of controlling fixed-wing and multi-rotor UAVs and UGVs among many other configurations. The PX4 SITL simulator can be run on any compatible computer (Windows, Linux, Mac), with relatively low resource requirements. The SITL itself runs the same flight controller algorithms that are used when flying actual UAV hardware. For it to simulate flight or driving, a simulation environment needs to be attached to it to provide information such as simulated world and sensor data and to resolve the dynamics of the vehicle. The simulation environment used in this project is Gazebo [50], which integrates seamlessly with the PX4 SITL simulator. This configuration is able to simulate up to 10 (and possibly more) UAVs, UGVs or any other type of vehicle for which models exist. The simulations communicate with the system through the UDP (user datagram protocol) protocol, and for the purposes of the conflict management framework they behave exactly like real hardware would. Simulated and real vehicles can be freely combined.

The X500 is a UAV development kit, from which a complete UAV can be built that uses the Pixhawk 4 flight controller, running the PX4 autopilot software. This is a very common configuration and probably provides the best solution in terms of achievable functions and ease of expandability. The other hardware, "PX4 in a box" also uses a Pixhawk hardware as the flight controller; however, it only includes the most essential sensors, so that the unit is able to connect to the system and provide telemetry data. There is no frame or propulsion unit, so it is not flight capable; rather, it is built into a box, which can be mounted on vehicles or even carried by hand, enabling integration into the system for arbitrary users. After boot, the box's controller needs to be "tricked" into believing it is flying or driving, which is achieved using the GCS software. Figure 4 shows this hardware as used during system tests.

Figure 4. PX4-based real hardware: X500 (**left**) "PX4 in a box" (**right**).

The PX4-based users connect to the system via telemetry radios (the particular type of hardware is called SiK radio), sending telemetry data and commands as MAVLink messages via serial communication. This solution is one of the most widespread and available for open-source drones today. They operate on nominal 433 MHz using a frequency-hopping method to enable a number of these units to work simultaneously without interference. One end of the unit is connected to the vehicle and the other to the PC running the conflict management framework. These radios are primarily designed to work in pairs, and while multiple units could be connected for multi-point comms, in the authors' experience, it does not work suitably for this purpose. Considering this, during the operation of a production version of the conflict management system, these radios will be plugged in to the computers of the users who are operating the drones and not to the central conflict management computer. These telemetry radios are used by the operators to monitor the status of the UAV and manage missions. Routing and forwarding the communication to the system via UDP or TCP (transmission control protocol) can relatively easily solve this problem. For other type of radio systems, different interfaces need to be implemented, and this is part of the longer-term development plans.

3.1.3. Conflict Detection Methods

For the initial prototype of the system two conflict detection methods are implemented, both of the deterministic class, and they detect conflicts between pairs of users, initially only in two dimensions:

- Pairwise waypoint-based static area detection;
- Pairwise dynamic projected area detection.

In the first method, it is assumed that both users provide their mission definition, and as such the trajectories they are going to take are determined in advance. An "s" ratio is introduced, which represents the ratio of the mission trajectory already completed by the user. This "s" ratio in theory simplifies the detection as the ratios where conflicts could potentially arise during the mission of the users can be predicted in advance, as soon as a user registers (connects) to the system. It is then only important to know what the current "s" ratios of the vehicles are (presuming the mission definition also includes velocity targets) and determine if a possible conflict is active or not based on the ratios. Figure 5 shows an example visualisation of the detection of the conflict areas using this algorithm. A uniform safety buffer of 5 m was used on both sides of the defined mission trajectory (middle black and grey lines).

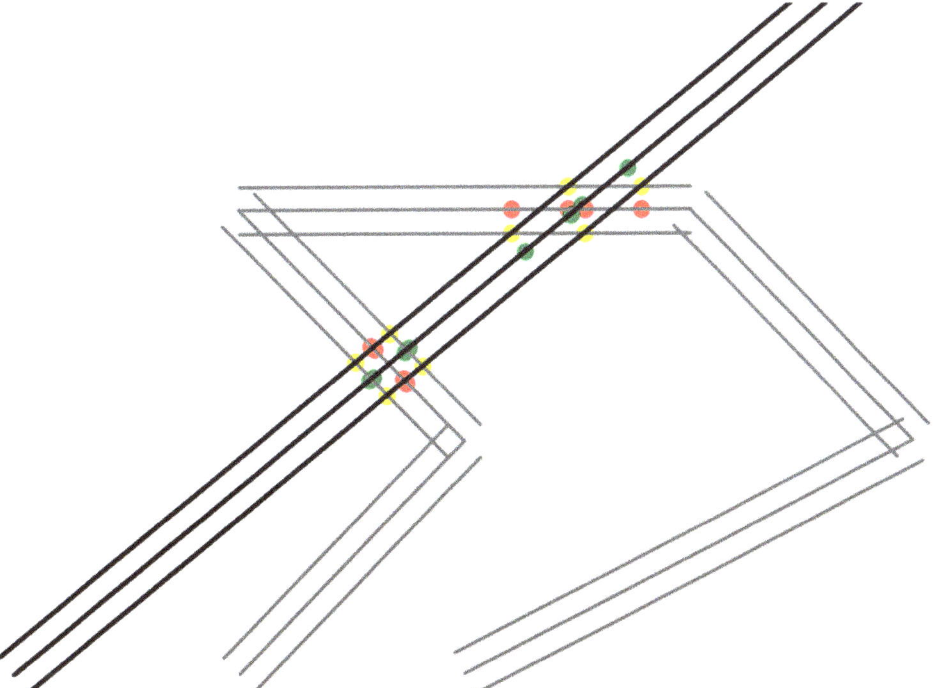

Figure 5. Pairwise waypoint-based static area detection method example; trajectories with safety offset (black and grey), determined conflict areas (yellow), "s" ratios along trajectory when conflict can be active (green for black trajectory and orange and red for grey trajectory).

While the algorithm in theory simplifies the detection and frees up resources from checking for conflicts where they simply cannot arise, the practical implementation showed that there are issues with the robust and accurate determination of the vehicles' "s" ratio. Most of the autopilot software solutions give some form of feedback regarding what the vehicle is doing; however, in the experience of the author, the current mission waypoint information is not always robust enough. Predicting which waypoint is the target without receiving it via telemetry also fails when the trajectories become more complex, such as self-intersecting trajectories, repeating waypoints, etc. Furthermore, any deviation from the flight path, including auto-land or return-to-home emergency functions, means that during these deviatory manoeuvres the system does not provide conflict management capabilities. Based on these results we decided not to pursue this form of conflict detection at the moment but to revisit it at a later stage of the research project to find potential solutions to the issues.

The pairwise dynamic projected area detection method works by calculating the predicted position of the user based on the current position and velocity, received as telemetry data. The predictions are updated at every cycle of the conflict management algorithm. The prediction time and a safety buffer are set for each user in each conflict pair. This allows the system to customise each conflict; for example, for users that operate with higher uncertainty a larger or differently shaped buffer area could be used, and slower responding users could have longer prediction times. Figure 6 shows an example of this methodology during testing in the first version of the GUI. Note the axes in Figure 6 are based on the UTM33N (EPSG:32633) coordinate system, and as such they include the full coordinate values to the points. Coordinate systems need to be converted often when using the framework system, as for example, Smart records coordinates in UTM33, but DJI and PX4 systems report telemetry in WGS84. In the first iteration it decided to display vehicles in UTM33N.

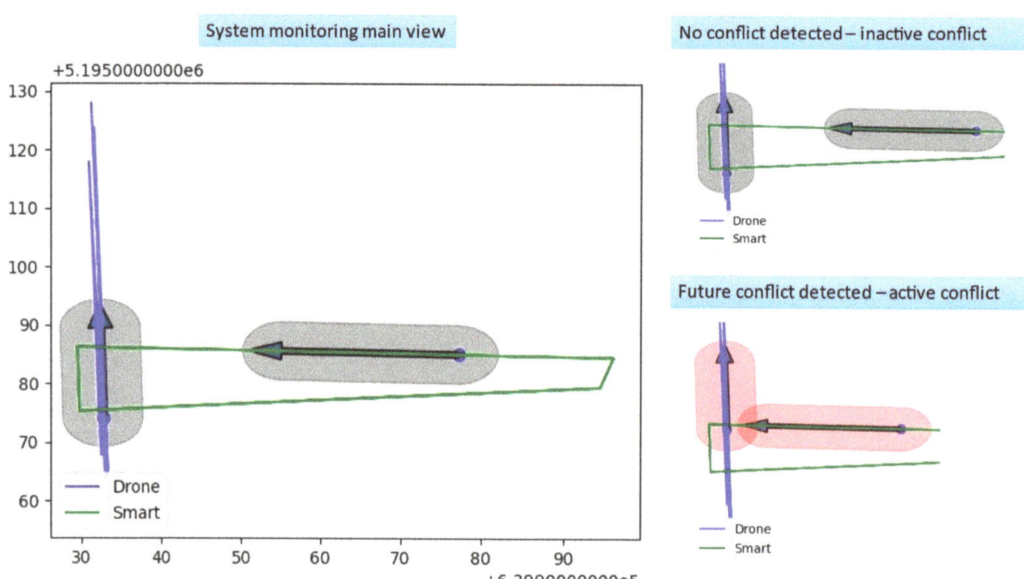

Figure 6. First version GUI display of pairwise dynamic projected area detection method; lines show user's mission (if available), points show the current position and arrows show predicted position. The grey (inactive) and red (active) areas show the conflict area with safety buffer applied).

Using the current and the predicted position and the safety buffer, a polygon is generated using the Shapely package of python. Shapely then can also be used to efficiently calculate intersections between the polygons of the two users. If the intersection area exists, then the conflict is set to be active, and it needs to be resolved. The method is very robust and quick, and as such it was set as the default algorithm for future development activities. An important aspect is the tuning of the prediction time and safety buffer parameters, which was performed using SITL simulations to yield an acceptable performance for the prototype system.

The system is intended to be used to develop and evaluate further methodologies, which will be modularly added in future steps of the development process.

3.1.4. Conflict Resolution Methods

Once the conflict detection method has established that a given conflict is active, then the allocated resolver method must provide a resolution for it. Detectors and resolvers are designed so they can be mixed within the framework and different combinations tested.

The baseline resolver that has been implemented is the stop-and-go resolution method. In this method one from the pair of vehicles in a conflict is stopped by the system, while the other is left to carry on with its mission. As one user can be involved in many conflicts, if a user receives a stop command from any of its conflicts, it must stop. Only those users who receive "carry on" commands (or rather, do not receive stop commands from any of its conflicts) can carry on with their mission.

To prevent impossible situations and deadlocks in the system, a priority assignment algorithm is created. The priority assignment considers first how a user is participating in the system. If the user is cooperative and it can receive and act on control commands from the system, it is assigned lower priority than a non-cooperative user. For each conflict pair, the user with the highest priority is given the carry-on command, and the lower priority user is sent the stop command. In the case of two non-cooperative users coming into conflict, no commands are sent, as the system has no means of affecting these users. In this case, it is up to the non-cooperative users to resolve the conflict to the best of their abilities. Non-cooperative users need to be dealt with at a different level, for example, by introducing policies that define mandatory standard protocols for vehicles, so that in theory they all become cooperative users. In the case where two cooperative users come into conflict, they will be allocated priority based on a set of rules. In the first implementation, the user's ID (unique for each user assigned when it registers to the system) is used to allocate priority, but this method can be used to give priority to less manoeuvrable or sensitive users, etc. It just needs to ensure that priority values are always absolute, and no deadlocks arise in multiple conflict pairs, in which all vehicles are given the stop command, and they all wait for each other forever.

The stop-and-go algorithm, despite its simplicity, works relatively well; however, it does not work for some cases, including coming into conflict head on, as one user would stop right on the trajectory of the second, possibly preventing its safe passage. For the second iteration of the system an improved algorithm is used, in which, in the case of head-on conflict, an evasive command is given to the lower priority vehicle before it is sent the stop command, so it moves out of the way. The robustness of this algorithm still needs to be improved, especially to make sure the lower priority vehicle is commanded to a position that is safe; this can be difficult to guarantee when multiple conflicts are active or the environment is confined.

When multiple vehicles are in conflict, how the multiple conflict is treated depends on the assigned conflict detection and resolution algorithms. Essentially, two types of algorithms can be defined, one which works between a pair of vehicles and one which resolves the detected conflict for an arbitrary number of vehicles. Obviously, when the pairwise algorithm is used, multiple algorithms are running parallel at the same time. When treating a very large number of vehicles, the management of the computational requirements needs to be considered. The number of conflict algorithms to be run in parallel scales as the factorial of the number of vehicles inolved, when pairwise treatment is chosen. The utilised stop-and-go algorithm for both versions of the management framework relies on pairwise detection and resolution, and due to the relatively low number of vehicles, computing capacity is of no concern.

Figure 7 shows the resolution process for a multi-vehicle conflict using the pairwise stop-and-go management algorithm. While all algorithms being developed have the capability to deal with multi-vehicle conflicts, as seen in the figure, the key focus at this stage of the research is the development of robust methodologies and testing with one pair of vehicles in conflict, as it is a lot clearer, easier and safer to track and control how a single pair of vehicles are behaving, especially during real hardware tests. In the following, discussion will be limited to a single pair of vehicles in conflict for this reason.

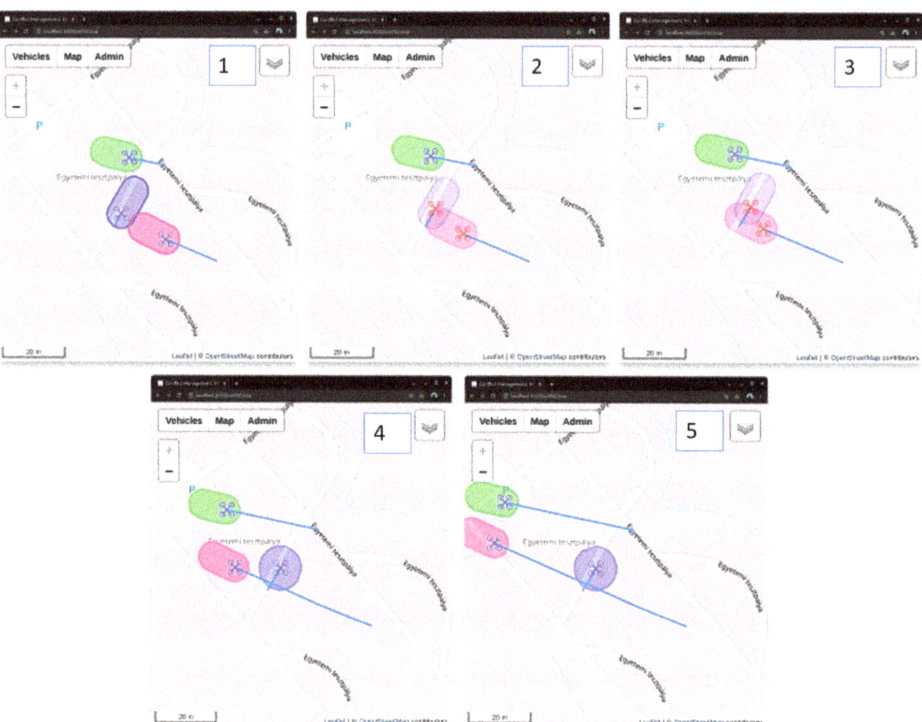

Figure 7. Multi-vehicle conflict resolution **1,2**: Conflict is detected between blue and purple, 3rd vehicle (green) unaffected; **3**: Lower priority (blue) vehicle is given command to stop and moves out of impact path; **4,5**: Conflict is resolved, vehicles not stopped carry on, commanded (blue) continues mission after safety delay. Non-English text are road section designations.

It is also worth mentioning that as the system does not deal with low-level flight control, as it only sends commands to the users. During testing it was observed that for some flight controllers, sending a pause or stop mission command would result in the UAV first stopping, then reversing to the position where it received the stop command. While it is debatable whether this is a feature or issue, it can be overcome, but the more important conclusion is that extensive testing is required for each type of user that is integrated into the system to understand how to best control them.

3.1.5. Lessons Learnt from Testing the First Version

The first version of the system was tested using primarily SITL simulators, but a successful real hardware test was also performed to validate the concept. During these tests some shortcomings were identified, which are presented and briefly discussed here.

Probably the greatest issue identified was the fragility of the software implementation. While the conflicts were managed safely when the system operated as intended, there were too many occasions when the operation of the system stopped. The primary culprit was the communication system. It was too easy for the telemetry radios to drop the connection between a user and the system, even if just momentarily. Since the connection to the users was established during initialization, this meant that the whole system had to be restarted to reconnect. The implementation of the MAVLink connection used (MAVSDK-python) also seemed to be a fragile type of link, as when the connection failed it often took manual intervention and effort to reset everything and initiate the connection again. This is clearly not acceptable for an automated system. Because the tools used in this case are open source, it would be possible to develop the communication link; however, that would go against the principles of the system. It was decided that a solution is required where instead of fixing the link's fragility, the system would manage the connection to make sure it can always recover after a failure.

Improving the robustness of the system's other aspects was also a priority. One of the biggest issues was the lack of persistent storage of data. Whenever the system stopped for any reason, the only way to fix it was essentially to reset it and start again. This is particularly problematic for the missions, as the missions are also started during the initialization step. This means that if the users were already executing their missions when it was restarted, they would first return to the first waypoint to start the mission from the beginning. This has undesirable consequences, as most controllers send the vehicle on the shortest possible direct route when the return command is received, irrespective of the conflicts that could be caused. This also makes testing difficult. When investigating specific conflict situations, the return activity throws off the timing required for the situation to arise, because it is not known from which point of its trajectory the vehicle will start its return when the system is restarted. While there was an attempt to mitigate these issues by not always restarting the missions when the system restarted, it was decided that a robust solution would be required rather than a workaround.

Another issue identified was the restriction of user interaction possibilities. When the system connects to MAVLink capable users it blocks the ports necessary to run GCS software. As such, the user loses its ability to monitor and control the vehicle's mission. This is a most undesirable outcome, as the system's role is only to manage conflicts; it must not take away the mission management possibility (and responsibility) from the users. While this could potentially be fixed by implementing GCS functionalities in the system and forcing users to use the system for their mission management, this also goes against the principle of relying on COTS components.

Among the issues there are also minor elements, such as the low quality of GUI implementation. As can be seen in Figure 6, the first iteration uses a Matplotlib plot to display the system status and conflicts. While it would be possible to develop this further, it would inherently be restrictive, as it can only be displayed on the computer running the system. Additionally, developing new GUIs when there are other possibilities that could be implemented and would probably work well is a waste of resources and goes against the LEAN principles. A final issue that also arises partly from the GUI implementation is that users cannot be added or removed from the system during runtime—the system needs to be stopped, and new users added in the code. While a production system would need to have automatic user management capabilities, for the prototype it would be useful if the system operator (supervisor) could add or remove users as required.

To summarise, the following shortcomings had to be addressed in the second version:

- Fragile connections;
- Difficult timing of scenarios for testing due to resetting;
- Persistent data storage;
- GCS use is prohibited;
- Low-quality GUI;
- Local monitoring capability only;
- Dynamic user management.

3.2. Second Version of Software Implementation

To address the shortcomings identified, a significant redesign of the system was required. This section discusses the second implementation of the conflict management system, focusing particularly on the features that solve the problems identified.

3.2.1. System Architecture

The principal change in version 2 was a change in the system architecture from a locally running code to a networked web-based application. The framework used was the python-based Django web framework (version 3.2). Django is a very popular framework for web application development with powerful capabilities and a large database of expansion modules. The default implementation of Django is a synchronous service, which means that when long-running tasks are initiated, the service temporarily becomes unresponsive until the task is resolved. This is not desirable for the conflict management, as the various activities need to be run parallel (asynchronously). In order to achieve this the Celery (version 5.1.2)-distributed task queue was integrated into the system. The application also relies on a PostgreSQL relational database server for persistent data storage and the RabbitMQ message passing interface for managing tasks. A high-level overview of the system architecture is shown in Figure 8, and the significant changes compared to the first version are discussed in the following section.

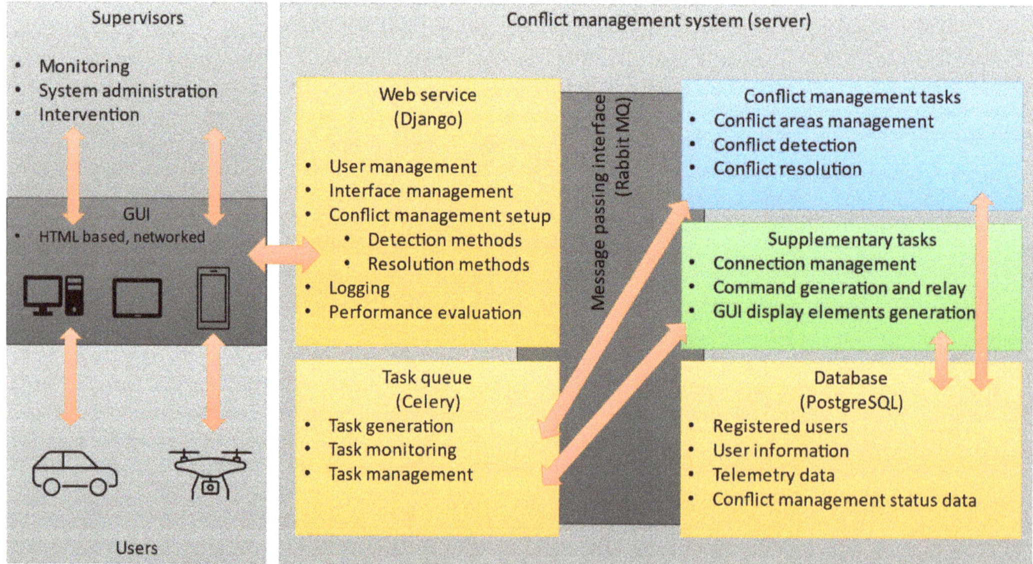

Figure 8. Conflict management system architecture—second version.

3.2.2. Communication Solutions

The introduction of the new architecture enables the implementation of robust communication solutions. In the current setup, each user is connected to the system by generating an individual connection manager task that is responsible for monitoring the health of the specific connection. The manager task is run periodically, and when the connection drops, it reinitiates the connection task. To facilitate this all tasks are designed to deliberately fail if they encounter non-desirable situations, implementing a fail-safe layer in the system. This is important, as tasks that are seemingly running but are stuck in a non-operative status can provide misleading information, which can result in dangerous situation during conflict management.

3.2.3. Data Persistence

The second version of the system uses a relational database to store data in a persistent and accessible manner. This is beneficial in two ways. First, it provides the system with an enhanced restart capability, as the relevant data are immediately available from the database even after a failure. Additionally, this approach enables the robust and parallel (asynchronous) handling of user data collection and the conflict detection and resolution tasks. Unlike the first version, where for each iteration of the conflict management loop, the system has to wait for each user to provide updated data, the data collection and usage run as separate tasks. A typical database system such as the one used is capable of handling in the order of 100,000 transactions per second with minimal tuning. Considering that the typical frequency of telemetry data acquisition from users is in the order of 20 Hz or less, the system should be capable of handling a large number of users without performance issues. In the case of a production system with a large number of users, the appropriate IT architecture planning (distributed databases, optimised transactions, etc.) solutions must be implemented. From the point of view of this research, these steps are not relevant, as it is known that solutions exist for this purpose, and it was not investigated further.

The database also has useful built-in utility functions, for example, timestamping the last modification of a table in the db. This is used, for example, to check the "freshness" of the collected telemetry data. From the system's standpoint, whether a user's connection is alive or not can be determined from the time the last telemetry data were written to the database. In the current system 3 s is set as the threshold. If the data are not "fresh", then the system treats the user as not connected, i.e., non-cooperative, until the connection manager reconnects the user. Figure 9 shows the second iteration of the GUI, displaying a user with timed-out telemetry messages and showing the indication of an active conflict zone.

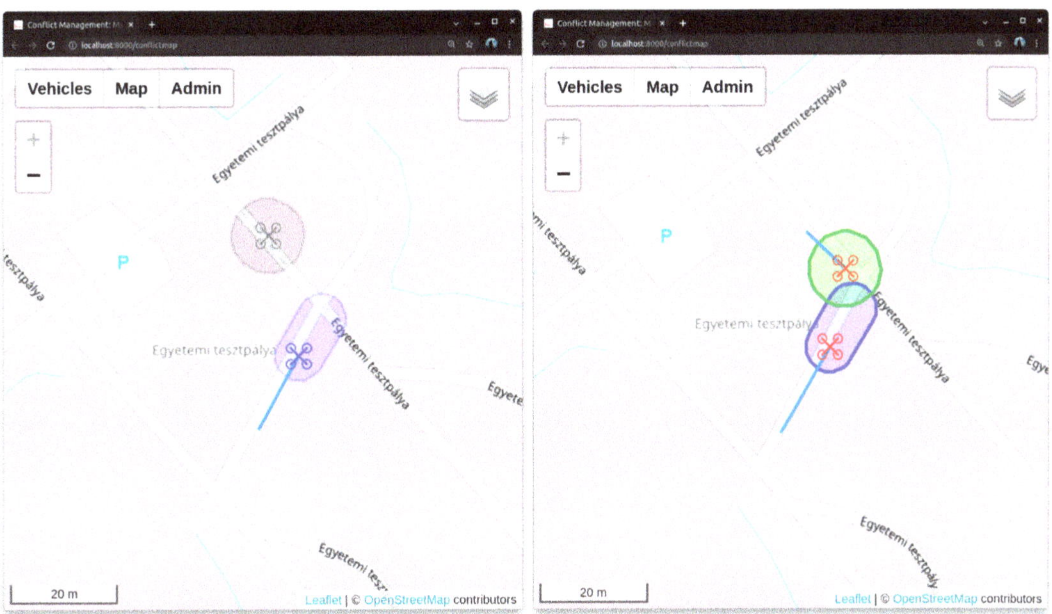

Figure 9. Version 2 GUI: indication of timed out telemetry (**left**), conflict area marked as active (**right**). Non-English text are road and area designations.

The colour coding of the current GUI is the following:

- Each vehicle is allocated a unique RGB colour during vehicle registration. (Red is not preferred, however can be allocated—it is purely a cosmetic question.) This colour is used to represent the conflict area when the conflict is not active.
- Vehicle icons by default are coloured in blue, as it was decided that the vehicle icon colour should represent the status of the vehicle.
- Trace of vehicle paths are marked with blue polylines.
- Vehicles with timed-out telemetry (lost connection or vehicle malfunction) are seen at their last reported location and last conflict area with greyed out colour.
- Vehicles in conflict are shown with flashing, bold line contoured coloured areas and flashing icon alternating between red and blue colour.

3.2.4. COTS Solutions Integration

Since the system is to be added as an additional security layer, not one replacing operator control, changes have been made to the way the users are connected. The PX4-based units are still connected through the SiK telemetry radios; however, the system no longer uses the connection COM port directly. The MAVLink-router package is used to run a forwarding service for each user, which in effect creates a mirrored copy of the connection port using the UDP protocol. This enables the computer running the system to connect to a user and a GCS at the same time. Or it can just receive UDP communication from a completely different computer, which connects to the user and its own GCS via the telemetry radio, as would be the case for an operating user. This way, the interface for connecting a simulated vehicle or a real hardware becomes identical as well. The same strategy was developed for DJI drones (using Windows or Mobile SDK—Software Development Kit), and it is currently in the process of being implemented. Integrating DJI SDK and MAVLink-based vehicles accounts for a very large segment of the current UAV solutions. MAVLink is a viable option for UGVs as well, either connecting to the vehicle itself or deploying the "PX4 in a box" solution. The use of pure position data from an arbitrary vehicle architecture is also a valid approach; however, this would provide only one-way communication, and thus would result in a non-cooperative user.

3.2.5. Graphical User Interface and Human Interaction

The GUI for the second version was significantly upgraded. Instead of using a single computer-based application and graphics, the system GUI is now served by the Django framework. It is an HTML response which can be viewed by any compatible device running an internet explorer (Edge, Firefox, Chrome, Safari, etc.). The base of the GUI is a Leaflet map. Leaflet is an open source javascript-based mapping library. It is simple to integrate into an HTML (HyperText Markup Language) view, and it is interactive and handles Geojson inputs. Geojson is a standard format for encoding various geographical features, such as points, paths, areas, etc. Plotting the users' positions, the conflict areas with the safety buffers, geofencing areas and other GIS features naturally lends itself to Geojson representation. From the conflict management related data stored in the database, the Django framework generates the Geojson features to be displayed on the Leaflet map, which is served to an arbitrary number of users via the web service.

The Leaflet map is completely customisable, the tilesets (background) of the map can be selected from different providers and the items displayed can be placed in layers that can be made hidden by the user and other similar features. Interacting with a map item (click or touch) can reveal additional information on the specific item in the form of a popup window or a separate HTML view. While this high level of customisation is beneficial, we was decided to deliberately limit the functionalities of the GUI to the conflict management specifics. A typical GCS has many options for setting up the vehicle, the mission, emergency functions and similar. It is our intention to keep these features separate, so the individual operators have to rely on their GCS and only use the conflict management system for its core purpose. This is important for maintaining the concept of relying on

COTS solutions. However, integrating the conflict management system into existing GCS is an option that will be considered in the future.

3.2.6. Improved Stop-and-Go Resolution Algorithm

For the second iteration of the system, the stop-and-go algorithm was also improved based on the lessons learnt from the first iteration. This chapter describes the algorithm used, highlighting the improvement and the reasons behind it for the second iteration.

In the first iteration of the methodology, as soon as the conflict was detected between two users, the system issued a "stop mission" command. This command is fairly universally present in the implementation of various flight controllers and their SDK mechanisms; however, it presents some significant drawbacks. First of all, the functionality might not exist for all COTS systems. Second, the implementation might not be done in the same way for different systems. The implementation in the PX4 flight controller stack used extensively in the research project suffers from a drawback; when the command is issued, the vehicle decelerates, stops, then reverses to the point where the vehicle has acknowledged the stop command. This is an undesirable behaviour, as the stopping point cannot be set deterministically due to the unknown latency in the communications systems. Furthermore, because the vehicle passes through the commanded stopping point, there is no guarantee that an impact is avoided in a conflict.

Upon investigating the specifics of the controller algorithm, the reason for this behaviour was found. Figure 10 shows the controller algorithm used by the flight control software. It can be seen that the system relies on a single proportional constant to ensure the position hold capacity of the vehicle. It can also be seen that the output of the system is a velocity control signal, which is saturated (limited) to prevent excessive accelerations. While it would be possible to include a derivative (and if necessary, an integral) constant to prevent overrun and reversing in the system, one of the key principles of the conflict management system is to not modify the individual control algorithms of the vehicles.

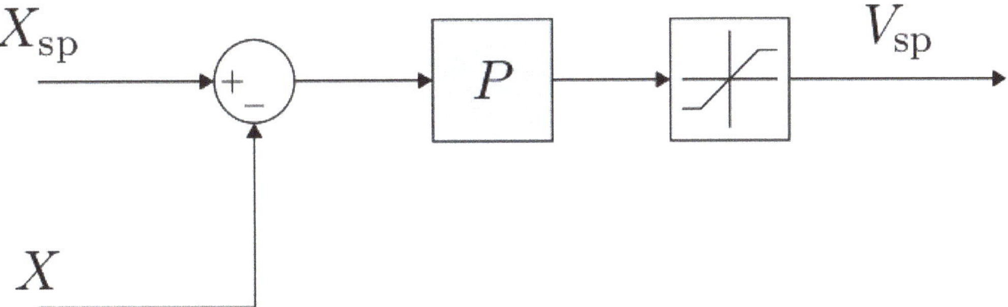

Figure 10. PX4 controller algorithm for position control (source: [51]).

The solution, which is adopted in the second iteration of the stop-and-go algorithm, is to use a "set position" command, instead of the "stop mission" command. When using this command, the vehicle dynamics are considered to issue a command where the accelerations can be kept below the saturation level of the controller. While this does not completely eliminate overswings while acquiring positions, in practice these overswings are reduced to the order of 10–20 cm, and as such they are no longer of concern.

Using set position commands also mitigates another big drawback of the stop-and-go algorithm, specifically, that it cannot handle head-on or shallow angle conflicts, such as if one vehicle is stopped in the path of the other one. Although it avoided immediate impact, this did not actually resolve the conflict. Using set positions allows the command to define the stopping point, so that the vehicle would stop outside the predicted conflict area of the other vehicle given the go signal. Due to the vehicle dynamics this stopping area needs to be chosen within a fairly narrow angle cone, below 30 degrees half angle measured from the velocity vector of the commanded vehicle. The quoted angle can be even more restrictive based on velocity, conflict prediction time and safety buffer size. In the second iteration, based on the test scenarios performed, to achieve acceptable behaviour, the safety buffer was set to a 4.5 m radius, and the conflict detection time was set to 4.0 s. The position target for stopping is derived based on 3.0 s of resolution time and the current vehicle velocity, which results in a distance that can be mapped within the cone defined by the dynamics. These specific settings do not necessarily provide an optimum behaviour, but rather they are settings that were chosen based on a trial-and-error process and seem to provide an acceptable qualitative performance for the algorithm in the test scenarios. For quantitative evaluation, please refer to the discussion section of this paper.

A final modification of the stop-and-go methodology involved the conflict detection algorithm. When the vehicle is commanded to stop, its conflict area will gradually reduce as its speed drops (refer to Figure 6, Figure 7 or Figure 9 for conflict area visualizations). In many cases during the tests in the first iteration of the system, this drop resulted in a situation where a stopped vehicle's conflict zone no longer intersected the conflict zone of the other. In this case, the algorithm assumes that the conflict is resolved, and the stopped vehicle is given the go signal, effectively sending it in front of the other vehicle, creating a significant impact risk. In the second iteration this issue is resolved by increasing the safety buffer of stopped or slow-moving (below 0.25 m/s velocity) vehicles to 1.5 times the defined conflict buffer distances. This allows the stopped or almost stopped vehicles to remain in conflict and stopped.

4. Discussion

This chapter summarises and discusses the insights gained from the testing of the prototype system and proposes further development directions. There has been extensive testing involving the SITL simulator, and it was confirmed that the robustness of the system and the communications layer has been increased significantly. Real hardware tests also showed that the second version provides a superior performance. The two performance criteria used in the case of prototype testing were qualitative evaluations of system functionality (which is effectively safety for the vehicles) and the robustness of the system. Regarding robustness, the first implemented method required resetting the system and users when non-desirable behaviour occurred; this involves returning the vehicles to the starting positions, restarting the management algorithm and disconnecting and reconnecting the communication systems (telemetry radios) of the vehicles. This can up to 1–5 min per reset. Note that the average endurance for drones is in the order of 30 min, so even a few resets take up significant portion of their available flight time. Recalling drones to replace batteries during the tests or demonstration would add an undesirable amount of waiting time, and in extreme cases could prevent the successful completion of all scenarios during a test day. Due to the persistent data storage and differing conflict resolution algorithm, the second iteration does not require resetting; furthermore, in the case of communication loss, it automatically attempts to reconnect to the users. In terms of the other criteria, the system functionality, the first implemented version of the stop-and-go algorithm showed a number of mismanaged conflicts, e.g., the lower priority vehicle did not stop at a discovered conflict, or first stopped and continued its mission. Using the second algorithm eliminated these system errors, and in the scenarios tested, it provided appropriate resolutions for all conflicts. Quantitative evaluation of system safety, management efficiency (time lost due to conflict management) and robustness is being carried out in the current, follow-up stage of the research project.

The current state of the research project was also presented at a live demo conducted in the ZalaZone Proving Grounds in Hungary. The demo was aimed at demonstrating the core principles of the conflict management system. Unfortunately, during the demo, the allocated space was rather limited, and only a single intersection of roads was made available, so the scope of the demonstration was also limited. Refer to Figure 11 for photographs taken at the event. In the demo, the UAV has received a repeating mission to fly between the two sides of the intersection. The ground vehicle had the PX4-in-a-box unit mounted to the roof and was operated by a driver. A manned ground vehicle seemed like the appropriate choice for such limited space, as the intention was to demonstrate that the system can react to a wide variety of situations that can arise. Relying on a driver means that the driver can also improvise in the situations and demonstrate the safe workings of the system. The demo was successfully presented, and while the system's performance was excellent, some areas for further improvement were identified.

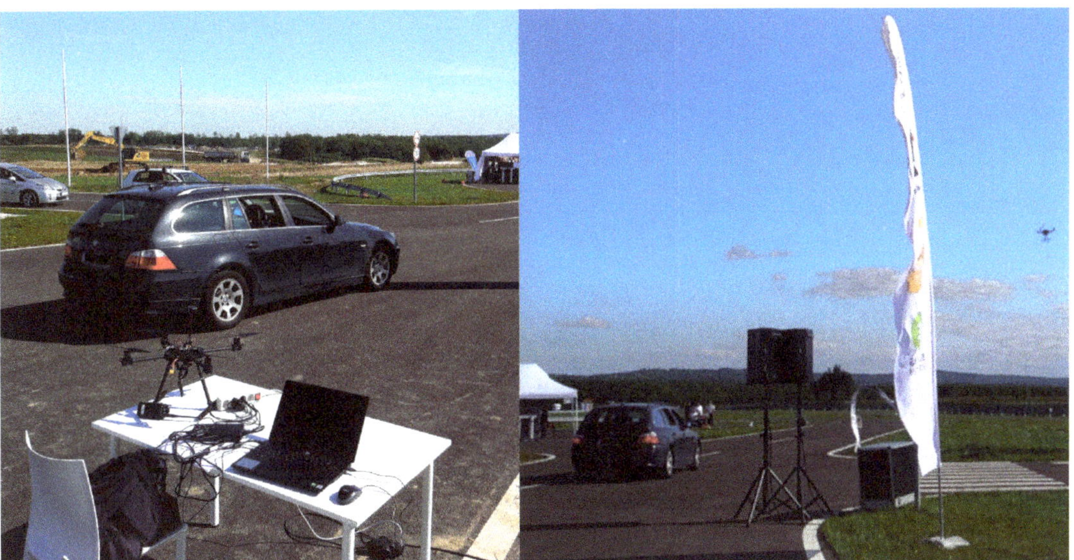

Figure 11. ZalaZone demo setup: ground control station and system users (**left**), system in operation (**right**).

Among the identified areas to improve was the aforementioned case, when head-on conflicts occur. In this case, the stopped vehicle needs to perform an evasive manoeuvre before stopping to ensure that separation is maintained between the users. To do that safely, in addition to the system users' positions, information about the environment is also required. As such the next important step is the integration of a GIS system.

Another direction is the closer integration of DJI products. The Windows SDK was successfully tested, and basic automatic operation was demonstrated. There are significant limitations in this SDK version, and based on the lack of updates it seems that it might be on the road to obsolescence. As such the Mobile SDK (Android and iOS) is under consideration. Mobile development can also unlock new opportunities, as it would enable smartphones to join the system as users. However, mobile development is a separate niche of programming, and the costs and benefits need to be weighed before effort is invested in building up this technology skillset in the research group.

As the framework's performance is acceptable, the next primary direction of the research group is to develop the conflict management algorithms. For the conflict detection algorithm, the most important thing is to include information from vehicle trajectories (mission plans), if available. Knowing the trajectories can potentially help to resolve some detected conflict situations more efficiently.

Regarding efficiency, another key development direction is a method to objectively evaluate the system's safety, performance and efficiency measures. While intuitively, a simple resolution algorithm such as stop-and-go seems to be less efficient than more advanced ones, this might be based on the inconvenience experienced as a human driver when the vehicle needs to be stopped or the route changed. For an autonomous system, however, convenience is not a factor that matters. Unless specific criteria are defined, it is impossible to compare methods objectively. Once the measurement method is developed, SITL simulations can be used to objectively compare detection and resolution methods. Furthermore, optimisation tools can be used to determine the optimum parameter settings for a specific type of algorithm.

In the follow-up studies since the completion of the first stage of the project the development of objective quantitative performance criteria is one of the key points of research. The following list highlights the criteria that are being used in the development for the system:

- Safety measures:

  - Number of users in the system: this is used to calculate the per-user performance measures to compare management of areas with varying traffic intensity;
  - Number of conflicts detected: the number of times the conflict detection methods turn the conflict areas active;
  - Number of resolution commands issued;
  - Number of mismanaged conflicts;
  - Number of impacts between users;
  - Number of near impacts between users.

- Performance measures:

  - Time and distance travelled by individual users without other users or conflicts;
  - Time and distance travelled by individual users while other users are present, and the conflict is being managed by the system;
  - Highest level of accelerations and decelerations commanded by the system;
  - Total change in altitude commanded by the system (only applicable to UAVs);
  - Time and distance spent off-road or parked (UGV);
  - Time spent landed, if commanded (UAV).

When evaluating the performance criteria, the following approach is being developed:

1. Simulate individual users, alone in the conflict management area, to evaluate how they would perform their mission without interactions from other users or the conflict management system.
2. Simulate (or when sufficient confidence in the system is achieved, live test) the conflict area with all users involved and log the individual positions, velocities, accelerations and commands received.
3. Post-process the logged information to evaluate the performance measures listed above.

The three-stage process described enables the objective evaluation of the performance measures, as it essentially compares the theoretical possible unrestricted operation of all users to the actual performance they can achieve while multiple users are around, and the conflict is managed. The best management solutions provide the highest level of safety, so no impacts or near impacts and detect all conflicts, while at the same time being the least intrusive, so issuing a minimum number of resolution commands and imposing minimum additional time and distance (detours). Quantitatively evaluating these measures enables the objective comparison and development of different conflict management methods and strategies, which is the ultimate aim of the research framework presented here.

5. Conclusions

In conclusion, in this research a system framework was developed and its functionality successfully demonstrated. The framework implements short- and mid-term tactical conflict management for a wide user base, including unmanned aerial and ground vehicles. The framework has gone through two iterations, the second of which significantly improved key issues that were identified during development and testing of the first. The final, second version of the framework achieves a technology and solution independent implementation of the stop-and-go conflict management algorithm. This system is capable of handling a wide range of unmanned and aerial ground vehicles without introducing any modifications to the communications or control systems of any particular vehicle. The framework is implemented as a modular suite of software algorithms, with simulation software integration. The framework is accessible as a web service, where an arbitrary number of users and supervisors can connect to the system using standard web browsers from various devices to monitor activities and perform administrative tasks. The system is designed at the moment to handle up to 250 users (vehicles). The web service framework relies natively on database persistent information storage, which improves the robustness and recovery characteristics of the system, including reconnecting communication systems when the connection is interrupted. The stop-and-go algorithm was also improved to consider system dynamics and head-on conflict cases.

The framework's performance is considered acceptable, and future research focus is based on conflict management methods development and evaluation.

Author Contributions: Conceptualization, D.R. and D.S.; methodology, D.S.; software, D.S.; validation, D.R. and D.S.; formal analysis, D.S.; investigation, D.R. and D.S.; resources, D.R. and D.S.; data curation, D.S.; writing—original draft preparation, D.S.; writing—review and editing, D.S.; visualization, D.S.; supervision, D.S.; project administration, D.R. and D.S.; funding acquisition, D.R. and D.S. All authors have read and agreed to the published version of the manuscript.

Funding: The research reported in this paper and carried out at the Budapest University Technology and Economics was supported by the National Research Development and Innovation Fund (TKP2020 National Challenges Subprogram, Grant No. BME-NC) based on the charter of bolster issued by the National Research Development and Innovation Office under the auspices of the Ministry for Innovation and Technology. In addition the research was supported by the National Research, Development and Innovation Office through the project "National Laboratory for Autonomous Systems" under Grant NKFIH-869/2020. Funder NKFIH Grant Nos.: TKP2020 BME-NC and NKFIH-869/2020.

Institutional Review Board Statement: Not applicable.

Informed Consent Statement: Not applicable.

Acknowledgments: The authors would like to acknowledge Gábor Jandó and Károly Veres for their contribution to the extensive testing and operation of the system during demo events.

Conflicts of Interest: The authors declare no conflict of interest.

References

1. Poponak, N.; Porat, M.; Hallam, C.; Jankowski, S.; Samuelson, A.; Nannizzi, M. Drones Flying into the Mainstream. 2016. Available online: https://www.goldmansachs.com/insights/pages/drones-flying-into-the-mainstream.html (accessed on 22 October 2021).
2. MarketsandMarkets. Unmanned Aerial Vehicle (UAV) Market by Point of Sale, Systems, Platform (Civil & Commercial, and Defense & Governement), Function, End Use, Application, Type, Mode of Operation, MTOW, Range, and Region—Global Forecast to 2026. In *Markets and Markets*; MarketsandMarkets Research Private Ltd.: Hadapsar, India, 2021. Available online: https://www.marketsandmarkets.com/Market-Reports/unmanned-aerial-vehicles-uav-market-662.html?gclid=CjwK CAiA78aNBhAlEiwA7B76p3r10jhv4HYaPU5K6iYWi-9Uq1S_SlE_W5bEvNb2U6_unbt5QHiLZhoC_FoQAvD_BwE (accessed on 22 October 2021).
3. *Global Drone Market Report 2021–2026*; Research and Markets: Dublin, Ireland, 2021.
4. *European Aviation in 2040, Challenges of Growth*; EUROCONTROL: Brussels, Belgium, 2018. Available online: https://www.euro control.int/sites/default/files/content/documents/official-documents/reports/challenges-of-growth-2018.pdf (accessed on 22 October 2021).
5. European Automobile Manufacturers Association. *Vehicles in Use Europe 2019*; ACEA Report: Brussels, Belgium, 2019.
6. Litman, A.T. Autonomous Vehicle Implementation Predictions, Implications for Transport Planning. 2021. Available online: https://www.vtpi.org/avip.pdf (accessed on 22 October 2021).
7. On-Road Automated Driving (ORAD) committee, Taxonomy and Definitions for Terms Related to Driving Automation Systems for On-Road Motor Vehicles. *SAE Int.* **2021**. [CrossRef]
8. *Unmanned Aircraft Systems Traffic Management (UTM)—A Common Framework with Core Principles for Global Harmonization*, 3rd ed.; International Civil Aviation Organization: Montreal, QC, Canada, 2020.

9. Bradford, S.; Kopardekar, P. *Uncrewed Aircraft Systems (UAS) Traffic Management (UTM), UTM Pilot Program (UPP), UPP Phase 2 Final Report, Version 1.0*; Federal Aviation Administration and National Aeronautics and Space Administration Joint Publication: City, WA, USA, 29 July 2021. Available online: https://www.faa.gov/uas/research_development/traffic_management/utm_pilo t_program/media/UTM_Pilot_Program_Phase_2_Progress_Report.pdf (accessed on 22 October 2021).

10. ASTM F3411-19. *Standard Specification for Remote ID and Tracking*; ASTM International: Conshohocken, PA, USA, 2019. [CrossRef]

11. *Initial View on Principles for the U-Space Architecture, European Union*; EUROCONTROL: Brussels, Belgium, 2019.

12. SESAR, GOF U-Space, European Union, SESAR JU GOF U-Space Project: Final Demo with Piloted Air Taxi Flight Successfully Completed, Helsinki. 2019. Available online: https://www.sesarju.eu/news/sesar-gulf-finland-u-space-project-first-demos-suc cessfully-completed (accessed on 22 October 2021).

13. Federal Office of Civil Aviation. Swiss U-Space ConOps, U-Space Program Management, Reference: FOCA muo/042.2-00002/00001/00005/00021/00003. 2019. Available online: https://susi.swiss/wp-content/uploads/2020/04/Swiss-U-Sp ace-ConOps-v.1.1.pdf (accessed on 22 October 2021).

14. Innovation Hub. *Beyond Visual Line of Sight in Non-Segregated Airspace, Fundamental Principles & Terminology*; UK Civil Aviation Authority: London, UK, 2020.

15. Eliot, L. Pairing Self-Driving Cars with Autonomous Drones for Faster Fast Food Delivery. 2019. Available online: https://ww w.forbes.com/sites/lanceeliot/2019/06/21/self-driving-cars-paired-with-autonomous-drones-for-faster-fast-food-delivery/ (accessed on 22 October 2021).

16. Nogar, S.M. Autonomous Landing of a UAV on a Moving Ground Vehicle in a GPS Denied Environment. In Proceedings of the 2020 IEEE International Symposium on Safety, Security, and Rescue Robotics (SSRR), Abu Dhabi, United Arab Emirates, 4–6 November 2020; pp. 77–83. [CrossRef]

17. Hament, B.; Oh, P. Unmanned aerial and ground vehicle (UAV-UGV) system prototype for civil infrastructure missions. In Proceedings of the 2018 IEEE International Conference on Consumer Electronics (ICCE), Las Vegas, NV, USA, 12–14 January 2018; pp. 1–4. [CrossRef]

18. Cocchioni, F.; Pierfelice, V.; Benini, A.; Mancini, A.; Frontoni, E.; Zingaretti, P.; Ippoliti, G.; Longhi, S. Unmanned Ground and Aerial Vehicles in extended range indoor and outdoor missions. In Proceedings of the 2014 International Conference on Unmanned Aircraft Systems (ICUAS), Orlando, FL, USA, 27–30 May 2014; pp. 374–382. [CrossRef]

19. Waslander, S.L. Unmanned Aerial and Ground Vehicle Teams: Recent Work and Open Problems. In *Autonomous Control Systems and Vehicles, Intelligent Systems, Control and Automation: Science and Engineering*; Nonami, K., Kartidjo, M., Yoon, K.J., Budiyono, A., Eds.; Springer: Tokyo, Japan, 2013; Volume 65. [CrossRef]

20. Logan, M.J.; Bird, E.; Hernandez, L.; Menard, M. Operational Considerations of Small UAS in Urban Canyons. In Proceedings of the AIAA SciTech 2020 Forum, Orlando, FL, USA, 6–10 January 2020.

21. Low, K.H. Framework for urban Traffic Management of Unmanned Aircraft System (uTM-UAS), Drone Enable. In Proceedings of the ICAO Unmanned Aircraft Systems (UAS) Industry Symposium (UAS2017), Montreal, QC, Canada, 22 September 2017.

22. Krozel, J.; Mueller, T.; Hunter, G. Free Flight Conflict Detection and Resolution Analysis Paper AIAA-96-3763. In Proceedings of the AIAA Guidance, Navigation, and Control Conference, San Diego, CA, USA, 29–31 July 1996.

23. Lijima, Y.; Hagiwara, H.; Kasai, H. Results of Collision Avoidance Maneuver Experiments Using a Knowledge-Based Autonomous Piloting System. *J. Navig.* **1991**, *44*, 194–204.

24. Coenen, F.P.; Smeaton, G.P.; Bole, A.G. Knowledge-Based Collision Avoidance. *J. Navig.* **1989**, *42*, 107–116. [CrossRef]

25. Radio Technical Committee on Aeronautics (RTCA), Minimum Performance Standards—Airborne Ground Proximity Warning Equipment, Document No. RTCA/DO-161A, WA, USA. 1976. Available online: https://my.rtca.org/NC__Product?id=a1B3600 0001IcnOEAS (accessed on 22 October 2021).

26. Bilimoria, K.D.; Sridhar, B.; Chatterji, G.B. Effects of Conflict Resolution Maneuvers and Traffic Density of Free Flight. In Proceedings of the AIAA Guidance, Navigation, and Control Conference, San Diego, CA, USA, 29–30 July 1996.

27. Krozel, J.; Peters, M. Conflict Detection and Resolution for Free Flight. *Air Traffic Control. Q.* **1997**, *5*, 181–212. [CrossRef]

28. Havel, K.; Husarcik, J. A Theory of the Tactical Conflict Prediction of a Pair of Aircraft. *J. Navig.* **1989**, *42*, 417–429. [CrossRef]

29. Radio Technical Committee on Aeronautics (RTCA), Minimum Performance Specifications For TCAS Airborne Equipment, Document No. RTCA/DO-185, Washington. 1983. Available online: https://my.rtca.org/NC__Product?id=a1B360000011cmZEAS (accessed on 22 October 2021).

30. Ford, R.L. The Conflict Resolution Process for TCAS II and Some Simulation Results. *J. Navig.* **1987**, *40*, 283–303. [CrossRef]

31. Tomlin, C.; Pappas, G.J.; Sastry, S. Conflict Resolution for Air Traffic Management: A Case Study in Multi-Agent Hybrid Systems. *IEEE Trans. Autom. Control.* **1998**, *43*, 509–521. [CrossRef]

32. Shepard, T.; Dean, T.; Powley, W.; Akl, Y. A Conflict Prediction Algorithm Using Intent Information. In Proceedings of the 36th Annual Air Traffic Control Association Conference Proceedings, Arlington, VA, USA, 29 September–3 October 1991.

33. Shewchun, M.; Feron, E. Linear Matrix Inequalities for Analysis of Free Flight Conflict Problems. In Proceedings of the IEEE Conference on Decision and Control, San Diego, CA, USA, 12 December 1997.

34. Ratcliffe, S. Automatic Conflict Detection Logic for Future Air Traffic Control. *J. Navig.* **1989**, *42*, 92–106. [CrossRef]

35. Paielli, R.A.; Erzberger, H. Conflict Probability Estimation for Free Flight. *J. Guid. Control. Dyn.* **1997**, *20*, 588–596. [CrossRef]

36. Carpenter, B.; Kuchar, J. Probability-Based Collision Alerting Logic for Closely-Spaced Parallel Approach, Paper AIAA-97-0222. In Proceedings of the 35th AIAA Aerospace Sciences Meeting and Exhibit, Reno, NV, USA, 6–10 January 1997.

37. Bakker, G.J.; Blom, H.A.P. Air Traffic Collision Risk Modeling. In Proceedings of the 32nd IEEE Conference on Decision and Control, San Antonio, TX, USA, 15–17 December 1993; Volume 2.
38. Williams, P.R. Aircraft Collision Avoidance using Statistical Decision Theory. In *Sensors and Sensor Systems for Guidance and Navigation II*; International Society for Optics and Photonics: Bellingham, WA, USA, 1992.
39. Durand, N.; Alliot, J.; Chansou, O. Optimal Resolution of En Route Conflicts. *Air Traffic Control. Q.* **1995**, *3*, 139–161. [CrossRef]
40. Glover, W.; Lygeros, J.A. *Stochastic Hybrid Model for Air Traffic Control Simulation*; Springer: Berlin/Heidelberg, Germany, 2004. [CrossRef]
41. Visintini, A.L.; Glover, W.; Lygeros, J.; Maciejowski, J. Monte Carlo optimization for conflict resolution in air traffic control. *IEEE Trans. Intell. Transp. Syst.* **2006**, *7*, 470–482. [CrossRef]
42. Palme, R.; Siket, Z.; Gati, B.; Rohacs, J. Non-cooperative target classification, with use of their measured motion kinematics. In Proceedings of the 12th Mini Conference on Vehicle System Dynamics, Identification and Anomalies, Budapest, Hungary, 8–10 November 2010; Zobory, I., Ed.; BME Budapest: Budapest, Hungary, 2012; pp. 369–384, ISBN 978 963 313 058 2.
43. Szalay, Z.; Ficzere, D.; Tihanyi, V.; Magyar, F.; Soós, G.; Varga, P. 5G-Enabled Autonomous Driving Demonstration with a V2X Scenario-in-the-Loop Approach. *Sensors* **2020**, *20*, 7344. [CrossRef] [PubMed]
44. Szalay, Z. Next Generation X-in-the-Loop Validation Methodology for Automated Vehicle Systems. *IEEE Access* **2021**, *9*, 35616–35632. [CrossRef]
45. Tihanyi, V.; Rövid, A.; Remeli, V.; Vincze, Z.; Csonthó, M.; Pethő, Z.; Szalai, M.; Varga, B.; Khalil, A.; Szalay, Z. Towards Cooperative Perception Services for ITS: Digital Twin in the Automotive Edge Cloud. *Energies* **2021**, *14*, 5930. [CrossRef]
46. Szalay, Z.; Nyerges, Á.; Hamar, Z.; Hesz, M. Technical Specification Methodology for an Automotive Proving Ground Dedicated to Connected and Automated Vehicles. *Period. Polytech. Transp. Eng.* **2017**, *45*, 168–174. [CrossRef]
47. Szalay, Z.; Hamar, Z.; Nyerges, Á. Novel design concept for an automotive proving ground supporting multilevel CAV development. *Int. J. Veh. Des.* **2019**, *80*, 1–22. [CrossRef]
48. Totalcar, Akik Megtanítják a Smartot Önvezetni. Available online: https://totalcar.hu/magazin/2019/05/10/akik_megtanitjak_a_smartot_onvezetni/ (accessed on 22 October 2021).
49. Meier, L.; Honegger, D.; Pollefeys, M. PX4: A node-based multithreaded open source robotics framework for deeply embedded platforms. In Proceedings of the 2015 IEEE International Conference on Robotics and Automation (ICRA), Seattle, WA, USA, 26–30 May 2015; pp. 6235–6240. [CrossRef]
50. Koenig, N.; Howard, A. Design and use paradigms for gazebo, an open-source multi-robot simulator. In Proceedings of the 2004 IEEE/RSJ International Conference on Intelligent Robots and Systems (IROS) (IEEE Cat. No. 04CH37566), Sendai, Japan, 28 September–2 October 2004; Volume 3, pp. 2149–2154.
51. PX4 Documentation. Controller Diagrams, Multicopter Position Controller. Available online: https://docs.px4.io/v1.12/en/flight_stack/controller_diagrams.html (accessed on 22 October 2021).

Article

Phenomenological Modelling of Camera Performance for Road Marking Detection

Hexuan Li [1,*], Kanuric Tarik [1], Sadegh Arefnezhad [1], Zoltan Ferenc Magosi [1], Christoph Wellershaus [1], Darko Babic [2], Dario Babic [2], Viktor Tihanyi [3], Arno Eichberger [1] and Marcel Carsten Baunach [4]

[1] Institute of Automotive Engineering, TU Graz, 8010 Graz, Austria; tarik.kanuric@tugraz.at (K.T.); s.arefnezhad@tugraz.at (S.A.); zoltan.magosi@tugraz.at (Z.F.M.); christoph.wellershaus@tugraz.at (C.W.); arno.eichberger@tugraz.at (A.E.)

[2] Faculty of Transport and Traffic Sciences, University of Zagreb, 10000 Zagreb, Croatia; darko.babic@fpz.unizg.hr (D.B.); o.babic@fpz.unizg.hr (D.B.)

[3] Department of Automotive Technologies, Budapest University of Technology and Economics, 1111 Budapest, Hungary; viktor.tihanyi@auto.bme.hu

[4] Institute of Technical Informatics, TU Graz, 8010 Graz, Austria; baunach@tugraz.at

* Correspondence: hexuan.li@tugraz.at

Abstract: With the development of autonomous driving technology, the requirements for machine perception have increased significantly. In particular, camera-based lane detection plays an essential role in autonomous vehicle trajectory planning. However, lane detection is subject to high complexity, and it is sensitive to illumination variation, appearance, and age of lane marking. In addition, the sheer infinite number of test cases for highly automated vehicles requires an increasing portion of test and validation to be performed in simulation and X-in-the-loop testing. To model the complexity of camera-based lane detection, physical models are often used, which consider the optical properties of the imager as well as image processing itself. This complexity results in high efforts for the simulation in terms of modelling as well as computational costs. This paper presents a Phenomenological Lane Detection Model (PLDM) to simulate camera performance. The innovation of the approach is the modelling technique using Multi-Layer Perceptron (MLP), which is a class of Neural Network (NN). In order to prepare input data for our neural network model, massive driving tests have been performed on the M86 highway road in Hungary. The model's inputs include vehicle dynamics signals (such as speed and acceleration, etc.). In addition, the difference between the reference output from the digital-twin map of the highway and camera lane detection results is considered as the target of the NN. The network consists of four hidden layers, and scaled conjugate gradient backpropagation is used for training the network. The results demonstrate that PLDM can sufficiently replicate camera detection performance in the simulation. The modelling approach improves the realism of camera sensor simulation as well as computational effort for X-in-the-loop applications and thereby supports safety validation of camera-based functionality in automated driving, which decreases the energy consumption of vehicles.

Keywords: lane detection; simulation and modelling; multi-layer perceptron

Citation: Li, H.; Tarik, K.; Arefnezhad, S.; Magosi, Z.F.; Wellershaus, C.; Babic, D.; Babic, D.; Tihanyi, V.; Eichberger, A.; Baunach, M.C. Phenomenological Modelling of Camera Performance for Road Marking Detection. *Energies* **2022**, *15*, 194. https://doi.org/10.3390/en15010194

Academic Editor: Rui Xiong

Received: 12 October 2021
Accepted: 20 December 2021
Published: 28 December 2021

Publisher's Note: MDPI stays neutral with regard to jurisdictional claims in published maps and institutional affiliations.

1. Introduction

The traffic safety problem is severe with an increasing number of vehicles on the road. According to [1], approximately 11 percent of road accidents result from lane departures caused by inattentive, distracted, or drowsy drivers. According to statistics from [2], in 2015 nearly 13,000 people died in single-vehicle run-off-road, head-on, and sideswipe crashes where a passenger vehicle left the lane without warning. Lane Keeping Assist (LKA) and lane departure warnings are designed to reduce potential risk and improve driving safety. They support more effective driving tasks that maintain safe lateral vehicle control. The study that investigated the safety potential of Lane Keeping Assist systems shows that the possibility to avoid fatal accidents is between 16.4% and 29.2%, depending on the capability of the system [3]. For passenger vehicles, these values even went up to 23.2% to 40.9%

Nowadays, almost every installed system relies on vision-based technologies to detect and trace lane marking. For most conventional methods [4–6], the lane edge is detected in the region of interest by image filtering and thresholding. With the development of artificial intelligence, the convolutional NN-based approach has stimulated a promising research direction for the extraction of lane marking from acquired images [7–9]. In contrast, Kim et al. [10] uses an MLP in the fully connected layer to manually extract the Region of Interest as the input of convolutional NN and directly outputs lane marking candidates. This approach ultimately outputs the detected lane marking by fitting a function. Thus, the camera's computational performance and the algorithm's detection efficiency affect the accuracy of the detection results. An appropriate lane marking detection model is required to analyze and validate vision-based lane marking detection systems. This model is developed based on the ground truth of the digital twin maps, which provides an excellent setting for detecting and reading a list of lane marking points to validate the performance of the lane marking model.

Meanwhile, Kalra et al. [11] and Shladover et al. [12] demonstrated that using Autonomous Driving Systems (ADS) statistically results in fewer collisions. However, hundreds of millions of kilometres of test drives should be conducted to verify the robustness of ADS algorithms and software. Furthermore, ADS are subject to different research challenges (technical, non-technical, social, and policy) [13]. In particular, different driving scenarios related to traffic and humans bring new system requirements to ADS [14]. These cases induce that certification of an automated system can only be achieved with the support of modelling and simulation [15]. More specifically, to realistically capture the complexity and diversity of the real world in a virtual environment, models that combine virtual scenarios, flexible simulations, and real measurement data should be considered [16,17].

In order to accommodate different requirements encountered during the vehicle development process, various camera model types with distinct detection performance are developed, as demonstrated in the prior studies. For example, Schlager et al. [18] defined low-fidelity sensor modules for input and output using object lists, which are filtered according to the sensor specific Field of View (FOV). In [19], and an error-free camera model is introduced, which can correctly recognize all objects within the FOV. Based on this sensor model, a more refined sensor model is proposed in [19,20], which supports arbitrarily shaped FOVs. In order to standardize the modelling process, a modular architecture was proposed [21], which defines the filtering process for input objects lists according to different sensor effects and occlusion situations [20]. A significant advantage of the described model is that it only considers detection results within the FOV of the sensor, which results in lower computing complexity.

However, due to the low-fidelity provided by the model, the detection performance of a specific sensor cannot be accurately replicated. Therefore, a stochastic model for errors in the position measurement is constructed based on an ideal sensor in [21] where the variation is a random Gaussian white noise. The real detection behaviour is still not reflected by a random error distribution. In order to improve the reality of sensor simulation and approximate the distribution of given measurements or a dataset, non-parametric machine learning approaches can be used. It estimates the outputs and ensures that the shape of the distribution will be learned from the data automatically [22–24]. Furthermore, the details of the perception function are usually not accessible to the developer of the automated driving system, i.e., the vehicle manufacturer. The measurement process of a comprehensive physical model is also computationally expensive. Accordingly, a statistical model of the perception process is proposed. Examples of statistical models can be found in [25,26]. In these models, the measurement and reference data drive the construction of the sensor model, where errors are calculated between data and the probability functions map the errors to reference data as the outputs of the model [27]. This approach can implicitly depict several sources of error. In contrast to previous techniques, the resulting sensor output distribution is no longer limited to a specific set of distributions. This statistical model was also employed in [28] as a lane marking detection model, where a direct relationship between sensing distance and error was developed by measuring errors of a real camera system [22]. These models only take the measurement error of the camera and ignore the impact of environmental and vehicle dynamic movement on the results. Hence, it is impossible to predict the output correctly based on the vehicle's current status.

In order to enhance the fidelity of camera simulation, a complex camera model is proposed that mimics the physics of imaging processes in [29,30] optical situations (e.g., optical distortion, blur, and vignetting) and additionally the image processing modules (e.g., signal amplification, objects or features identification, and detection) are modelled. In [31], an optical model was presented to validate the functional and safety limits of camera-based ADAS, which is based on the real, measured lens used in the product. In addition, Carlson et al. [32] proposed an efficient, automatic, and physically based augmentation pipeline to vary sensor effects to augment camera simulation performance. As more or changing requirements emerge, the model must be updated with optical characterization models, which results in increasing effort. Therefore, the main design paradigm of the model presents a barrier to allowing iterative development cycles.

Additionally, a semi-physical approach combining geometric and stochastic approaches to simulate dedicated short-range communication was developed in [33] and calibrated for different environmental conditions with on-road measurements.

This paper aims to remove the drawbacks and limitations of these previous research studies by fitting lane marking detection errors. It is based on statistical models using real-time vehicle measurement data collected in real-world tests. As the camera sensing algorithm is highly confidential, it is impossible to determine primary factors driving the detection error from extensive vehicle data. Therefore, feature selection is introduced, removing the data containing redundant or irrelevant features without losing informative features. In this study, the lane detection error model is constructed from the MLP. One of the main advantages of the MLP is the capability of simulating both linear and nonlinear relationships between the parameters. Meanwhile, the trained MLP is applied to estimate the output from new input data in the virtual simulation environment.

The structure of the subsequent sections of this paper is as follows: The problem is defined in Section 2. Section 3 represents the method for data collection and ground truth definition. Section 4 describes the methodology and structure of the designed MLP for lane marking detection using vehicle-based data. Experimental results are presented and discussed in Section 5. Finally, a conclusion is provided in Section 6.

2. Problem Definition

Numerical models of cameras can be used for simulation and digital twin-based testing for automated vehicles. In prior studies [28,34,35], varieties of sensor models with a distinct performance and detail profile were introduced that can replicate the performance of real cameras in simulation. These camera models can be adapted to accommodate specific simulation requirements. Three camera models that are frequently utilized in a simulation scenario can be categorized as follows:

- Ideal Sensor Model: This model provides the most accurate detection results from the geometric space of sensor coverage. This kind of model is frequently employed in multibody simulation software. However, the ideal sensor model is not able to measure and estimate perception errors. Hence, reliability is reduced during the simulation.
- Physical Sensor Model: This model is more numerically complicated and often produces higher accuracy. Since the model parameters correspond to the physical imaging process of the sensors, the output can be used to replicate physical effects and principles correctly. However, developing a physical sensor model requires knowledge about the physical characteristics and internal imaging algorithm. In our study, a MO-

BILEYE camera series 630 [36] is used, which includes complicated and confidential perception algorithms that are difficult to be simulated in software.

- Phenomenological Sensor Model: It simulates sensor performance, whereas phenomenological output effects are modelled without consideration for internal processes or algorithms of a camera, but with an emphasis on reproducing the real effects that are the difference between camera outputs and reference data. The phenomenological sensor model places greater emphasis on physical effects to establish the relationship between input and output of the camera model. While using this model, it is possible to map the realistic behaviour of lane detection more quickly and efficiently. Moreover, the camera modelling framework avoids complex algorithms.

Camera recognition is mainly responsible for detecting road marking. For the current study, our test vehicle is equipped with a MOBILEYE camera series 630, which employs a third-degree polynomial to estimate detected lane markings. Thus, the stored output of the image processing unit is four coefficients $C \in \mathbb{R}^4$, $C = [C_0, C_1, C_2, C_3]$ for each detected lane marking, the polynomial function is presented in Equation (1).

$$Y_{Cam}(X_{Cam}) = \sum_{i=0}^{3} C_i \cdot X_{Cam} \tag{1}$$

The measurement coordinate is relative to the camera system, where X_{Cam} points in a forwards direction and Y_{Cam} points to the right side illustrated in Figure 1.

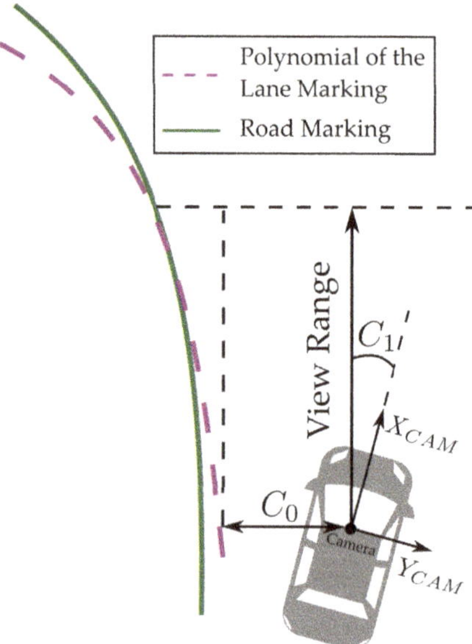

Figure 1. Illustration of lane marking detection.

These coefficients are explained in Table 1. Since our test scenarios are primarily focused on straight segments of the highway, C_2 and C_3 are ignored. However, C_0 is the lateral distance to the detection lane marking at the height of the camera. C_1 indicates the vehicle heading relative to the lane heading and the road markings on the measurement section are symmetrical, implying that C_1 values for the left and right lane markings are identical. As a result, this paper will only focus on C_0 and C_1 estimation, as shown in Figure 1.

Table 1. Lane detection coefficients from MOBILEYE camera.

Parameters	Definition
C_0: Lane position	Lateral distance from the centerline of the host vehicle to the left/right lane marking
C_1: Heading angle	The vehicle heading relative to the lane heading
C_2: Curvature	The curvature of the lane ahead
C_3: Curvature derivative	Curvature rate

Vision-based lane detection is influenced by different factors that contain external environmental parameters [37] (e.g., lane line reflectivity, appearance, and lighting conditions, etc.) as well as vehicle dynamic performance [38] (e.g., speed and heading angle, a departure from the road centerline, etc.), resulting in discrepancies between detection results and the ground truth. This phenomenon can be observed by comparing two different road markings in Figure 1. According to the guide to the expression of uncertainty in measurement, the detection result of the camera can be stated as the best reference quantity plus the measurement uncertainty [39], where uncertainty can be treated as the detection error, and it is estimated by using an NN-based approach in this paper. Finally, a phenomenological camera model is proposed to approximate real-world camera detection performances.

3. Experimental Setup

3.1. Data Collection

The highway was publicly closed during data collection. Lane detection data were collected using the MOBILEYE 630 system installed in the test car. The MOBILEYE camera provides real-time image processing to recognize various road objects such as lane markings, pedestrians, and so on. For this study, the data related to the type of detected longitudinal marking (continuous or dashed), polynomial coefficient of lane marking, and view range were recorded. Meanwhile, the six-degree-of-freedom inertial measurement system of the GENESYS Automotive Dynamic Motion Analyzer (ADMA) for motion analysis is combined with the NOVATEL RTK-GPS receiver to provide a highly accurate vehicle kinematic data. Figure 2 shows the measurement setup used for data collection.

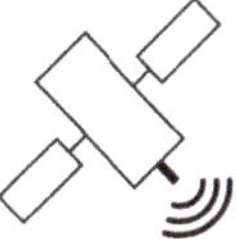

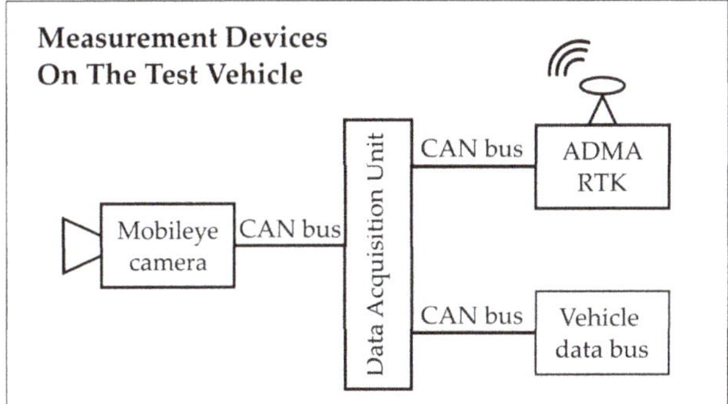

Figure 2. Measurement setup for measuring vehicle.

3.2. Ground Truth Definition

ADMA-RTK combination is a strap-down inertial measurement system. The extended-Kalman filter used in the ADMA can estimate several important sensor errors in order to enhance system performance. Depending on the capability of the GPS receiver, the position accuracy range down to 1 cm. Meanwhile, six inertial sensors provide high accuracy data [40]. Due to the accurate performance of this combination, it is used as a reference system. The measurements and data collection were conducted on the M86 highway in Hungary, see Figure 3. The construction of a close road section facilitates the development and testing of connected and autonomous vehicles. The total length of the test road section is 3.4 km [41].

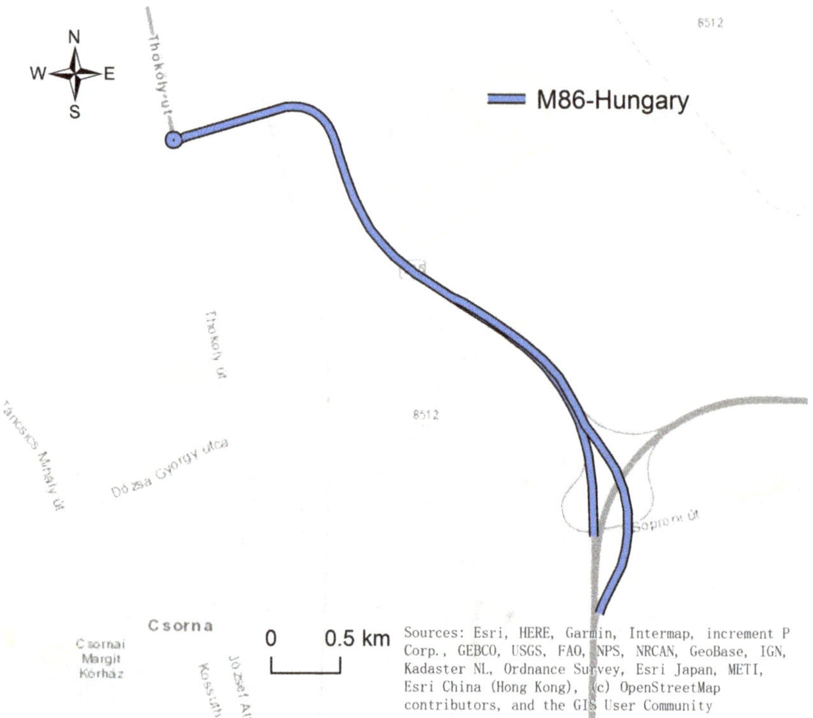

Figure 3. M86 freeway located near Csorna (Hungary) on route E65 (GNSS coordinates: 47.625778, 17.270162).

In order to perfectly duplicate the real-world test scenario in the simulation environment, the M86 road was converted into an Ultra-High-Definition (UHD) map, a digital twin of reality that accurately represents every detail of the test environment. The production workflow that was applied for the production of the UHD map was presented in [41]. A digital twin-based M86 map was explicitly produced for testing and validating ADAS/AD driving functions with an absolute precision of +/−2 cm as a quality reference source. The extreme high precision of the lane marking data in this map will be used as the ground truth for comparison with the camera detection output. Additionally, this map will also be used for further virtual testing to duplicate simulation results.

4. Methodology

The camera modelling approach and process are presented in Figure 4. The test vehicle collects information from mounted experimental equipment, such as vehicle dynamic data, GPS sensor data and camera data introduced in Section 3. These data will be used for target determination and feature selection. Depending on modelling requirements, sensor detection results contain essential information about the modelling target, which facilitates the calculation of the differences between measured and reference data. This error represents both the camera's performance and uncertainty in lane detection and the target of the model. In order to improve the performance of the camera model and decrease the training time of NN, data extraction and input features selection are applied, which contribute most to the prediction variable or output used in this case. The selected features based on the ReliefF algorithm are used as input for MLP. The relationships between each input feature and target are evaluated using ReliefF, which is a feature weighting method designed for multi-class, noisy, and incomplete dataset classification issue [42,43]. Once inputs and targets are determined, the MLP-based approach is applied for modelling.

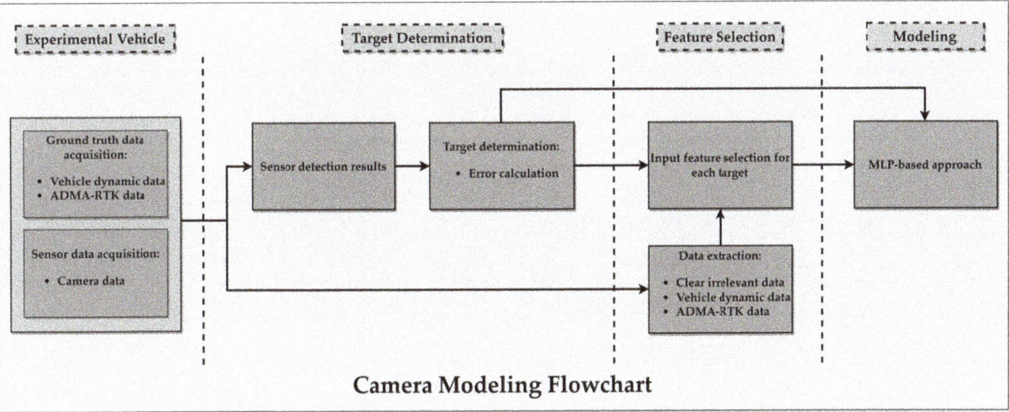

Figure 4. Schematic representation of necessary components for camera model.

4.1. Target Determination

As previously discussed in Section 2, our model primarily focuses on straight highway segments. Therefore, C_0 Lane Position Error (C_0-LPE) and C_1 Heading Angel Error (C_1-HAE) are considered as our targets of MLP. Reference data for each target were taken from M86 road marking coordinates and ADMA-RTK reference system, respectively. In addition, detection data for each target were taken from the MOBILEYE camera.

C_0-LPE is calculated as the difference between M86 road markings coordinates and detection data. The calculation process is defined in the next steps:

- Replicating trajectory of GPS data on M86 road map;
- For each timestamp of trajectory data, the test car is positioned on the road, and C_0 is calculated for each side of the road, resulting in C_0 Left and C_0 Right;
- The difference for each side of the road is calculated independently, resulting in C_0-LPE Left and C_0-LPE Right;
- Combining results into a two-dimensional vector provides us with C_0-LPE as the target of MLP.

C_1-HAE is calculated as the difference between the heading angle data provided by the reference system and the detection output of the camera. The calculation is conducted in the same way as mentioned for C_0-LPE, resulting in a one-dimension vector as a target of the MLP. As detailed in the next section, different input features were selected for each target.

4.2. Feature Selection

Various features were collected from different experimental devices and electronic controllers during the measurement process. However, a mass of data often contains many irrelevant or redundant features. In this study, the ADMA reference system provides details on the available data. Some features (ambient temperature, GPS receiver states, altitude, etc.) were discarded because they did not significantly impact vehicle dynamics or camera model functionality. Feature selection aims to maximize information associated with the target, carried out by the extracted features from raw data. Additionally, considering that different features have different update cycles, time synchronization is also required during data processing to align all features on the same timeline. The time synchronization process is as follows:

- The test car's ADMA-RTK-based trajectory data are selected as a base timeline. Each timestamp from it will be used as a reference point.
- Features will be checked with respect to whether the their timestamp aligns with a reference point within an offset interval from -0.02 s to 0.02 s. They will be saved in a database aligning values with the reference timestamp.
- The process is repeated until it proceeds through all reference points.

In order to further refine and reduce the parameters input to the predictive model, features should be selected from extracted data, which minimizes the number of input features. The benefit of the process is to reduce training time, lower the risk of overfitting, and improve the model's performance. The primary notions and applications of ReliefF are to rate the quality of features based on their ability in order to distinguish samples that are close to one another. The final weight assigned to each feature is calculated. According to ReliefF results, the final features with the greatest relevance to each target are selected and shown in Table 2, while the corresponding arguments are illustrated in Figure 5. Each set of input chosen features has a defined target. These features are used as inputs to the corresponding MLP model.

Table 2. Final Feature Selection for each target.

Features	C_0-**LPE**	C_1-**HAE**	Description
d_L	✓	✓	The distance between real trajectory of the vehicle and center line of the road
a_Y	✓	✓	The lateral acceleration of vehicle
a_Z	-	✓	The vertical acceleration of vehicle
θ	✓	✓	Pitch angle
ϕ	-	✓	Roll angle
$\dot{\theta}_Y$	✓	-	Pitch rate
$\dot{\psi}_Z$	✓	-	Yaw rate

All features are connected to the body coordinate system, described in [44].

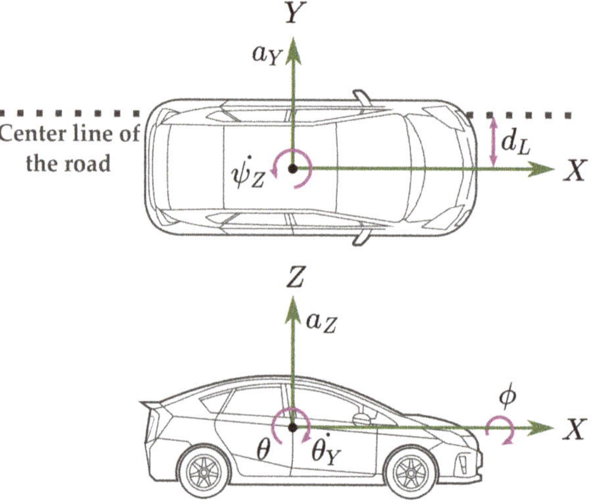

Figure 5. Illustration of the selected input feature variables on side and top views of the vehicle.

4.3. Neural Network Modelling

MLP is used here to estimate the performance of the camera. It is widely used in different fields, such as system modelling, anomaly detection, and classification applications to solve complex problems in a variety of computer applications [45–49]. Additionally, the MLP approach has been preferred as a method for state estimation and simulation implementation [50]. MLP is useful in research for its ability to solve problems stochastically. Therefore, it is employed here to estimate C_0-LPE and C_1-HAE.

A typical architecture of MLP has one input layer, one or more hidden layers, and one output layer. The working principle uses the connecting layers, which are components of neurons, to transfer normalized input data to the output. The number of layers in the network and the number of neurons in each layer are typically determined empirically. The architecture of the used MLP is presented in Figure 6. This architecture is used for both prediction models, including the estimation of heading angle and lateral position errors. The data mapping process from input data to the output data is presented in Equations (2)–(4). Furthermore, the arithmetic process in a node is illustrated in Figure 7.

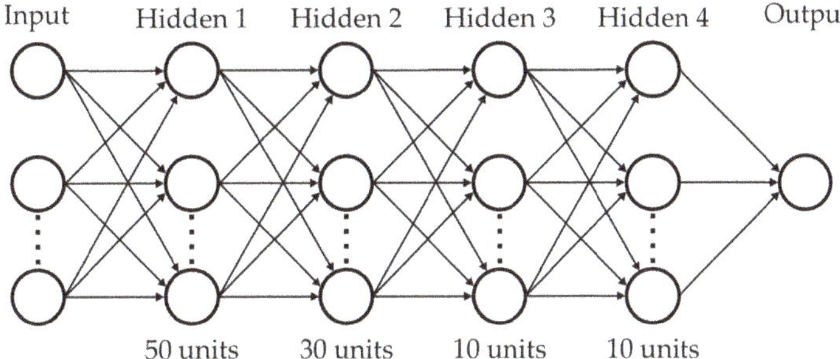

Figure 6. Proposed MLP architecture.

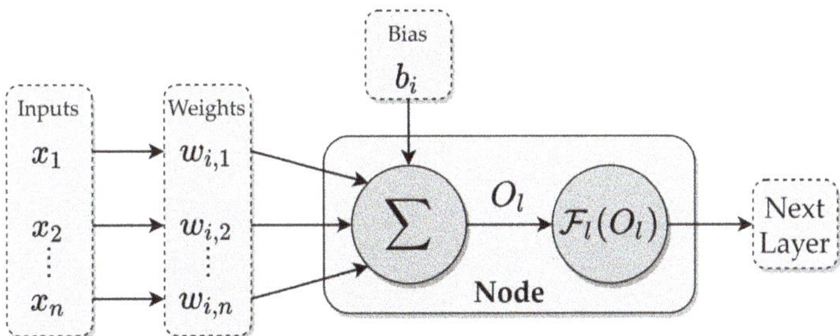

Figure 7. Data processing in a neural network node.

$$O_l = \sum_{j=1}^{n} (w_{i,j,l} x_j + b_{i,l}) \quad l = 1, 2, ..., m \tag{2}$$

Each hidden layer contains the associated coefficient weights and bias. The inputs of each node are calculated from the previous layer or are the initial input of the network, then the results of the mathematical operation can be provided by Equation (2), where x is the normalized input variable, w is the weight of each input, i is the input counter, b is the bias of this node, n is the number of input variables, and k and m are the counters of the hidden layer and the number of neural network nodes, respectively.

$$\mathcal{F}_l(O_l) = \frac{2}{1 + e^{-2O_l}} - 1 \tag{3}$$

Subsequently, the results of O_l are applied to function F_l. Here, the hyperbolic tangent sigmoid function is used as an activation function calculated by Equation (3), which defines the output of that node given an input or set of inputs.

$$\widehat{y} = \left(\sum_{l=1}^{m} w_{out,l} \mathcal{F}_l \right) + b_{out} \tag{4}$$

Finally, multiple nodes and hidden layers build up the MLP, as shown in the Figure 6. The output of each node is forwarded to the next layer to continue the same operations. The output layer of the entire network is defined by Equation (4), where output $\widehat{y}$ calculates the weighted sum of the signals provided by the hidden layer. The coefficients associated with them are grouped into matrices $w_{out,l}$ and b_{out}.

The most critical step in MLP modelling is training. Backpropagation is the most often used training algorithm, which is described as a process for adjusting the network parameters (weights and biases) to minimize the error function between the estimated and real outputs. In comparison to other back-propagation algorithms, a supervised learning algorithm called Scaled Conjugate Gradient (SCG) was selected [51].

The number of layers in the network and the number of neurons in each layer are typically determined empirically. By comparing training result performances, four hidden layers were decided to be utilized in the MLP model, with the number of neurons distributed as 50, 30, 10, and 10 in each layer. The architecture of hidden layers and the number of neurons in each layer are used for both prediction models (C_0-LPE and C_1-HAE estimation). As shown in Section 4.2, the number of inputs for each target is shown in Table 2.

After the definition of the MLP architecture, its performance is evaluated using three different metrics: Mean Squared Error (MSE), Root Mean Square Error (RMSE), and correlation coefficient (R^2). MSE is used to represent the average squared difference between the estimated values and the actual value (see Equation (5)). On the other hand, RMSE is a typical metric for regression models and is used to quantify the model's prediction error, with a larger error resulting in a higher value (see Equation (6)). Finally, R^2 represents the proportion of real output dynamics that could be caught by the MLP model. R^2 varies between 0 and 1. A higher number indicates that the model is more accurate in its predictions (see Equation (7)).

$$MSE = \frac{1}{n} \sum_{i=1}^{n} (y_i - \widehat{y}_i)^2 \tag{5}$$

$$RMSE = \left(\frac{1}{n} \sum_{i=1}^{n} (y_i - \widehat{y}_i)^2 \right)^{\frac{1}{2}} \tag{6}$$

$$R^2 = 1 - \frac{\sum_{i=1}^{n} (y_i - \widehat{y}_i)^2}{\sum_{i=1}^{n} (y_i - \overline{y_i})^2} \tag{7}$$

In Equations (5)–(7), y is the target (observe) value, $\overline{y_i}$ is the average value of the target, and n is the number of the MLP output data samples.

5. Results and Discussion

Real-world collected road test data are utilized to train the network model in order to evaluate the accuracy of PLDM better. This section explains MLP training results. Moreover, in order to verify the accuracy of model predictions and the validity of the approach, the employed MLP model results are compared with five other algorithms. Finally, this model will be deployed in the vehicle simulation software CarMaker from IPG Automotive GmbH [52].

5.1. Training Results

In order to train the MLP model, the data gathered by the various devices are synchronized, and 9010 samples were selected from collected data to optimize the model. Input data are randomly separated into three sets: training (70%), validation (15%), and test (15%). The configuration of the MLP model is discussed in Section 4.3 and presented in Table 3. Supervised training is performed on the model using the training set. The validation set is also used to mitigate the issue of overfitting. Finally, the test set is used to evaluate model performance on unseen data.

Table 3. MLP network model configuration for C_0 and C_1 estimation.

Hyper Parameter	MLP Configuration
Learning rate	Adaptive
Hidden layer	4
Hidden units for each layer	[50 30 10 10]
Training function	SCG
Activation function	Hyperbolic tangent sigmoid

As discussed in Sections 4.1 and 4.2, the MLP model is used to estimate C_0-LPE, which consists of five features and a two-dimension target. The combination of five input features and a one-dimension target is used to estimate C_1-HAE. The regression graphs obtained as results of the MLP training are given in Figure 8. The models are evaluated for the test set after convergence, and regression accuracy can achieve 94.0% and 95.5%, respectively, for C_0-LPE and C_1-HAE estimation in order to further evaluate the estimation model, where evaluation metrics (MSE, RMSE, and R^2) measure regression performance. As a result, MLP training performance is provided in Table 4. The results show good agreement between actual and estimated values, and the prediction errors of the results are within an acceptable range. The models are more consistent with the trend of real values in terms of the predicted value. Therefore, the training of the network has been successfully provided.

Table 4. Performance evaluation of MLP for C_0-LPE and C_1-HAE estimation.

Metrics	C_0-LPE	C_1-HAE
MSE	0.085 m^2	0.008 rad^2
RMSE	0.092 m	0.089 rad
R^2	95.5%	94.0%

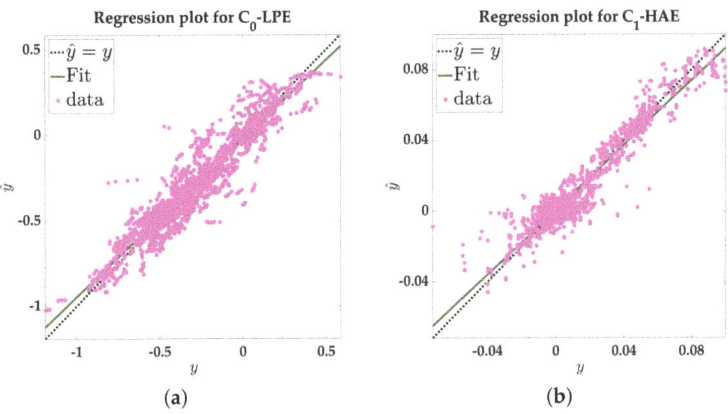

Figure 8. MLP training regression graph. (a) Training regression result for C_0-LPE (b) Training regression result for C_1-HAE.

5.2. Comparing with Other Approaches

The proposed method can be used for C_0-LPE and C_1-HAE predictions and for comparing the effectiveness and accuracy of this method. Moreover, five other machine learning methods are introduced in [53,54], and these algorithms are categorized and introduced as follows:

- Support Vector Machine (SVM): It is a widely utilized soft computing method in various fields. The fundamental idea is to fit data in specific areas by using non-linear mappings and to apply linear methods in function space, which has been applied for a regression problem and demonstrates superior generalization performance [55].
- Linear Regression (LR): It attempts to model the connection between two variables by fitting a linear equation to the observed data. One is the explanatory variable, and the other is the dependent variable. This algorithm is a fundamental regression method introduced in [56].
- Gaussian Regression of Process (GPR): It combines the structural properties of Bayesian NN with the nonparametric flexibility of Gaussian processes [57]. This model considers the input-dependent signal and noise correlations between various response variables. It performs well on small datasets and can also be used to measure prediction uncertainty.
- Ensemble Boosting (EB): The idea of an EB is presented in [58], and it fits a wide range of regression problems, and the architecture is the generation of sequential hypotheses, where each hypothesis tries to improve the previous one. General bias errors are eliminated throughout the sequencing process, and good predictive models are generated.
- Stepwise regression (SR): It is the iterative process of building a regression model by selecting independent variables to be used in a final model, which is introduced and applied in [59]. It entails gradually increasing or decreasing the number of putative explanatory factors and evaluating statistical significance after each cycle.

Finally, these five algorithms and MLP model are compared with performance metrics, as shown in Table 5. In comparison, the suggested MLP achieves outstanding results while outperforming alternative approaches. Additionally, the GPR demonstrated a rather good regression result, with an accuracy of 83% and 86% for C_0-LPE and C_1-HAE estimation, respectively; this is probably because the GPR kernel can extract sequential data from complex temporal structures. All three models, SVM, LR, and SR showed comparable performance and underfitting for C_0-LPE. Furthermore, two other neural network models based on data-driven approaches are introduced in [23,24], which include Mixture Density Network (MDN) and deep Gaussian Process (GP). MDN outputs a Gaussian mixture through a multilayer perceptron. Each Gaussian distribution is assigned a corresponding weight, which predicts the entire probability distribution. Deep GP is a deep belief network based on GP mappings. The data are modelled as the outputs of a multivariate GP. Both models can accurately represent uncertainty between camera detection and measurement results, but they do not produce an accurate estimate compared to MLP. Therefore, driven by the goal of the digital twin, MLP can more accurately represent the behaviour of sensors in real environments and still show substantial advantages.

Table 5. Performance comparison between several regression algorithms.

Output	Metrics	Regression Algorithm					
		MLP	SVM	LR	GPR	EB	SR
C_0-LPE estimation	MSE	0.085	0.077	0.075	0.022	0.035	0.075
	RMSE	0.092	0.278	0.274	0.15	0.187	0.274
	R2	95.50%	40%	42%	83%	73%	42.40%
C_1-HAE estimation	MSE	0.008	0.023	0.022	0.012	0.012	0.021
	RMSE	0.089	0.151	0.148	0.11	0.11	0.148
	R2	94.00%	73%	74%	86%	86%	74.30%

5.3. Virtual Validation in CarMaker

In this section, a test run is randomly selected from the test set samples carried out in the co-simulation based on the Carmaker-Simulink software, which provides a multi-body simulation environment that includes vehicle dynamics control and sensor modules. These modules can support custom modifications. Thus, PLDM replaces the default camera model in CarMaker and tests detection performance in a virtual environment. As illustrated in Figure 9, the entire model is integrated into the co-simulation platform. Realistic reproduction of the virtual scenario is produced using the digital twin-based M86 map. Subsequently, at each time step, the ideal CarMaker object sensor detects the object from the map and provides precise information feedback in list format. In particular, test run data from previous tests conducted in a real-world environment, such as vehicle dynamics and positioning information, are stored in an external file that could be utilized as input for free movement in CarMaker. This module is mainly responsible for real measurement playback, and necessary data are transmitted to the MLP-based error estimation module, where the estimator predicts the corresponding error values for C_0-LPE and C_{C_1}-HAE, respectively, based on the current vehicle state. Due to the fact that the MLP model was trained on prior training data successfully, ground-truth lane marking data are manipulated according to the model's output. In this case, two polynomial coefficients (C_0-LPE and C_{C_1}-HAE) of lane marking detection can be determined.

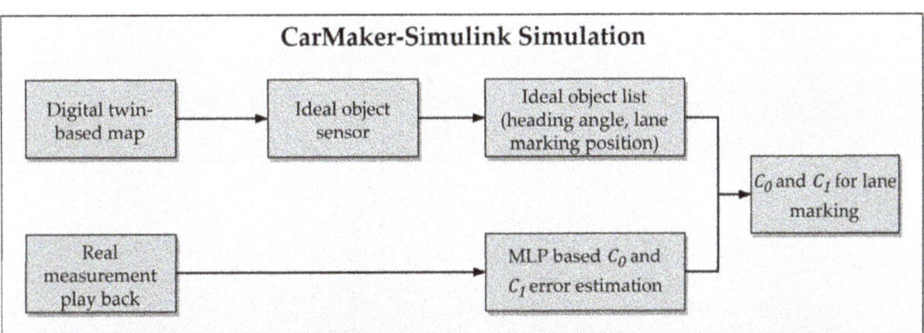

Figure 9. The procedure of the phenomenological lane detection model in simulation.

As shown in Figure 10, the estimated value and real value of the randomly selected samples are generally consistent with the trend of the sample change. Affected by real factors, the method to make predictions for certain samples still contains certain errors. A larger C_0 of the left lane estimation error greater than the right lane estimation can be observed, probably because the left lane line is dashed and the right lane line is continuously solid. Namely, a dashed lane marking is usually more challenging to determine from the background in a captured image, as explained in [60]. Overall, the predicted peak and valley space corresponding to the estimated values still contains some errors compared to the actual values. However, the maximum error value of 0.05 m is still acceptable. The reason is that this method takes into account many more factors than traditional regression forecasting methods, and it is hard to avoid errors in the weights of some secondary factors. However, in terms of the overall trend, the effectiveness of the chosen model is proven.

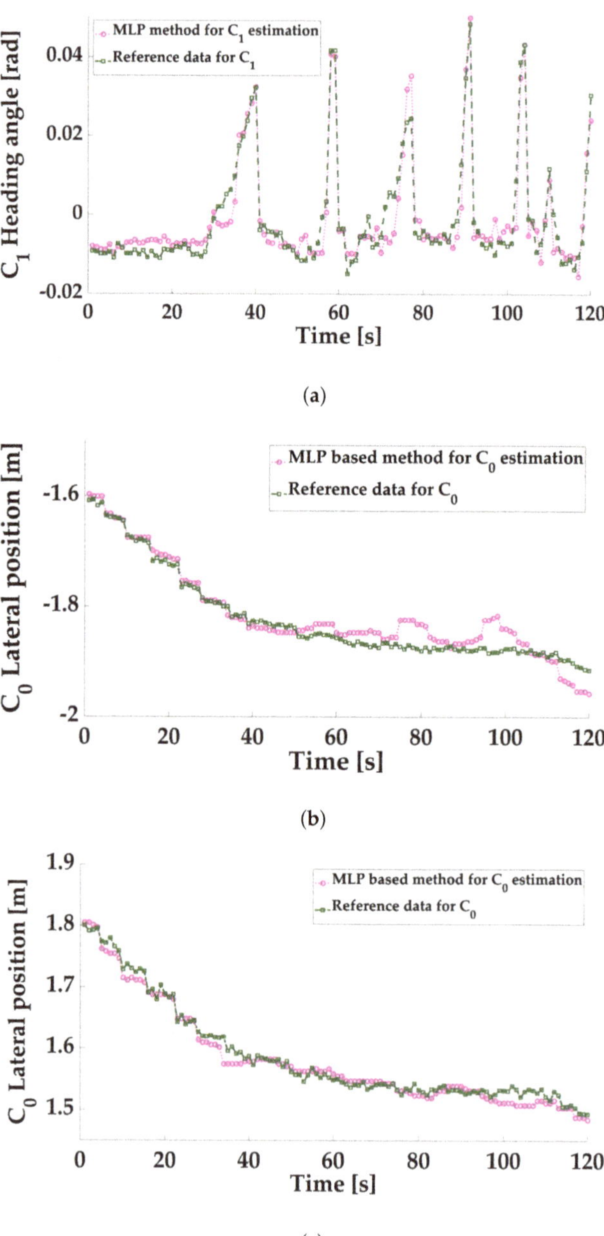

Figure 10. Simulation result for C_0 and C_1 estimation. (**a**) C_1 estimation comparison between camera detection data and MLP based output. (**b**) C_0 of left lane estimation comparison between camera detection data and MLP based output. (**c**) C_0 of right lane estimation comparison between camera detection data and MLP-based output.

6. Conclusions

In order to support efficient virtual test and validation of LKA systems, this study developed an MLP model to determine lane detection C_0-LPE and C_1-HAE estimation based on the relationship of vehicle dynamic data. This relationship is complex for an actual dataset derived from real-world measurements and requires an artificial intelligence method to create a reliable model to analyze the problem. This approach was divided into three parts. Firstly, the measurements and data collection were carried out for the testing procedure, and digital twin-based data were defined as ground truth. The second part was to extract data and select features from the actual collected data to find input data that had greater influences on the model, thus improving training efficiency. In the third part of the study, an MLP model was developed, and the selected features were used as inputs to train the model. The results also showed that MLP can produce higher accuracy than other regression approaches. Finally, the technique was employed to reproduce lane detection behaviour of an automotive camera system in a simulation platform. Combined with the analysis of the simulation results, we found that the best regression is achieved for a given non-linear dataset. Due to the fact that existing data and tests were conducted primarily on straight roads, lane marking detection on curved roads will be taken into account to refine the model further and improve our approach.

The model fits the detection error of the sensor output by using selected features, which enables fast and efficient sensor modelling. Compared to the physical model, this approach simplifies the modelling process by ignoring physical performance modelling of the camera components as well as the perception algorithm and focusing only on the inputs and outputs of the camera system, thus improving computational performance. Moreover, in contrast to the ideal models previously mentioned, ideal sensor models provide only ground truth information without any specific post-processing function. Therefore, physical effects do not influence these models. However, PLDM models based on the MLP approach can provide more details about sensor detection performance than an ideal model, enhancing the simulation's realism. Although there is a strong correlation between modelling complexity, training time, data composition and volume, modelling efficiency is improved, and this approach is generic. It can be applied to various sensors with low efforts after initial development.

Author Contributions: Conceptualization, H.L.; methodology, H.L. and S.A.; software, H.L. and K.T.; validation, H.L. and K.T.; investigation, H.L.; resources, D.B. (Darko Babic), D.B. (Dario Babic), Z.F.M. and V.T.; data curation, H.L., K.T. and C.W.; writing—original draft preparation, H.L. and K.T.; writing—review and editing, H.L., S.A. and A.E.; visualization, H.L.; supervision, A.E. and M.C.B. All authors have read and agreed to the published version of the manuscript.

Funding: Open Access Funding by the Graz University of Technology.

Data Availability Statement: Not applicable.

Acknowledgments: The author thanks those who have supported this research, to the Graz University of Technology also.

Conflicts of Interest: The authors declare no conflict of interest.

References

1. Administration, N.H.T.S. National motor vehicle crash causation survey: Report to congress. *Natl. Highw. Traffic Saf.* **2008**, *811*, 059.
2. Cicchino, J.B. Effects of lane departure warning on police-reported crash rates. *J. Saf. Res.* **2018**, *66*, 61–70. [CrossRef] [PubMed]
3. Eichberger, A.; Rohm, R.; Hirschberg, W.; Tomasch, E.; Steffan, H. RCS-TUG Study: Benefit potential investigation of traffic safety systems with respect to different vehicle categories. In Proceedings of the 24th International Technical Conference on the Enhanced Safety of Vehicles (ESV), Washington, DC, USA, 8–11 June 2011.
4. Jianwei, N.; Jie, L.; Mingliang, X.; Pei, L.; Zhao, X. Robust Lane Detection Using Two-stage Feature Extraction with Curve Fitting. *Pattern Recognit.* **2016**, *59*, 225–233.
5. Aly, M. Real time detection of lane markers in urban streets. In *2008 IEEE Intelligent Vehicles Symposium*; IEEE: Piscataway, NJ, USA, 2008; pp. 7–12.

6. Zhang, Y.; Lu, Z.; Zhang, X.; Xue, J.H.; Liao, Q. Deep Learning in Lane Marking Detection: A Survey. *IEEE Trans. Intell. Transp. Syst.* **2021**. [CrossRef]

7. Wang, Z.; Ren, W.; Qiu, Q. Lanenet: Real-time lane detection networks for autonomous driving. *arXiv* **2018**, arXiv:1807.01726.

8. Cireşan, D.C.; Giusti, A.; Gambardella, L.M.; Schmidhuber, J. Mitosis detection in breast cancer histology images with deep neural networks. In Proceedings of the International Conference on Medical Image Computing and Computer-Assisted Intervention, Nagoya, Japan, 22–26 September 2013; pp. 411–418.

9. Chng, Z.M.; Lew, J.M.H.; Lee, J.A. RONELD: Robust Neural Network Output Enhancement for Active Lane Detection. In Proceedings of the 2020 25th International Conference on Pattern Recognition (ICPR), Milan, Italy, 10–15 January 2021; pp. 6842–6849.

10. Kim, J.; Lee, M. Robust lane detection based on convolutional neural network and random sample consensus. In Proceedings of the International Conference on Neural Information Processing, Montreal, QC, Canada, 8–13 December 2014; pp. 454–461.

11. Kalra, N.; Paddock, S.M. Driving to safety: How many miles of driving would it take to demonstrate autonomous vehicle reliability? *Transp. Res. Part A Policy Pract.* **2016**, *94*, 182–193. [CrossRef]

12. Shladover, S.E. Connected and automated vehicle systems: Introduction and overview. *J. Intell. Transp. Syst.* **2018**, *22*, 190–200. [CrossRef]

13. Hussain, R.; Zeadally, S. Autonomous cars: Research results, issues, and future challenges. *IEEE Commun. Surv. Tutorials* **2018**, *21*, 1275–1313. [CrossRef]

14. Bardt, H. Autonomous Driving—A Challenge for the Automotive Industry. *Intereconomics* **2017**, *52*, 171–177. [CrossRef]

15. Bellem, H.; Klüver, M.; Schrauf, M.; Schöner, H.P.; Hecht, H.; Krems, J.F. Can we study autonomous driving comfort in moving-base driving simulators? A validation study. *Hum. Factors* **2017**, *59*, 442–456. [CrossRef]

16. Uricár, M.; Hurych, D.; Krizek, P.; Yogamani, S. Challenges in designing datasets and validation for autonomous driving. *arXiv* **2019**, arXiv:1901.09270.

17. Li, W.; Pan, C.; Zhang, R.; Ren, J.; Ma, Y.; Fang, J.; Yan, F.; Geng, Q.; Huang, X.; Gong, H.; et al. AADS: Augmented autonomous driving simulation using data-driven algorithms. *Sci. Robot.* **2019**, *4*. [CrossRef] [PubMed]

18. Schlager, B.; Muckenhuber, S.; Schmidt, S.; Holzer, H.; Rott, R.; Maier, F.M.; Saad, K.; Kirchengast, M.; Stettinger, G.; Watzenig, D.; et al. State-of-the-Art Sensor Models for Virtual Testing of Advanced Driver Assistance Systems/Autonomous Driving Functions. *SAE Int. J. Connect. Autom. Veh.* **2020**, *3*, 233–261. [CrossRef]

19. Stolz, M.; Nestlinger, G. Fast generic sensor models for testing highly automated vehicles in simulation. *EI* **2018**, *135*, 365–369. [CrossRef]

20. Muckenhuber, S.; Holzer, H.; Rübsam, J.; Stettinger, G. Object-based sensor model for virtual testing of ADAS/AD functions. In Proceedings of the 2019 IEEE International Conference on Connected Vehicles and Expo (ICCVE), Graz, Austria, 4–8 November 2019; pp. 1–6.

21. Hanke, T.; Hirsenkorn, N.; Dehlink, B.; Rauch, A.; Rasshofer, R.; Biebl, E. Generic architecture for simulation of ADAS sensors. In Proceedings of the 2015 16th International Radar Symposium (IRS), Dresden, Germany, 24–26 June 2016; pp. 125–130.

22. Genser, S.; Muckenhuber, S.; Solmaz, S.; Reckenzaun, J. Development and Experimental Validation of an Intelligent Camera Model for Automated Driving. *Sensors* **2021**, *21*, 7583. [CrossRef]

23. Yang, T.; Li, Y.; Ruichek, Y.; Yan, Z. Performance Modeling a Near-Infrared ToF LiDAR Under Fog: A Data-Driven Approach. *IEEE Trans. Intell. Transp. Syst.* **2021**. [CrossRef]

24. Fang, W.; Zhang, S.; Huang, H.; Dang, S.; Huang, Z.; Li, W.; Wang, Z.; Sun, T.; Li, H. Learn to Make Decision with Small Data for Autonomous Driving: Deep Gaussian Process and Feedback Control. *J. Adv. Transp.* **2020**, *2020*, 8495264. [CrossRef]

25. Hanke, T.; Hirsenkorn, N.; Dehlink, B.; Rauch, A.; Rasshofer, R.; Biebl, E. Classification of sensor errors for the statistical simulation of environmental perception in automated driving systems. In Proceedings of the 2016 IEEE 19th International Conference on Intelligent Transportation Systems (ITSC), Rio de Janeiro, Brazil, 1–4 November 2016; pp. 643–648.

26. Hirsenkorn, N.; Hanke, T.; Rauch, A.; Dehlink, B.; Rasshofer, R.; Biebl, E. A non-parametric approach for modeling sensor behavior. In Proceedings of the 2015 16th International Radar Symposium (IRS), Dresden, Germany, 24–26 June 2015; pp. 131–136.

27. Eder, T.; Hachicha, R.; Sellami, H.; van Driesten, C.; Biebl, E. Data Driven Radar Detection Models: A Comparison of Artificial Neural Networks and Non Parametric Density Estimators on Synthetically Generated Radar Data. In Proceedings of the 2019 Kleinheubach Conference, Miltenberg, Germany, 23–25 September 2019; pp. 1–4.

28. Höber, M.; Nalic, D.; Eichberger, A.; Samiee, S.; Magosi, Z.; Payerl, C. Phenomenological Modelling of Lane Detection Sensors for Validating Performance of Lane Keeping Assist Systems. In Proceedings of the 2020 IEEE Intelligent Vehicles Symposium (IV), Las Vegas, NV, USA, 23 June 2020; pp. 899–905.

29. Schneider, S.A.; Saad, K. Camera Behavior Models for ADAS and AD functions with Open Simulation Interface and Functional Mockup Interface. *Cent. Model Based Cyber Phys. Prod. Dev.* **2018**, *20*, 19–19.

30. Schneider, S.A.; Saad, K. Camera behavioral model and testbed setups for image-based ADAS functions. *EI* **2018**, *135*, 328–334. [CrossRef]

31. Wittpahl, C.; Zakour, H.B.; Lehmann, M.; Braun, A. Realistic image degradation with measured PSF. *Electron. Imaging* **2018**, *2018*, 149-1. [CrossRef]

32. Carlson, A.; Skinner, K.A.; Vasudevan, R.; Johnson-Roberson, M. Modeling camera effects to improve visual learning from synthetic data. In Proceedings of The European Conference on Computer Vision (ECCV) Workshops, Munich, Germany, 8–14 September 2018.
33. Eichberger, A.; Markovic, G.; Magosi, Z.; Rogic, B.; Lex, C.; Samiee, S. A Car2X sensor model for virtual development of automated driving. *Int. J. Adv. Robot. Syst.* **2017**, *14*, 1729881417725625. [CrossRef]
34. Bernsteiner, S.; Magosi, Z.; Lindvai-Soos, D.; Eichberger, A. Radarsensormodell für den virtuellen Entwicklungsprozess. *ATZelektronik* **2015**, *10*, 72–79. [CrossRef]
35. Ponn, T.; Müller, F.; Diermeyer, F. Systematic analysis of the sensor coverage of automated vehicles using phenomenological sensor models. In Proceedings of the 2019 IEEE Intelligent Vehicles Symposium (IV), Paris, France, 9–12 June 2019; pp. 1000–1006.
36. Mobileye. *LKA Common CAN Protocol;* Mobileye: Jerusalem, Israel, 2019.
37. Borkar, A.; Hayes, M.; Smith, M.T.; Pankanti, S. A layered approach to robust lane detection at night. In Proceedings of the 2009 IEEE Workshop on Computational Intelligence in Vehicles and Vehicular Systems, Nashville, TN, USA, 30 March–2 April 2009; pp. 51–57.
38. Li, Y.; Zhang, W.; Ji, X.; Ren, C.; Wu, J. Research on lane a compensation method based on multi-sensor fusion. *Sensors* **2019**, *19*, 1584. [CrossRef]
39. BIPM; IEC; ISO; IUPAC; IUPAP; OML. *Guide to the Expression of Uncertainty in Measurement;* BIPM: Tokyo, Japan, 1995.
40. Schneider, D.; Schick, B.; Huber, B.; Lategahn, H. Measuring Method for Function and Quality of Automated Lateral Control Based on High-precision Digital Grund Truth Maps. In *VDI/VW-Gemeinschaftstagung Fahrerassistenzsysteme und Automatisiertes Fahren 2018;* VDI: Düsseldorf, Germany, 2018; pp. 3–15.
41. Tihanyi, V.; Tettamanti, T.; Csonthó, M.; Eichberger, A.; Ficzere, D.; Gangel, K.; Hörmann, L.B.; Klaffenböck, M.A.; Knauder, C.; Luley, P. Motorway measurement campaign to support R&D activities in the field of automated driving technologies. *Sensors* **2021**, *21*, 2169. [PubMed]
42. Zhang, J.; Chen, M.; Zhao, S.; Hu, S.; Shi, Z.; Cao, Y. ReliefF-based EEG sensor selection methods for emotion recognition. *Sensors* **2016**, *16*, 1558. [CrossRef] [PubMed]
43. Palma-Mendoza, R.J.; Rodriguez, D.; De-Marcos, L. Distributed ReliefF-based feature selection in Spark. *Knowl. Inf. Syst.* **2018**, *57*, 1–20. [CrossRef]
44. GmbH, G.E. *Technical Documentation ADMA Version 1.0;* GeneSys Electronik GmbH: Offenburg, Germany, 2013.
45. Khatib, T.; Mohamed, A.; Sopian, K.; Mahmoud, M. Estimating ambient temperature for Malaysia using generalized regression neural network. *Int. J. Green Energy* **2012**, *9*, 195–201. [CrossRef]
46. Lee, D.; Yeo, H. A study on the rear-end collision warning system by considering different perception-reaction time using multi-layer perceptron neural network. In Proceedings of the 2015 IEEE Intelligent Vehicles Symposium (IV), Seoul, Korea, 28 June–1 July 2015; pp. 24–30.
47. Liu, B.; Zhao, Q.; Jin, Y.; Shen, J.; Li, C. Application of combined model of stepwise regression analysis and artificial neural network in data calibration of miniature air quality detector. *Sci. Rep.* **2021**, *11*, 1–12.
48. Bishop, C.M.; Roach, C. Fast curve fitting using neural networks. *Rev. Sci. Instruments* **1992**, *63*, 4450–4456. [CrossRef]
49. Li, Y.; Tang, G.; Du, J.; Zhou, N.; Zhao, Y.; Wu, T. Multilayer perceptron method to estimate real-world fuel consumption rate of light duty vehicles. *IEEE Access* **2019**, *7*, 63395–63402. [CrossRef]
50. Ceven, S.; Bayir, R. Implementation of Hardware-in-the-Loop Based Platform for Real-time Battery State of Charge Estimation on Li-Ion Batteries of Electric Vehicles using Multilayer Perceptron. *Int. J. Intell. Syst. Appl. Eng.* **2020**, *8*, 195–205. [CrossRef]
51. Møller, M.F. A scaled conjugate gradient algorithm for fast supervised learning. *Neural Netw.* **1993**, *6*, 525–533. [CrossRef]
52. IPG CarMaker. *Reference Manual (V 8.1.1).;* IPG Automotive GmbH: Karlsruhe, Germany, 2019.
53. Chandran, V.; Patil, C.K.; Karthick, A.; Ganeshaperumal, D.; Rahim, R.; Ghosh, A. State of charge estimation of lithium-ion battery for electric vehicles using machine learning algorithms. *World Electr. Veh. J.* **2021**, *12*, 38. [CrossRef]
54. Liao, X.; Li, Q.; Yang, X.; Zhang, W.; Li, W. Multiobjective optimization for crash safety design of vehicles using stepwise regression model. *Struct. Multidiscip. Optim.* **2008**, *35*, 561–569. [CrossRef]
55. Cherkassky, V.; Ma, Y. Practical selection of SVM parameters and noise estimation for SVM regression. *Neural Netw.* **2004**, *17*, 113–126. [CrossRef]
56. Montgomery, D.C.; Peck, E.A.; Vining, G.G. *Introduction to Linear Regression Analysis;* John Wiley & Sons: Hoboken, NJ, USA, 2021.
57. Quinonero-Candela, J.; Rasmussen, C.E. A unifying view of sparse approximate Gaussian process regression. *J. Mach. Learn. Res.* **2005**, *6*, 1939–1959.
58. Avnimelech, R.; Intrator, N. Boosting regression estimators. *Neural Comput.* **1999**, *11*, 499–520. [CrossRef]
59. Zhou, N.; Pierre, J.W.; Trudnowski, D. A stepwise regression method for estimating dominant electromechanical modes. *IEEE Trans. Power Syst.* **2011**, *27*, 1051–1059. [CrossRef]
60. Hoang, T.M.; Hong, H.G.; Vokhidov, H.; Park, K.R. Road lane detection by discriminating dashed and solid road lanes using a visible light camera sensor. *Sensors* **2016**, *16*, 1313. [CrossRef]

Article

Driver Monitoring of Automated Vehicles by Classification of Driver Drowsiness Using a Deep Convolutional Neural Network Trained by Scalograms of ECG Signals

Sadegh Arefnezhad [1,*], **Arno Eichberger** [1], **Matthias Frühwirth** [2], **Clemens Kaufmann** [3], **Maximilian Moser** [2] and **Ioana Victoria Koglbauer** [1]

[1] Institute of Automotive Engineering, Faculty of Mechanical Engineering and Economic Sciences, Graz University of Technology, 8010 Graz, Austria; arno.eichberger@tugraz.at (A.E.); koglbauer@tugraz.at (I.V.K.)

[2] Human Research Institute of Health Technology and Prevention Research, Franz-Pichler-Strasse 30, 8160 Weiz, Austria; matthias.fruehwirth@humanresearch.at (M.F.); maximilian.moser@humanresearch.at (M.M.)

[3] Apptec Ventures Factum, Slamastrasse 43, 1230 Vienna, Austria; clemens.kaufmann@factum.at

* Correspondence: sadegharefnezhad1@gmail.com; Tel.: +43-316-873-35270

Abstract: Driver drowsiness is one of the leading causes of traffic accidents. This paper proposes a new method for classifying driver drowsiness using deep convolution neural networks trained by wavelet scalogram images of electrocardiogram (ECG) signals. Three different classes were defined for drowsiness based on video observation of driving tests performed in a simulator for manual and automated modes. The Bayesian optimization method is employed to optimize the hyperparameters of the designed neural networks, such as the learning rate and the number of neurons in every layer. To assess the results of the deep network method, heart rate variability (HRV) data is derived from the ECG signals, some features are extracted from this data, and finally, random forest and k-nearest neighbors (KNN) classifiers are used as two traditional methods to classify the drowsiness levels. Results show that the trained deep network achieves balanced accuracies of about 77% and 79% in the manual and automated modes, respectively. However, the best obtained balanced accuracies using traditional methods are about 62% and 64%. We conclude that designed deep networks working with wavelet scalogram images of ECG signals significantly outperform KNN and random forest classifiers which are trained on HRV-based features.

Keywords: convolutional neural network; driver drowsiness; ECG signal; heart rate variability; wavelet scalogram

Citation: Arefnezhad, S.; Eichberger, A.; Frühwirth, M.; Kaufmann, C.; Moser, M.; Koglbauer, I.V. Driver Monitoring of Automated Vehicles by Classification of Driver Drowsiness Using a Deep Convolutional Neural Network Trained by Scalograms of ECG Signals. *Energies* **2022**, *15*, 480. https://doi.org/10.3390/en15020480

Academic Editors: Giovanni Lutzemberger and Aldo Sorniotti

Received: 4 November 2021
Accepted: 6 January 2022
Published: 10 January 2022

Publisher's Note: MDPI stays neutral with regard to jurisdictional claims in published maps and institutional affiliations.

1. Introduction

Drowsiness is defined as a transitional state fluctuating between alertness and sleep that increases the reaction time to critical situations and leads to impaired driving [1,2]. According to previous studies, driver drowsiness is one of the leading causes of traffic accidents. For example, the National Highway Transportation Safety Administration (NHTSA) reported that drowsy drivers were involved in about 800 fatal crashes in 2017 [3]. Another study announced that about 22–24% of crashes or near-crash risks are contributed by drowsy drivers [4]. The American Automobile Association (AAA) has also reported that about 24% of drivers acknowledged feeling extremely sleepy during driving at least once in the previous month [5].

Moreover, monitoring of driver alertness is an implicit requirement in the forthcoming SAE level of conditional automated driving (level 3) since handing over vehicle control to drowsy drivers is unsafe [6,7]. Various driver drowsiness detection systems (DDDS) have already been proposed in recent studies [8–11]. In our previous work [2], we developed a method for drowsiness classification only in manual driving mode and by using only vehicle-based data. As the drivers insert no input into the vehicle during automated driving tests, the proposed method in [2] cannot be used in automated driving. Moreover, vehicle-based data can be significantly affected by road geometry and the driving behavior of the specific driver. However, the proposed method in this paper uses the ECG data as inputs to the deep CNNs and can be applied in both manual and automated driving modes. Moreover, biosignals such as ECG can provide more accuracy to detect the onset of drowsiness than vehicle-based data [12,13]. This paper offers a new method using deep neural networks trained by wavelet scalograms of an electrocardiogram (ECG) signal.

1.1. Related Works

ECG signals present the heart's electrical activity over time that is typically recorded using attached electrodes to the chest [14]. Figure 1 shows the schematic representation of a standard ECG signal [15]. Heart rate variability (HRV) information is extracted by detecting the R-peaks in the ECG signals and evaluating the fluctuations of the time intervals between adjacent R-peaks [16]. HRV is well-known physiological information that presents the activity of the autonomic nervous system (ANS) [17], fluctuations markedly over a day, and the sleep-wake-cycle [18]. Therefore, it is assumed to be indicative not only of the sleep stages [19] but also of sleepiness as well. HRV has been frequently employed to design a DDDS. For example, Fujiwara et al. [20] developed a system based on eight extracted features from HRV data where multivariate statistical process control was used as an anomaly detection method in HRV data. Results showed that the proposed method detected 12 out of 13 drowsiness onsets and the false-positive rate of the anomaly detection system was about 1.7 times per hour. Huang et al. used machine learning with four different traditional classifiers (support vector machine, K-nearest neighbor, naïve Bayes, and logistic regression) for binary detection of drowsiness by training on time and frequency domain features from HRV data [17]. Results showed that the K-nearest neighbor achieved the best accuracy, which was about 75.5%.

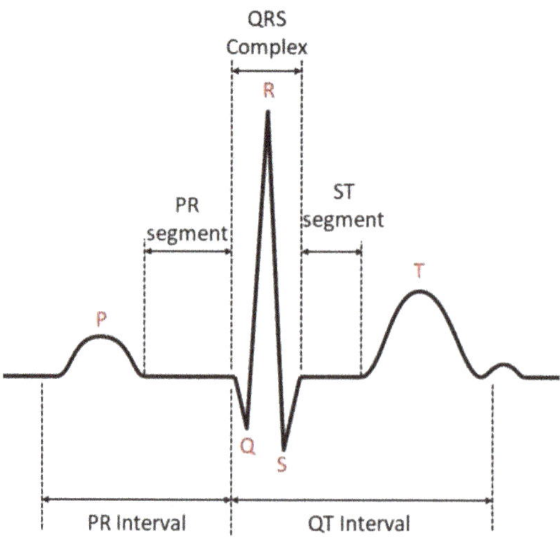

Figure 1. Schematic representation of a standard ECG signal.

To discriminate between the HRV dynamics in two states of fatigue (caused by sleep deprivation) and drowsiness (caused by monotonous driving), two different monitoring systems were proposed in [21] based on features from HRV and respiration signals. One of these systems is a binary classifier (alert/drowsy) for assessing the level of driver vigilance every minute. Another system detects the level of the driver's sleep deprivation in the first three minutes of driving. That study showed that the balanced accuracy of the drowsiness detection system which used only HRV-based features is about 65.5%. However, by adding the features from respiration signals, this system achieved a balanced accuracy of 78.5%, an improvement of about 13%. The balanced accuracy of the sleep deprivation system was also about 75%, and it detected 8 out 13 sleep-deprived drivers correctly. Another study conducted by Buendia et al. [22] investigated the relationship between the drowsiness levels rated with the Karolinska sleepiness scale (KSS) and heart rate dynamics. Results showed that the average heart rate decreased with increasing KSS (which means higher drowsiness levels), whereas heart rate variance increased in drowsy states. Patel et al. [23] also developed a neural network classifier to detect the early onset of driver drowsiness by analyzing the power of low- and high-frequency HRV sub-bands. The spectral image, plotted from the power spectral density of the HRV data, was the input given to the neural network that yielded an accuracy of 90%. In [24], Li and Chung used a wavelet transformation to extract features from HRV signals and compared them with fast Fourier transform (FFT)-based features. Receiver operation curves were used for feature selection and support vector machines as a classifier. The wavelet method outperformed the system designed using FFT. Classification results showed that the wavelet-based feature system achieved an overall accuracy of 95%. Furman et al. [25] reported that HRV activity in the very-low-frequency range (0.008–0.04 Hz) significantly and consistently decreases approximately five minutes before extreme signs of drowsiness can be observed.

1.2. Contribution of the Method

Previous studies commonly used hand-crafted techniques or dimensionality reduction methods to extract features from HRV data for driver drowsiness classification. Most commonly, heart rate variability data are derived by the detection of R-peaks in the ECG signal and processing the information of R-peak time points only. However, other segments of the ECG signals (see Figure 1) might also be associated with different levels of drowsiness. Furthermore, previous studies widely used traditional machine learning classifiers to classify driver drowsiness; however, deep neural networks are expected to outperform them if a large data set is available for training. In this study, we first employed the wavelet transformation to generate 2D scalogram images of the ECG signal, which capture time–frequency domain features. These images are inserted as input data to a deep convolutional neural network. Bayesian optimization is applied to optimize the hyperparameters of this network. To compare the results of this approach with previous methods, HRV data is also derived from ECG signals in a common way, and its extracted features are utilized to classify driver drowsiness using two traditional classifiers: K-nearest neighbors (KNN) and random forest.

The rest of this paper is structured as follows: Section 2 explains the experimental setup and the testing procedure that was used to collect the dataset. Section 3 describes the methodology for the classification of driver drowsiness. Section 4 presents the results of the proposed method, discusses the results, and compares them with the outcomes of other algorithms. Finally, Section 5 presents our conclusions and suggests future tasks to improve the proposed method.

2. Experimental Setup and Testing Procedure

This study utilizes the dataset collected during the WACHSens project, a joint project of the Human Research Institute Weiz, the Graz University of Technology, apptec Factum Vienna, and AVL U.K. The tests were performed in the automated driving simulator of Graz (ADSG) at the Institute of Automotive Engineering, Graz University of Technology. The driving simulator is presented in Figure 2. The following subsections explain the structure of the ADSG, simulated driving test procedure, and definition of ground truth for driver drowsiness.

Figure 2. Automated driving simulator of Graz (ADSG). To cancel the external noise and adjust the indoor temperature, ADSG is separated from its surrounding area using an insulating housing cube.

2.1. Driving Simulator

In the ADSG, the visual cues are simulated by eight LCD panels, covering 180 degrees of view and the rear screen, which the inner mirror observes. The side mirrors are also implemented in the LCDs covering the side windows. The acoustic cue is simulated by generating engine and wind noise applied at the car's sound system. Moreover, four bass shakers generate the vibration in the car chassis and the driver and passenger seats. Haptic feedback is provided by the Sensodrive™ simulator steering wheel (Weßling, Germany) [26], and an active brake pedal simulator, gas pedal, and gear-shift input are taken from the vehicle unmodified controls. The vehicle dynamics states are calculated by a full vehicle software AVL-VSM™ (Graz, Austria) [27], parametrized with a middle-class passenger car. The vehicle model calculates dynamics states as well as engine speed and torque for the acoustic simulation. Adaptive cruise control (ACC) and lane-keeping assist (LKA) systems are also implemented in this simulator for controlling the vehicle's longitudinal and lateral dynamics during tests on automated driving. The ADSG is surrounded by a noise- and temperature-insulating cube/box. Different features of this simulator were studied in our previous works [28,29].

2.2. Participants and Driving Tests Procedure

In this project, different types of physiological data were collected from 92 drivers. These drivers participated in manual and automated driving tests when they were in two different vigilance states: fatigued and rested. This procedure results in four different driving sessions for each participant: fatigued automated driving, fatigued manual driving, rested automated driving, and rested manual driving. In the rested condition, drivers were required to have a full night's sleep before performing the tests. For the fatigued condition, the drivers could choose one of the two following options: (1) extended wakefulness (being awake for at least 16 h continuously before starting the tests in the conditions fatigued automated and fatigued manual) and perform the tests at their usual bedtime, or (2) being sleep-restricted by sleeping a maximum of four hours in the night before the tests. The age and gender of participants were balanced across the sample as presented in Table 1. The Female-60+ group has only 12 participants since we could not hire more still active drivers from this group in the available time frame.

Several biosignals, namely, ECG, EEG, EOG, skin conductivity, and oronasal respiration, were collected using a g.Nautilus^{TM} device (Schiedlberg, Austria; research version) with a sampling frequency of 500 Hz. Facial-based data such as eyelid opening, pupil diameter, and gaze direction were also measured with a sampling frequency of 100 Hz using a SmartEye^{TM} (Gothenburg, Sweden) eye-tracker system installed on the car dashboard. In this study, only ECG signals are employed to classify the driver's drowsiness. The study was conducted according to the ethical guidelines of the Declaration of Helsinki and the General Data Protection Regulation of the European Union. The study protocol was approved by the Ethics Committee of the Medical University of Graz in vote 30-409 ex 17/18 dated 1 June 2018. Written informed consent was obtained from participants before the experiments, and they were compensated with EUR 50 after finishing the sessions. More details of the driving test procedure are described in a previous publication [2].

Table 1. Gender–age groups of the participants in the driving tests. SD: standard deviation.

Gender	Age Range	Mean of Age	SD of Age	Number of Participants
Female	20–40	25.2	5.3	16
Female	40–60	50.4	6.5	16
Female	60+	65.4	4.3	12
Male	20–40	24.7	3.7	16
Male	40–60	51.9	4.2	16
Male	60+	69.0	7.3	16
-	-	47.0	18.4	Total: 92

2.3. Ground Truth Definition for Driver Drowsiness

To monitor the participants' driving behavior, four cameras were placed in the ADSG that recorded different views of the driver and the test track (see Figure 3). Traffic psychologists thoroughly observed these videos and assigned labels to the driver's drowsiness level based on drowsiness signs such as yawning, long blinks, and head nodding. The driver's vigilance state is reported in four classes: alert (AL), moderately drowsy (MD), extremely drowsy (ED), and falling asleep event (SL). These drowsiness levels are collected with their corresponding SmartEye^{TM} video frame numbers to synchronize drowsiness level ratings with the recorded data channels (more details of data synchronization are explained in Section 3.1). Figure 4 shows an example of the defined ground truth for driver drowsiness in all four driving tests (all performed by the same driver). As that Figure shows, micro-sleep events (SL) were also reported by video observers. However, we merged the SL class with the extremely drowsy (ED) class since the overall number of SL samples was too small to be considered as a separate class for machine learning training. This figure also shows that even in the rested condition, some drivers showed signs of moderately and extremely drowsy states. More details of the ground truth definition for driver drowsiness using video observations are explained in our previous publication [30].

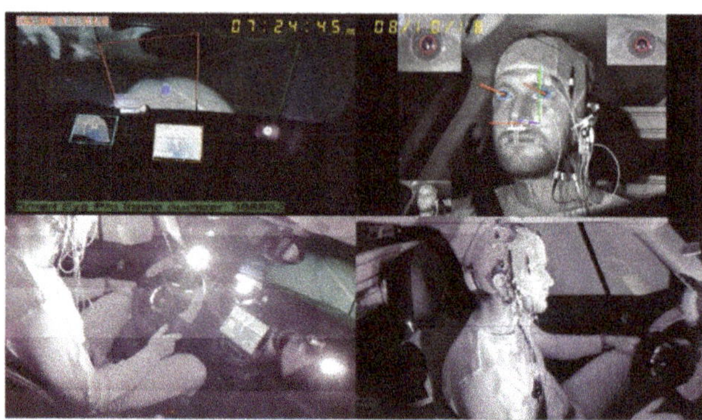

Figure 3. Four different views of the driver and the test track. These views were observed thoroughly by an expert to define a ground truth for driver drowsiness based on drowsiness signs into three classes (informed consent was obtained from the driver to publish his image in this paper; reprinted from our previous study [2], license no. 5218171384545).

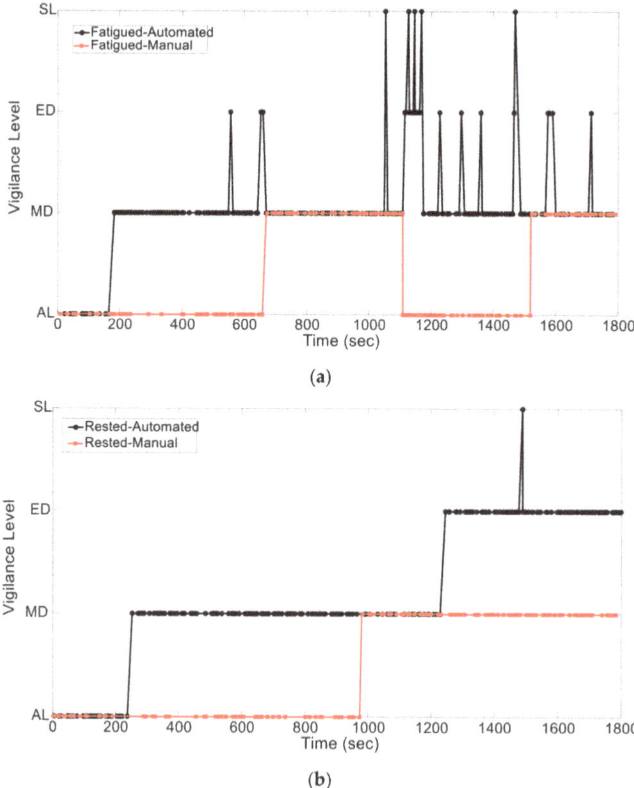

Figure 4. Reported ground truth for driver drowsiness by ratings of the driving test videos: (**a**) video observations in the fatigued automated and fatigued manual tests; and (**b**) video observations in the rested automated and rested manual tests. Four different levels for drivers' vigilance were reported: alert (AL), moderately drowsy (MD), extremely drowsy (ED), and sleep (SL). In this paper, we merge the SL level into the ED level.

3. Methodology

Two different methodologies are employed to classify driver drowsiness using ECG signals: (1) two traditional classifiers (random forest and KNN) trained by features extracted from HRV signals, and (2) one deep convolutional neural network (CNN) model trained by ECG wavelet scalogram images. The Bayesian optimization method is used to optimize the hyperparameters of the classifiers. Figure 5 shows the flowchart of these methods. The following subsections describe the structure of these methodologies.

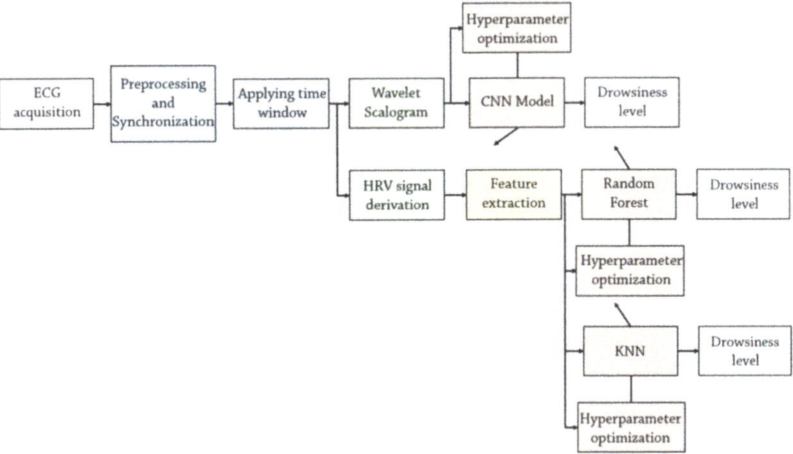

Figure 5. Two different approaches to classify driver drowsiness using ECG signals: wavelet scalograms or derived HRV features. The hyperparameter of KNN, random forest, and CNN model are optimized using the Bayesian optimization method.

3.1. Data Synchronization

Ground truth is defined based on the video observation and recorded using the frame number information of SmartEye^{TM} data collected with a sampling frequency of 100 Hz. Physiological signals were recorded with separate equipment at 500 Hz sampling frequency, but also fed into the central recording module and stored with the same sampling frequency of 100 Hz. The lower sample rate is not sufficient for the high-quality processing of physiological data. Therefore, we need to synchronize video and physiological data sources with the help of the respiration signal which is available at both sampling rates of 100 Hz and 500 Hz. The normalized cross-correlation between the two respiration signals is calculated at all possible lags. The delay between these two signals is calculated as the lag with the largest absolute value of normalized cross-correlation. Figure 6 shows an example of data synchronization where 500Hz-respiration data is shifted about 21.4 s forward to be synchronized with the 100Hz-respiration data. The same time shift is also applied to the ECG signals collected with the sampling frequency of 500 Hz to sync them with the video observations. In those data, offset correction was sufficient for an accuracy ±1 video frame.

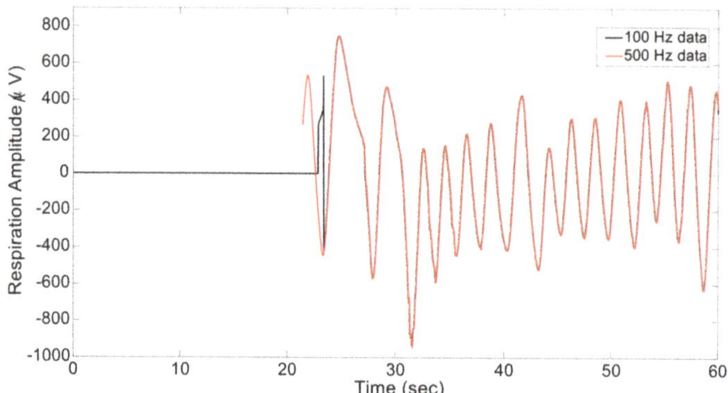

Figure 6. Data synchronization using respiration signals collected with two different frequencies of 100 Hz and 500 Hz. In this example, the 500 Hz data is shifted about 21.4 s forward to be synced with the 100 Hz data.

3.2. ECG Preprocessing

Generally, ECG signals are contaminated with different noise sources such as power line interference (50 Hz) [31] and baseline wander [32]. A second-order infinite impulse response (IIR) notch filter [33] is utilized here to remove the power line noise from ECG signals. Furthermore, a high pass filter with a passband frequency of 0.5 Hz is also employed to remove the low-frequency baseline wander noise. Figure 7 shows one part of the noisy and denoised ECG signals after removing the baseline wander and power line noise.

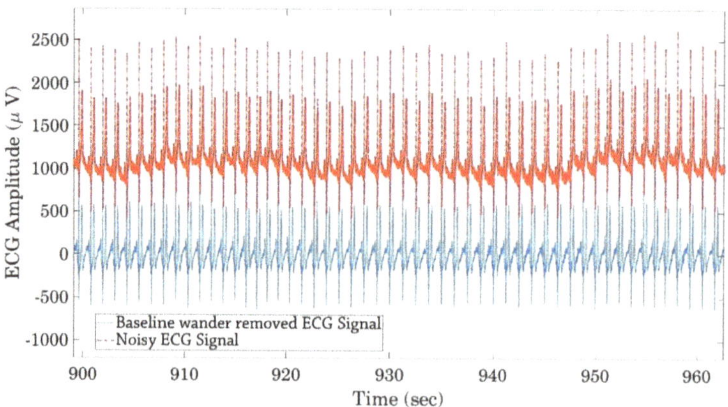

Figure 7. Noisy and denoised ECG signals after removing baseline wander and power line noise.

3.3. Driver Drowsiness Classification Using Scalograms of ECG Signals

This section describes the proposed method for driver drowsiness classification using deep neural networks trained by wavelet scalogram images of the ECG signals.

3.3.1. Wavelet Scalogram

Wavelet analysis calculates the correlation (similarity) between an input signal and a given wavelet function $\psi(t)$. Unlike Fourier transform, wavelet analysis provides a multi-resolution time–frequency output under the assumption that low frequencies maintain the same characteristics for the whole duration in the input signal. In contrast, high frequencies are assumed to appear at different time points as short events.

Therefore, the wavelet function is scaled and translated by two parameters $s \in \mathcal{R}^+$ and $u \in \mathcal{R}$ to generate a wavelet filter bank, $\psi_{u,s}$ as presented by Equation (1) [34].

$$\psi_{u,s} = \frac{1}{\sqrt{s}} \psi \left(\frac{t-u}{s} \right) \tag{1}$$

By using this transformed wavelet, continuous wavelet transform (CWT) of the input signal $x(t)$ at the time u and scale s can be calculated as:

$$X_{WT}(u,s) = \int_{-\infty}^{\infty} x(t) \psi_{u,s}^*(t) \mathrm{d}t \tag{2}$$

where $\psi^*(t)$ is the complex conjugate of $\psi(t)$ and $X_{WT}(u,s)$ provides the frequency contents of $x(t)$ corresponding to the time u and scale s. By using the two parameters of u and s, it is possible to investigate the input $x(t)$ in two domains of time and frequency simultaneously, whereby the resolution of time and frequency depends on the scale parameter s. CWT provides a time–frequency decomposition of $x(t)$ in the time–frequency plane. This method can be more beneficial than other methods such as short-time Fourier transform (STFT) when investigating the non-stationary signals since it provides a higher time resolution in the higher frequencies by reducing frequency resolution, and a higher frequency resolution in lower frequencies by reducing time resolution. In contrast, the time and frequency resolutions are constant in STFT. The scalogram if $x(t)$ in any positive scale is calculated as the norm of $X_{WT}(u,s)$:

$$S(u,s) = ||X_{WT}, (u,s)|| = \left(\int_{-\infty}^{\infty} |X_{WT}(u,s)|^2 \, \mathrm{d}u \right)^{\frac{1}{2}} \tag{3}$$

This equation calculates the energy of X_{WT} at a scale s. Therefore, we can find the significant scales (which correspond to frequencies) in the signal using the scalogram.

The wavelet scalogram is used to transform the time series ECG signal to the time–frequency domain. Here, the Morse wavelet [35] is employed to calculate the wavelet transformation of the ECG signals. A sliding window with a length of 10 s and an overlap of 5 s between two consecutive windows was employed, and the scalogram image of every window of the data is calculated. The resulting number of data windows in each level of driver drowsiness are provided in Table 2 for manual driving, and in Table 3 for automated driving. As these tables show, the generated data sets are imbalanced in both manual and automated modes. This fact must be taken into account in the structure of the deep and traditional classifiers. Moreover, the percentages of MD and ED classes are higher in the automated driving tests than in the manual tests. Thus, the drivers were generally drowsier in automated.

Figure 8 shows sample images of ECG signals and their corresponding scalogram images for all three drowsiness levels in a rested automated test. The generated images are resized to 224 × 224 pixels. To reduce the computational complexity of the training process of the deep network, the RGB scalogram images are also transformed to grayscale images as presented in Figure 9. These grayscaled images are used as input data to train the deep convolutional neural network.

Table 2. The number of data samples in each drowsiness class after applying time windows to generate the scalograms in the **manual** driving tests.

Class	Number of Samples	Percentage
AL	23,722	67.38%
MD	9371	26.62%
ED	2111	6.00%
Sum	35,204	100%

Table 3. The number of data samples in each drowsiness class after applying time windows to generate the scalograms in the **automated** driving tests.

Class	Number of Samples	Percentage
AL	19,508	56.33%
MD	10,699	30.89%
ED	4427	12.78%
Sum	34,634	100%

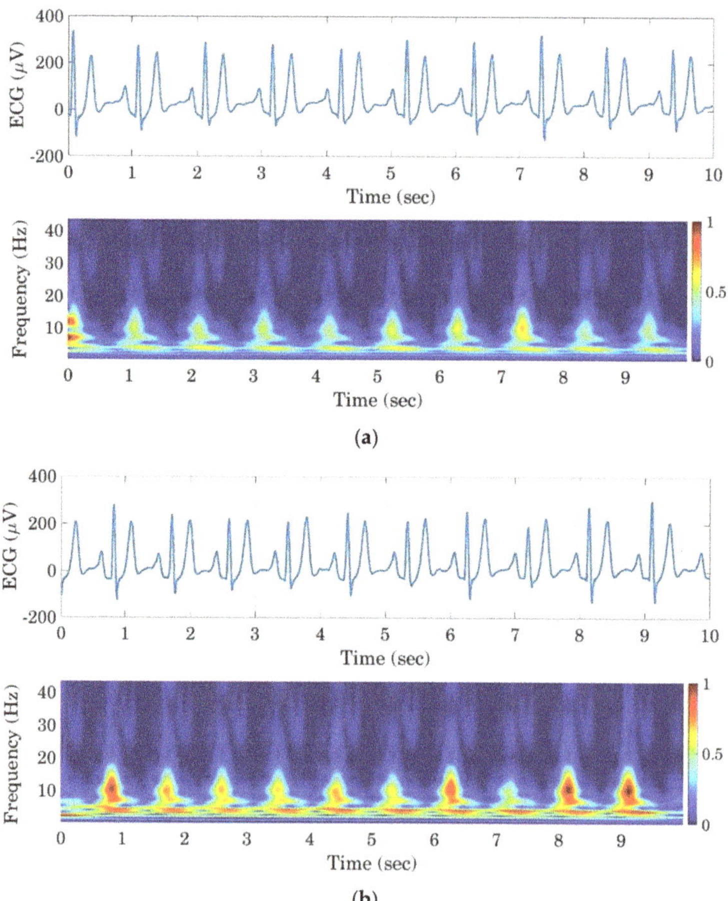

(a)

(b)

Figure 8. *Cont.*

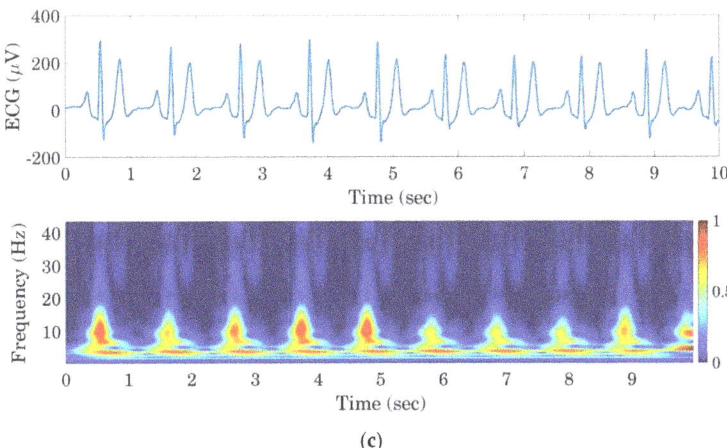

(c)

Figure 8. Examples of ECG signals segments and their corresponding scalograms for the AL (**a**), MD (**b**), and ED (**c**) classes.

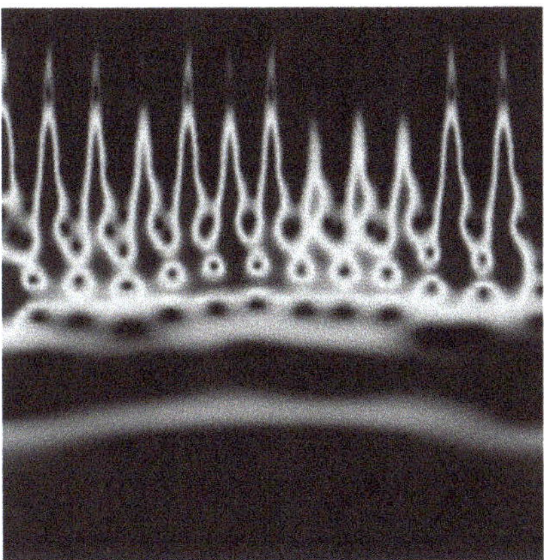

Figure 9. Sample of a grayscale resized image (224 × 224 pixels) of the ECG scalogram image.

3.3.2. Architecture of Deep CNN and Optimization of its Hyperparameters

Convolution neural networks (CNN) have been widely used to learn features from input images in different applications [36–39]. These networks help to capture the spatial dependencies in different parts of an input image by applying a convolution operation of some specific filters to input images [40]. This study used scalogram images of ECG signals to train a deep CNN and classify the driver drowsiness. As scalograms are time–frequency representations of an underlying time signal, temporal information is coded in the spatial features of the image. The input images are first normalized to have zero mean and unit variance. Then, the whole data set is split randomly into the train (80% of the input data), validation (10% of the input data), and test (10% of the input data) subsets in a way that the percentages of the drowsiness classes are approximately the same as presented in Tables 2 and 3 in each of the subsets.

The utilized deep CNN is composed of five convolutional blocks and one fully connected block in its hidden layer. Convolution and fully connected blocks are presented in Figure 10, where Conv, BN, ReLU, Max Pool, and FC are convolution layers, batch normalization layer, the ReLU activation function ($max(x, 0)$), and a fully connected layer.

respectively. The hidden layer is followed by the output layer that is constructed using an FC layer, a soft-max layer, and a weighted classification layer (weight). The number of neurons in the fully connected layer of the output layer is equal to the number of classes (here, three). The weight layer is used to mitigate the data imbalance issue. This layer contains one element per drowsiness class where every element is calculated using Equation (4):

$$W_i = \frac{N_c}{C_i \sum_{i=1}^{N_c} \frac{1}{C_i}} \tag{4}$$

where N_c is the number of classes (here, three), C_i is the number of data samples that belong to the i-th class, and finally, W_i is the weight of i-th class. By applying Equation (4) to the data samples that belong to the drowsiness classes in the manual and automated modes (presented in Tables 2 and 3), the corresponding weights for every class are computed. Table 4 provides these weights. As this table shows, the class weights of the ED class are highest in both manual and automated mode tests. By using these weights, misclassification errors of the MD and ED classes get more weight in comparison to the AL class. Therefore, if the network classifies an MD or ED sample into the AL class wrongly, it results in a large misclassification error that has a significant influence on the optimization process and thus will reduce the frequency of this kind of classification error.

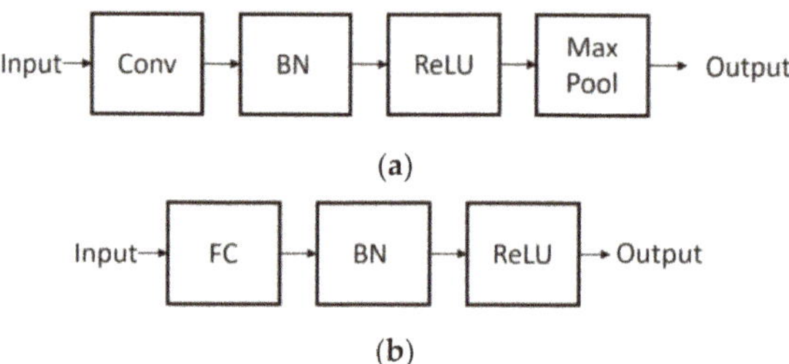

Figure 10. Convolutional (**a**) and fully connected (**b**) blocks are used to construct the deep CNN.

Table 4. Class weights of the different drowsiness classes used in the deep CNNs to alleviate the imbalanced data set issue.

Class	Manual	Automated
AL	0.203	0.415
MD	0.514	0.757
ED	2.283	1.828

Figure 11 presents the architecture of the deep CNN, where five convolution blocks are followed by one fully connected block. Moreover, one dropout layer is also added after convolution blocks to reduce the possibility of overfitting or getting stuck in local minima during the training process. The dropout layer temporarily eliminates some neurons with a predefined probability, along with all of their input and output connections [41].

Deep neural networks have several hyperparameters such as the learning rate, the regularization parameter, and the number of neurons of filters that can influence network performance. Finding a proper combination of these hyperparameters so that they provide the optimal performance of the deep network is a primary active task in the research field of deep learning [42,43]. In this study, the Bayesian optimization method [44,45] is applied to optimize the hyperparameters of the deep CNN. This method has the capability of reasoning about the iterations' performance before they are performed. Therefore, fewer iterations are needed to find the optimal hyperparameter combination than with other hyperparameter optimization methods [45]. Moreover, this method yields a better generalization on the tests data set [46]. Here, four different hyperparameters were considered to be optimized using the Bayesian optimization method, including learning rate, L2 regularization, dropout probability, the number of filters in convolution layers (Conv1 to Conv5 in Figure 11), and neurons in the fully connected layer in the hidden layers (FC1 in Figure 11). Here, it is assumed that the number of filters in Conv1 to Conv5 and the number of neurons in FC1 are equal, so only one hyperparameter is defined to find their optimal values. Table 5 presents the specified search space for each of these hyperparameters.

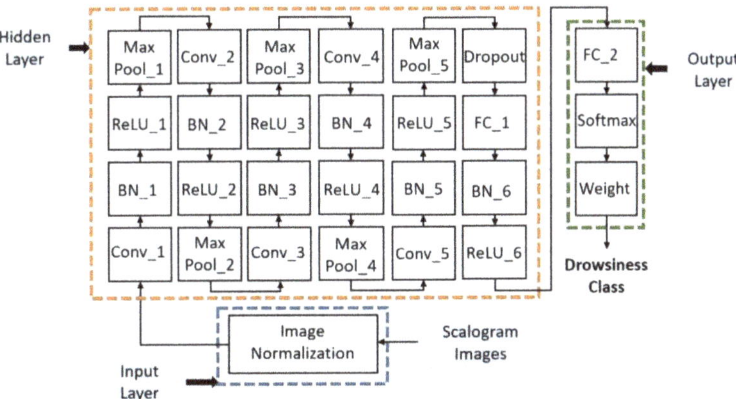

Figure 11. The architecture of the deep CNN used to classify driver drowsiness using ECG scalogram images.

Table 5. Used hyperparameters of the deep CNN to be optimized using a Bayesian optimizer.

Index	Hyperparameter	Search Space
H_1	Learning rate	$[5 \times 10^{-5}\text{–}0.001]$
H_2	Dropout probability	$[0.2\text{–}0.4]$
H_3	L2 regularization	$[10^{-8}\text{–}0.01]$
H_4	Number of filters in convolution layers (Conv1 to Conv5) and neurons in fully connected layer (FC1)	$[30\text{–}60]$

An adaptive moment estimation (ADAM) optimizer is employed to train the parameters of the designed deep CNNs (weights and biases). The maximum number of epochs is empirically chosen to be 15. A schedule for learning rate is utilized that multiplies the initial learning rate by 0.1 after 12 epochs to alleviate overfitting in the latter training epochs. The size of the mini-batch is defined to be constant and equals 16. The training process was conducted on a system with CPU and GPU types of Intel Core^{TM};i7-782HQ and NVIDIA^{TM} Quadro M2200, respectively.

3.4. Driver Drowsiness Classification Using Heart Rate Variability Data

This section describes the proposed method for driver drowsiness classification using feature extraction from HRV data.

3.4.1. Derivation of Heart Rate Variability Data from ECG Signals

The heart rate variability signal is derived from preprocessed ECG signals by applying an R-peak detection algorithm to detect heart rate. In this study, we used the automatic multiscale-based peak detection (AMPD) method [47] as an ECG R-peak detector, then RR Intervals (RRIs) that are defined as the time intervals between every two consecutive R-peaks are calculated. Figure 12 shows the detected R-peaks in a part of the ECG signal.

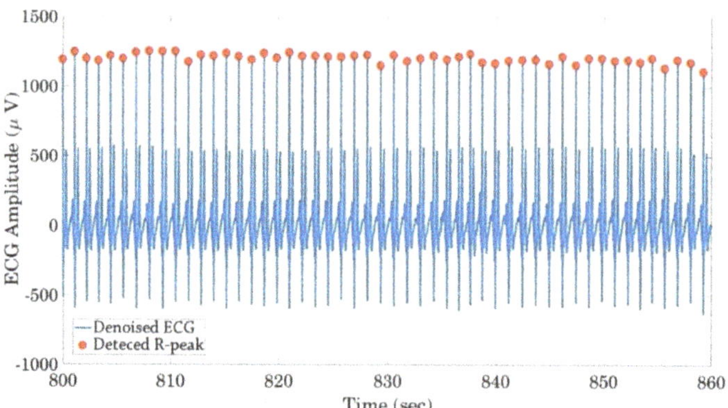

Figure 12. Detected R-peaks in the denoised ECG signal using the AMPD method.

3.4.2. Feature Extraction from HRV Data

Literature has proposed some features to be extracted from RR intervals for driver drowsiness detection [18], which conform to measures that are well established in clinical contexts [48]. Other HRV features are based on a visualization technique called the Poincaré plot. In this subsection, firstly, this plot is introduced, then those and other commonly extracted features from RR intervals are explained.

Poincaré plot: This plot is a type of recurrence plot to investigate the similarity in time series that can be used to analyze the nonlinear properties of HRV data [49]. Consider $X = [RR_t, RR_{t+1}, \ldots, RR_N]$ as a RR interval time series. The Poincaré plot first plots (RR_t, RR_{t+1}), then plots (RR_{t+1}, RR_{t+2}), then plots (RR_{t+2}, RR_{t+3}) and so on. This plot provides information about the short-term and long-term dynamics of the RR interval. An ellipse is fitted to the plotted data points and the minor and major semi-axes of the ellipse are associated with short-term and long-term HRV, respectively. Figure 13 shows the Poincaré plot for RR intervals collected in a rested automated driving test. The least-square method was employed to fit an ellipse on given RR intervals [50] and geometrical properties of this ellipse are extracted as features to describe the HRV dynamics.

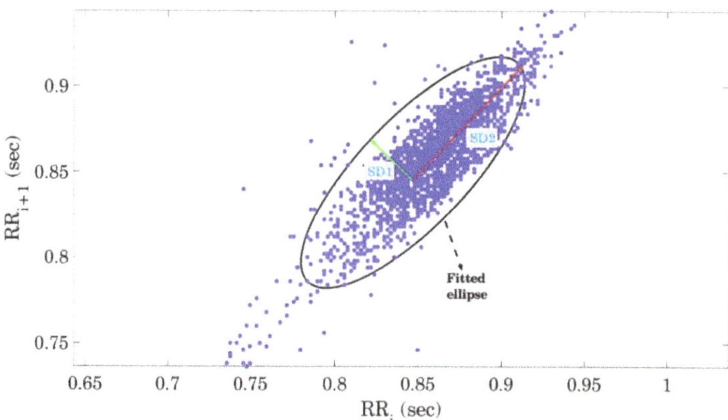

Figure 13. Poincaré plot and fitted ellipse for RR intervals during a rested automated driving test. Minor and major semi-axes of the fitted ellipse, SD1, SD2, and their ratio, are calculated as features to capture the dynamics of HRV data.

Three features are extracted from this plot:

1. SD1: SD1 is the standard deviation of the Poincaré plot perpendicular to the line of identity and the semi-minor axis (half of the shortest diameter) of the fitted ellipse, see Figure 13 (green vector). SD1 is an estimation of short-term HRV that describes parasympathetic activity since it represents the deviation of heart rate from the line-of-identity (constant heart rate).
2. SD2: SD2 is the standard deviation of the Poincaré plot along the line of identity and the semi-major axis (half of the largest diameter) of the fitted ellipse, see Figure 13 (red vector). SD2 is an estimation of long-term HRV that describes sympathetic and mixed activity since SD2 is along the line of identity.
3. SD1/SD2: SD1/SD2 is the ratio of SD1 to SD2 that describes the ratio of short-term to long-term HRV and the relationship between parasympathetic and sympathetic activity.

Other features that have been proposed by previous studies [18,51,52] are also extracted from RR intervals. These features are:

1. MeanRR: This feature presents the mean values of the time intervals between every two consecutive R-peaks. The MeanRR is calculated by Equation (5):

$$\text{MeanRR} = \frac{1}{N_R - 1} \sum_{i=1}^{N_R - 1} RR_{i+1} \tag{5}$$

where N_R is the number of heartbeats in the sliding windows and RR_{i+1} is equal to the time interval between R_i and R_{i+1}.

2. SDRR: This feature represents the standard deviation of RR intervals, calculated by Equation (6).

$$\text{SDRR} = \sqrt{\frac{1}{N_R - 1} \sum_{i=1}^{N_R-1} (RR_{i+1} - \text{MeanRR})^2} \tag{6}$$

3. RMSSD: This feature calculates the root mean square of consecutive RR intervals' differences, calculated by Equation (7). It reflects parasympathetic activity.

$$\text{RMSSD} = \sqrt{\frac{1}{N_R - 1} \sum_{i=1}^{N_R-1} (RR_{i+1} - RR_i)^2} \tag{7}$$

4. pRR50: This feature measures the ratio of the number of R-peaks that differ more than 50 ms from their next R-peak to the total number of RR intervals in every sliding window. Equation (8) calculates the pRR50.

$$\text{pRR50} = \frac{RR50_{count}}{N_R} \tag{8}$$

5. VLF: This feature presents the power in the very-low-frequency ranges of 0.003–0.04 Hz of the RR interval time series. To calculate this feature and the LF and HF, the PSD of the RR intervals is computed using the Lomb–Scargle periodogram method [53,54] in every sliding window.
6. LF: This feature presents the power in the low-frequency ranges of 0.04–0.15 Hz of the RR interval time series.
7. HF: This feature presents the power in the high-frequency ranges of 0.15–0.40 Hz of the RR interval time series and reflects parasympathetic activity.
8. LF/HF: This feature is the ratio of LF divided by HF and is also indicative of the sympathetic–parasympathetic balance.

Overall, eleven features are extracted from HRV data and are used to classify the driver's drowsiness. A window length of ten seconds, which was used in the ECG scalogram approach above, is considered extremely short for evaluation of HRV as it conforms nearly exclusively to the fast fluctuations of heart rate according to parasympathetic activity [55]. For exploratory purposes, we also applied longer sliding windows in comparison to the deep learning model. Two additional sliding windows are employed: (1) 60 s with 30 s overlap, and (2) 40 s with 20 s overlap. These longer windows help to provide a better estimation of mid-range dynamics of HRV data for classifiers than we can expect from short windows that are used in the deep learning method. The results of these sliding methods are compared together in Section 4.1.

The following subsection explains the two classifiers (KNN and random forest) used for drowsiness classification.

3.4.3. Classify Driver Drowsiness Using Traditional Classifiers

The KNN and random forest are employed to classify the driver drowsiness using extracted features from HRV data. Each one of these classifiers has two different hyperparameters. The KNN hyperparameters are the number of neighbors for every sample (numNei) and the function used to measure the distance between samples (distance) [56]. The random forest hyperparameters are also the minimum of leaf size (minLS) and number of predictors to sample at each node (numPTS) [57]. These hyperparameters are also optimized using the Bayesian optimization method to find the optimal set. Moreover, the issue of the imbalanced data set is removed by using the uniform prior probability of every class for the KNN [58] and random under-sampling boosting (RUSBoost) [59] for the random forest classifier.

4. Results and Discussion

To evaluate the generated classifiers, confusion matrices were calculated for the test dataset. These matrices provide four different values that are computed for every drowsiness level:

1. True-negative (TN): The number of samples that do not belong to a specific class (for example, AL) and are also classified in any of the two other classes (MD or ED) by the classifier.
2. True-positive (TP): The number of samples that belong to a specific class (for example, AL) and are correctly classified in that class.
3. False-negative (FN): The number of samples that belong to a specific class (for example, AL) but are wrongly classified in any of the two other classes (MD or ED).
4. False-positive (FP): The number of samples that do not belong to the specific class (for example, AL) and are wrongly classified in that class.

These four values are used to calculate five different metrics for every level of driver drowsiness:

1. Specificity (true negative rate): The specificity is TN divided by the sum of TN and FP. It can be interpreted as the probability of a sample not being classified in a class if it does not belong there
2. Sensitivity (true positive rate): The sensitivity is TP divided by the sum of TP and FN.
3. Precision (positive predictive value): The precision is TP divided by the sum of TP and FP.
4. F1-score: The F1-score is the harmonic mean of precision and sensitivity.
5. Balanced accuracy: The balanced accuracy is equal to the average of the accuracies of the three classes. The accuracy of every class is also equal to the ratio of TP of every class to the number of samples that belong to the corresponding class based on the actual labels.

The following subsections present the results of the two proposed methods for driver drowsiness classification.

4.1. Results of Driver Drowsiness Classification Using Heart Rate Variability Data

To evaluate the performance of the KNN and random forest classifiers, confusion matrices of these classifiers trained by HRV-based features with three different sliding windows of 10 s, 40 s, and 60 s are provided in Figures 14–16, respectively. In these Figures, the diagonal elements (in gray) provide the number of the sliding windows that are correctly classified in different classes of drowsiness, according to the ground truth classification from the video observations. Accordingly, the percentage numbers written in these elements show correct classification accuracy for every specific drowsiness level. Non-diagonal cells also present the number of samples that are misclassified. As these figures present, the classification accuracy of the MD class is lower than two other classes in the manual mode. Furthermore, random forest performs better than KNN for drowsiness classification in manual and automated modes regardless of the used sliding window. Balanced accuracies for every classifier in every driving mode are provided in Table 6. These accuracies are calculated as the average TP accuracies in confusion matrices that are shown in grey elements in Figure 16c,d. Therefore, the best balanced accuracy that is achieved using traditional methods in manual mode is the average of 64.7%, 56.2%, and 56.4%. The best balanced accuracy in the automated mode that is achieved by the same methods is also the average of 63.4%, 63.5%, and 66.6%. According to this table, the best balanced accuracies in the automated and manual modes are respectively 63.8% and 62.1%, which are obtained using the random forest classifier and the sliding window of 60 s with 30 s overlap.

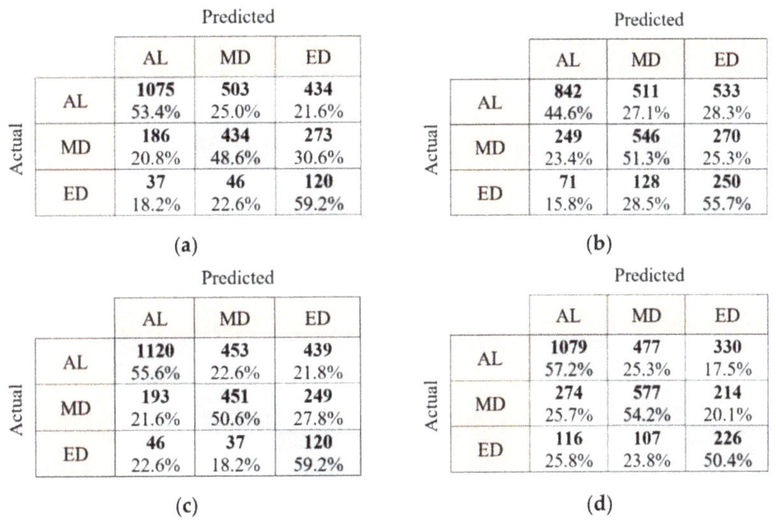

Figure 14. Confusion matrices of **KNN** classifier in the manual tests (**a**), **KNN** classifier in automated tests (**b**), **random forest** in the manual tests (**c**), and **random forest** in the automated tests (**d**) for driver drowsiness classification. The length of the sliding window for feature extraction is 10 s with a 5 s overlap. AL: alert, MD: moderately drowsy, and ED: extremely drowsy.

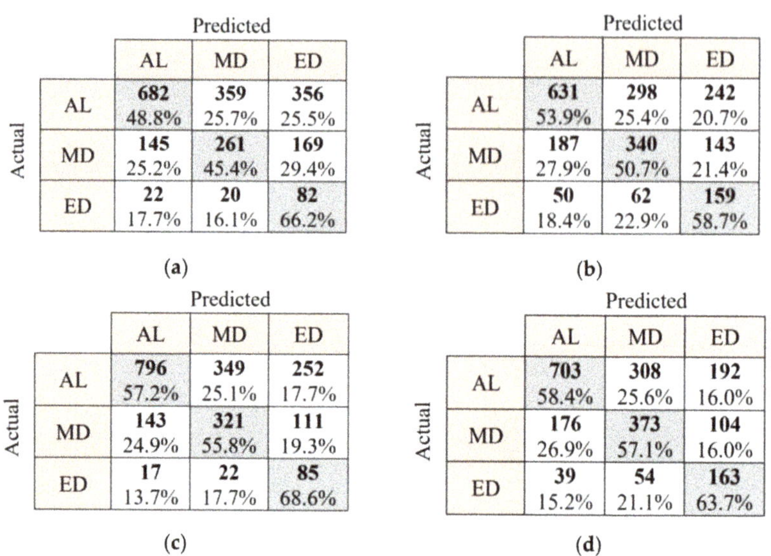

Figure 15 (a) — KNN classifier, manual tests

Actual \ Predicted	AL	MD	ED
AL	**682** 48.8%	359 25.7%	356 25.5%
MD	145 25.2%	**261** 45.4%	169 29.4%
ED	22 17.7%	20 16.1%	**82** 66.2%

(a)

Figure 15 (b) — KNN classifier, automated tests

Actual \ Predicted	AL	MD	ED
AL	**631** 53.9%	298 25.4%	242 20.7%
MD	187 27.9%	**340** 50.7%	143 21.4%
ED	50 18.4%	62 22.9%	**159** 58.7%

(b)

Figure 15 (c) — random forest, manual tests

Actual \ Predicted	AL	MD	ED
AL	**796** 57.2%	349 25.1%	252 17.7%
MD	143 24.9%	**321** 55.8%	111 19.3%
ED	17 13.7%	22 17.7%	**85** 68.6%

(c)

Figure 15 (d) — random forest, automated tests

Actual \ Predicted	AL	MD	ED
AL	**703** 58.4%	308 25.6%	192 16.0%
MD	176 26.9%	**373** 57.1%	104 16.0%
ED	39 15.2%	54 21.1%	**163** 63.7%

(d)

Figure 15. Confusion matrices of **KNN** classifier in the manual tests (**a**), **KNN** classifier in automated tests (**b**), **random forest** in the manual tests (**c**), and **random forest** in the automated tests (**d**) for driver drowsiness classification. The length of the sliding window for feature extraction is 40 s with a 20 s overlap. AL: alert, MD: moderately drowsy, and ED: extremely drowsy.

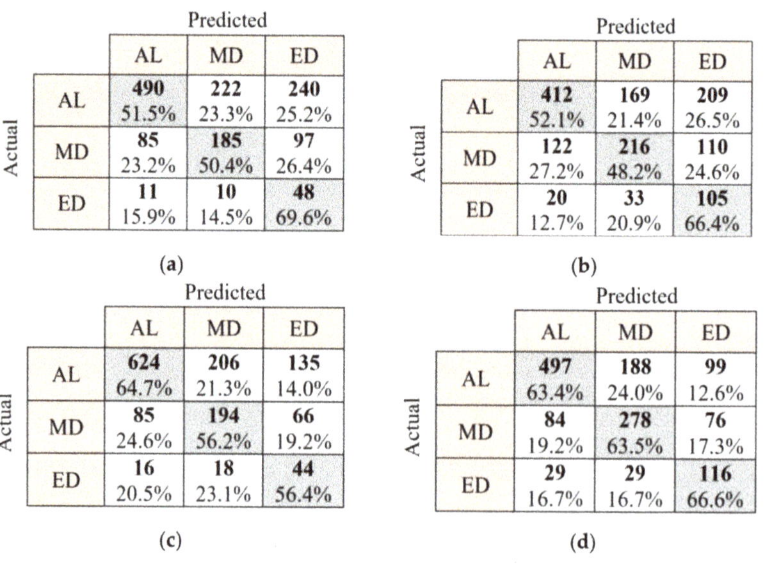

Figure 16 (a) — KNN classifier, manual tests

Actual \ Predicted	AL	MD	ED
AL	**490** 51.5%	222 23.3%	240 25.2%
MD	85 23.2%	**185** 50.4%	97 26.4%
ED	11 15.9%	10 14.5%	**48** 69.6%

(a)

Figure 16 (b) — KNN classifier, automated tests

Actual \ Predicted	AL	MD	ED
AL	**412** 52.1%	169 21.4%	209 26.5%
MD	122 27.2%	**216** 48.2%	110 24.6%
ED	20 12.7%	33 20.9%	**105** 66.4%

(b)

Figure 16 (c) — random forest, manual tests

Actual \ Predicted	AL	MD	ED
AL	**624** 64.7%	206 21.3%	135 14.0%
MD	85 24.6%	**194** 56.2%	66 19.2%
ED	16 20.5%	18 23.1%	**44** 56.4%

(c)

Figure 16 (d) — random forest, automated tests

Actual \ Predicted	AL	MD	ED
AL	**497** 63.4%	188 24.0%	99 12.6%
MD	84 19.2%	**278** 63.5%	76 17.3%
ED	29 16.7%	29 16.7%	**116** 66.6%

(d)

Figure 16. Confusion matrices of **KNN** classifier in the manual tests (**a**), **KNN** classifier in automated tests (**b**), **random forest** in the manual tests (**c**), and **random forest** in the automated tests (**d**) for driver drowsiness classification. The length of the sliding window for feature extraction is 60 s with a 30 s overlap. AL: alert, MD: moderately drowsy, and ED: extremely drowsy.

Table 6. Balanced accuracies of the KNN and random forest classifiers in the manual and automated driving modes with two different sliding windows. B.Acc.: balanced accuracy, 10/5: sliding window with the length of 10 s and 5 s overlap, 40/20: sliding window with the length of 40 s and 20 s overlap, and 60/30: sliding window with the length of 60 s and 30 s overlap.

Classifier	Driving Mode	Sliding Window	B.Acc. %
KNN	Automated	10/5	50.5%
KNN	Automated	40/20	53.5%
KNN	Automated	60/30	52.5%
Random Forest	Automated	10/5	53.9%
Random Forest	Automated	40/20	58.7%
Random Forest	**Automated**	**60/30**	**63.8%**
KNN	Manual	10/5	53.7%
KNN	Manual	40/20	48.9%
KNN	Manual	60/30	52.1%
Random Forest	Manual	10/5	55.1%
Random Forest	Manual	40/20	57.3%
Random Forest	**Manual**	**60/30**	**62.1%**

For the sake of brevity, classification metrics including specificity, sensitivity, precision, and F1-score are shown only for the best classifier (random forest trained by HRV-based features with a 60 s sliding window) in the manual and automated modes. Table 7 presents these metrics. As this table shows, the precision value for the ED class is low. This has occurred because the number of TP is low for this class. According to this table, the AL class has the maximum F1-score in both manual and automated modes. Therefore, the accuracy of the random forest for the AL class is higher than two other classes, and accordingly, the false alarm of this is reduced using this classifier.

Table 7. Classification metrics for the **random forest** classifier trained by HRV-based features in the **manual** and **automated** driving modes. Spe.: specificity; Sen.: sensitivity; Pre.: precision; and F1S: F1-score.

Mode	Class	Spe.	Sen.	Pre.	F1S.
Manual	AL	0.76	0.95	0.86	0.90
Manual	MD	0.78	0.56	0.46	0.51
Manual	ED	0.85	0.56	0.18	0.27
Automated	AL	0.81	0.63	0.81	0.71
Automated	MD	0.77	0.63	0.56	0.59
Automated	ED	0.86	0.67	0.40	0.50

4.2. Results of Driver Drowsiness Classification Using Scalogram of ECG Signals

As presented in Section 3.3.2, four different hyperparameters of deep CNN are considered to be optimized using the Bayesian optimizer. Table 8 shows the optimized hyperparameters of deep CNNs in the manual and automated driving modes. As this Table shows, the number of filters in the convolution layers and neurons in the fully connected layer (presented by the hyperparameter H_4) is higher in the automated driving mode. Therefore, the computational cost is higher in the automated tests to classify driver drowsiness using the proposed deep CNNs. The L2 regularization value (presented by the hyperparameter H_3) is much higher in manual tests than in automated tests. Thus, deep CNN needs larger parameters to classify the driver drowsiness in the manual tests. The dropout probability (presented by the hyperparameter H_2) of the trained deep CNN for the ECG signals in the automated tests is higher than the designed deep CNN for the manual tests. The number of neurons is also higher for the deep CNN trained by the ECG signals for automated driving. Therefore, its network is wider than another one. Consequently, the dropout probability of the network trained by the ECG signals of the automated tests should also be higher to turn off more neurons and avoid overfitting.

Table 8. Optimized values of hyperparameters in different driving modes and by inputting ECG scalogram to the deep CNNs. Hyperparameters H_1 to H_4 are defined in Table 4.

Driving Mode	H_1	H_2	H_3	H_4
Manual	0.0001	0.202	0.0015	42
Automated	0.0002	0.321	2.74×10^{-8}	60

In comparison with other widely used deep CNNs that are implemented in embedded systems for real-time face recognition or object detection, our developed deep CNNs have much fewer parameters. Table 9 compares the number of parameters of four different frequently used deep networks in real-time applications (AlexNet [60], VGG16 [61], ResNet18 [62], and GoogLeNet [63]) with our developed networks.

Table 9. Comparison of the approximate number (app. no.) of parameters in the developed deep CNN with other deep networks that were implemented in real-time applications in previous works. ECG-automated and ECG-manual in this table are deep networks and driving tests in automated and manual modes are used to train them, respectively.

Deep Network	App. No. Parameters (Million)
ECG-automated	2.2
ECG-manual	0.6
AlexNet	62
VGG16	138
ResNet18	11
GoogLeNet	6.5

Confusion matrices of the trained deep CNNs using ECG signals of the manual and automated tests are provided in Figure 17 to evaluate their classification performance. As this Figure shows, the MD class and AL class have the lowest and highest classification accuracy in both manual and automated driving modes, respectively. Therefore, reducing the number of classes from three to two can increase classification accuracy. However, it will not be possible to capture the transition between the AL to ED states in the case of binary classification. The balanced accuracy of the deep CNNs in both manual and automated modes is also provided in Table 10. These accuracies are calculated as the average TP accuracies in confusion matrices that are shown in grey elements in Figure 17a,b. Therefore, the balanced accuracy in manual mode is the average of 81.2%, 78.6%, and 79.1%. The balanced accuracy in the automated mode is also the average of 82.2%, 73.8%, and 82.0%. According to Table 10, the balanced accuracy of the deep CNN in the manual and automated driving modes are respectively about 77% and 79%. By comparing Table 10 with Table 6, deep CNNs significantly outperform the random forest and KNN methods in both manual and automated modes. Therefore, the input ECG scalograms are more informative than HRV-based features regarding driver drowsiness levels.

Classification metrics of the deep CNNs in the manual and automated modes are provided in Table 11. Comparing Table 11 with Table 7 shows that the F1-scores of all classes in both driving modes except the AL class in manual mode are improved by using the deep CNN method. According to this table, the precision value for the ED class is also lower than other classes since the numbers of the data samples of this class are much lower than the MD and AL classes.

<table>
<tr><td colspan="5" align="center">Predicted</td></tr>
<tr><td></td><td></td><td>AL</td><td>MD</td><td>ED</td></tr>
<tr><td rowspan="6">Actual</td><td rowspan="2">AL</td><td>**1927**</td><td>**385**</td><td>**60**</td></tr>
<tr><td>81.2%</td><td>16.2%</td><td>2.6%</td></tr>
<tr><td rowspan="2">MD</td><td>**111**</td><td>**736**</td><td>**90**</td></tr>
<tr><td>11.8%</td><td>78.6%</td><td>9.6%</td></tr>
<tr><td rowspan="2">ED</td><td>**7**</td><td>**37**</td><td>**167**</td></tr>
<tr><td>3.4%</td><td>17.5%</td><td>79.1%</td></tr>
</table>

(a)

<table>
<tr><td colspan="5" align="center">Predicted</td></tr>
<tr><td></td><td></td><td>AL</td><td>MD</td><td>ED</td></tr>
<tr><td rowspan="6">Actual</td><td rowspan="2">AL</td><td>**1604**</td><td>**279**</td><td>**68**</td></tr>
<tr><td>82.2%</td><td>14.3%</td><td>3.5%</td></tr>
<tr><td rowspan="2">MD</td><td>**143**</td><td>**790**</td><td>**137**</td></tr>
<tr><td>13.4%</td><td>73.8%</td><td>12.8%</td></tr>
<tr><td rowspan="2">ED</td><td>**13**</td><td>**67**</td><td>**363**</td></tr>
<tr><td>2.9%</td><td>15.1%</td><td>82.0%</td></tr>
</table>

(b)

Figure 17. Confusion matrices of **deep CNN** for driver drowsiness classification in the manual (**a**) and automated (**b**) driving tests.

Table 10. Balanced accuracy of the deep CNN in the manual and automated driving modes. B.Acc.: balanced accuracy.

Classifier	Driving Mode	B.Acc. %
Deep CNN	Manual	76.63
Deep CNN	Automated	79.33

Table 11. Classification metrics for the **deep CNN** trained by ECG scalogram images in the **manual** and **automated** driving tests. Spe.: specificity; Sen.: sensitivity; Pre.: precision; and F1S: F1-score.

Mode	Class	Spe.	Sen.	Pre.	F1S.
Manual	AL	0.90	0.81	0.94	0.87
Manual	MD	0.83	0.78	0.62	0.69
Manual	ED	0.95	0.79	0.53	0.63
Automated	AL	0.90	0.82	0.91	0.86
Automated	MD	0.85	0.74	0.69	0.71
Automated	ED	0.93	0.82	0.64	0.72

5. Conclusions

Two different methodologies were proposed in this paper to classify driver drowsiness using ECG signals. In the first methodology, R-peaks are firstly detected from ECG signals to obtain the HRV data. Then, eleven features are extracted from HRV data, and finally, random forest and KNN are used to classify drowsiness into three classes: alert, moderately drowsy, and extremely drowsy. In another method, a deep CNN was used to classify the drowsiness to the same classes when wavelet scalogram images of the ECG signals were inputs to this network. Results showed that the classification with deep CNN on ECG scalograms was more accurate than the random forest and KNN classifiers on HRV in both manual and automated driving modes. It is noteworthy that the length of ECG signals for the scalograms was only 10 s. For direct comparison, we also calculated HRV features on 10 s windows, though we are aware that this time frame captures fast, mostly respiratory, fluctuation only. Time frames from at least 1–2 min or even longer are necessary for a good agreement to usual short-term HRV measures [55,64]. We also computed longer time frames of 40 s and 60 s to verify the hypothesis that these longer windows capture more relevant information. Indeed, the classification accuracy of both KNN and RF classifiers increases with the duration of the time window used for HRV calculation.

In contrast, the deep CNN on ECG scalograms performs better already based on 10 s windows only. We conclude that the time–frequency content of the entire ECG signal captures information about the autonomous state of an individual beyond the RRI signal, which is the only information used for classical HRV parameters. Further research is suggested to understand which feature of an ECG exactly codes relevant information.

The following tasks are also suggested to improve the designed driver drowsiness classification system:

1. Applying a quality assessment method to the ECG signals in every sliding window can help to remove noisy data and increase classification accuracy. Moreover, a quality index can be derived for each sliding window to specify its influence on the reported drowsiness level in a predefined time interval (e.g., 1 min).
2. In this study, ECG signals are collected using attached electrodes to the driver's chest. Non-invasive sensors such as smart watches can also be used to design a non-disturbing system for drivers. However, the accuracy of these devices should be compared with accurate medical sensors, and differences in information gain according to the different characteristics of an ECG to an optical pulse signal need to be evaluated.
3. The proposed methods in this study developed generic driver drowsiness classification systems that consider no driver-specific differences. Only two hours of data is available for every driver which might not be sufficient to train a driver-specific deep network. To build a driver-specific system, transfer learning [65] can be employed. Using this method, the trained deep CNNs can be fine-tuned for a specific driver using a shuffled portion of their ECG data as the training set. Then, the fine-tuned deep CNN can be used to evaluate the drowsiness for the specific driver in the unseen test set or in real time. This approach can also reduce the amount of data needed from each driver to build a driver-specific system.
4. In this study, data from signal segments were treated as independent from each other by random selection and the sequential time information was ignored. Though this is presumably an advantage for the ability of a practical system to react fast, the transition from wakefulness to drowsiness might also be considered a continuous slower process. Therefore, it should be evaluated if outcomes of the deep network profit from the inclusion of sequential information of training segments.

Author Contributions: Conceptualization, S.A., A.E. and M.M.; methodology, S.A. and C.K.; validation, I.V.K. and A.E.; resources, M.F., C.K., M.M. and I.K.; data curation, M.F., A.E. and I.V.K.; writing—original draft preparation, S.A.; supervision, A.E.; project administration, A.E. and M.F.; funding acquisition, M.M., A.E. and M.F. All authors have read and agreed to the published version of the manuscript.

Funding: This project called WACHSens was carried out by the Human Research Institut für Gesundheitstechnologie und Präventionsforschung GmbH, Graz University of Technology, AVL Powertrain UK Limited, and Factum apptec ventures GmbH. It was funded by the Austrian Research Promotion Agency (FFG) via the Future Mobility Program (grant no. 860875).

Institutional Review Board Statement: The study was conducted in accordance with the Declaration of Helsinki, and approved by the Ethics Committee of the Medical University of Graz in vote 30-409 ex 17/18 dated 1 June 2018.

Informed Consent Statement: Written informed consent was obtained from participants before the experiments.

Data Availability Statement: Not applicable.

Acknowledgments: The authors are indebted to all drivers who participated in the experiment, the experimenters, and the many people who helped set up the tests.

Conflicts of Interest: The authors declare no conflict of interest.

References

1. Baulk, S.D.; Reyner, L.A.; Horne, J.A. Driver sleepiness—Evaluation of reaction time measurement as a secondary task. *Sleep* **2001**, *24*, 695–698. [CrossRef]
2. Arefnezhad, S.; Samiee, S.; Eichberger, A.; Frühwirth, M.; Kaufmann, C.; Klotz, E. Applying deep neural networks for multi-level classification of driver drowsiness using Vehicle-based measures. *Expert Syst. Appl.* **2020**, *162*, 113778. [CrossRef]

3. National Highway Traffic Safety Administration. Traffic Safety Facts: 2017 Fatal Motor Vehicle Crashes: Overview DOT HS 812 603, 1200 New Jersey Avenue SE., Washington, 2018. Available online: https://crashstats.nhtsa.dot.gov/Api/Public/ViewPublication/812603 (accessed on 14 April 2021).

4. Klauer, S.; Neale, V.; Dingus, T.; Sudweeks, J.; Ramsey, D.J. *The Prevalence of Driver Fatigue in an Urban Driving Environment Results from the 100-Car Naturalistic Driving Study*; Virginia Tech Transportation Institute: Blacksburg, VI, USA, 2005.

5. AAA Foundation for Traffic Safety. *2019 Traffic Safety Culture Index*; AAA Foundation for Traffic Safety: Washington, DC, USA, 2020.

6. Hirz, M.; Walzel, B. Sensor and object recognition technologies for self-driving cars. *Comput. Aided Des. Appl.* **2018**, *15*, 501–508. [CrossRef]

7. Inagaki, T.; Sheridan, T.B. A critique of the SAE conditional driving automation definition, and analyses of options for improvement. *Cogn Tech. Work.* **2019**, *21*, 569–578. [CrossRef]

8. Mou, L.; Zhou, C.; Xie, P.; Zhao, P.; Jain, R.C.; Gao, W.; Yin, B. Isotropic Self-supervised Learning for Driver Drowsiness Detection With Attention-based Multimodal Fusion. *IEEE Trans. Multimed.* **2021**, *1*. [CrossRef]

9. Zhu, M.; Chen, J.; Li, H.; Liang, F.; Han, L.; Zhang, Z. Vehicle driver drowsiness detection method using wearable EEG based on convolution neural network. *Neural Comput. Applic.* **2021**, *33*, 13965–13980. [CrossRef] [PubMed]

10. Dua, M.; Singla, R.; Raj, S.; Jangra, A. Deep CNN models-based ensemble approach to driver drowsiness detection. *Neural Comput. Applic.* **2021**, *33*, 3155–3168. [CrossRef]

11. Bakheet, S.; Al-Hamadi, A. A Framework for Instantaneous Driver Drowsiness Detection Based on Improved HOG Features and Naïve Bayesian Classification. *Brain Sci.* **2021**, *11*, 240. [CrossRef]

12. Jung, S.-J.; Shin, H.-S.; Chung, W.-Y. Driver fatigue and drowsiness monitoring system with embedded electrocardiogram sensor on steering wheel. *IET Intell. Transp. Syst.* **2014**, *8*, 43–50. [CrossRef]

13. Gromer, M.; Salb, D.; Walzer, T.; Madrid, N.M.; Seepold, R. ECG sensor for detection of driver's drowsiness. *Procedia Comput. Sci.* **2019**, *159*, 1938–1946. [CrossRef]

14. Rodríguez, R.; Mexicano, A.; Bila, J.; Cervantes, S.; Ponce, R. Feature Extraction of Electrocardiogram Signals by Applying Adaptive Threshold and Principal Component Analysis. *J. Appl. Res. Technol.* **2015**, *13*, 261–269. [CrossRef]

15. Thomas, J.; Rose, C.; Charpillet, F. A Multi-HMM Approach to ECG Segmentation. In Proceedings of the 18th IEEE International Conference on Tools with Artificial Intelligence (ICTAI'06), Arlington, VA, USA, 13 November 2006.

16. Kiran kumar, C.; Manaswini, M.; Maruthy, K.N.; Siva Kumar, A.V.; Mahesh kumar, K. Association of Heart rate variability measured by RR interval from ECG and pulse to pulse interval from Photoplethysmography. *Clin. Epidemiol. Glob. Health* **2021**, *10*, 100698. [CrossRef]

17. Huang, S.; Li, J.; Zhang, P.; Zhang, W. Detection of mental fatigue state with wearable ECG devices. *Int. J. Med. Inform.* **2018**, *119*, 39–46. [CrossRef] [PubMed]

18. Moser, M.; Frühwirth, M.; Penter, R.; Winker, R. Why life oscillates—From a topographical towards a functional chronobiology. *Cancer Causes Control* **2006**, *17*, 591–599. [CrossRef]

19. Moser, M.; Frühwirth, M.; Kenner, T. The symphony of life. Importance, interaction, and visualization of biological rhythms. *IEEE Eng. Med. Biol. Mag.* **2008**, *27*, 29–37. [CrossRef]

20. Fujiwara, K.; Abe, E.; Kamata, K.; Nakayama, C.; Suzuki, Y.; Yamakawa, T.; Hiraoka, T.; Kano, M.; Sumi, Y.; Masuda, F.; et al. Heart Rate Variability-Based Driver Drowsiness Detection and Its Validation With EEG. *IEEE Trans. Biomed. Eng.* **2019**, *66*, 1769–1778. [CrossRef]

21. Vicente, J.; Laguna, P.; Bartra, A.; Bailón, R. Drowsiness detection using heart rate variability. *Med. Biol. Eng. Comput.* **2016**, *54*, 927–937. [CrossRef] [PubMed]

22. Buendia, R.; Forcolin, F.; Karlsson, J.; Sjöqvist, B.A.; Anund, A.; Candefjord, S. Deriving heart rate variability indices from cardiac monitoring—An indicator of driver sleepiness. *Traffic Inj. Prev.* **2019**, *20*, 249–254. [CrossRef]

23. Patel, M.; Lal, S.K.L.; Kavanagh, D.; Rossiter, P. Applying neural network analysis on heart rate variability data to assess driver fatigue. *Expert Syst. Appl.* **2011**, *38*, 7235–7242. [CrossRef]

24. Li, G.; Chung, W.-Y. Detection of driver drowsiness using wavelet analysis of heart rate variability and a support vector machine classifier. *Sensors* **2013**, *13*, 16494–16511. [CrossRef] [PubMed]

25. Furman, G.D.; Baharav, A.; Cahan, C.; Akselrod, S. Early detection of falling asleep at the wheel: A Heart Rate Variability approach. *Comput. Cardiol.* **2008**, *35*, 1109–1112. [CrossRef]

26. SENSODRIVE GmbH. SENSODRIVE GmbH—Robotic und Force-Feedback. Available online: https://www.sensodrive.de/products/force-feedback-steering-wheels.php (accessed on 14 June 2021).

27. AVL. AVL VSM™ Vehicle Simulation; Release 2020 R1: Highlights of the Latest Release of Our Vehicle Dynamics Simulation Tool. Available online: https://www.avl.com/-/avl-vsm-vehicle-simulation (accessed on 14 June 2021).

28. Schinko, C.; Peer, M.; Hammer, D.; Pirstinger, M.; Lex, C.; Koglbauer, I.; Eichberger, A.; Holzinger, J.; Eggeling, E.; Fellner, D.W.; et al. Building a Driving Simulator with Parallax Barrier Displays. In Proceedings of the 11th Joint Conference on Computer Vision, Imaging and Computer Graphics Theory and Applications, Rome, Italy, 27–29 February 2016; pp. 281–289.

29. Lex, C.; Hammer, D.; Pirstinger, M.; Peer, M.; Samiee, S.; Schinko, C.; Ullrich, T.; Battel, M.; Holzinger, J.; Koglbauer, I.; et al. *Multidisciplinary Development of a Driving Simulator with Autostereoscopic Visualization for the Integrated Development of Driver Assistance Systems*; Graz University of Technology: Graz, Austria, 2015.

30. Kaufmann, C.; Frühwirth, M.; Messerschmidt, D.; Moser, M.; Eichberger, A.; Arefnezhad, S. Driving and tiredness: Results of the behaviour observation of a simulator study with special focus on automated driving. *ToTS* **2020**, *11*, 51–63. [CrossRef]
31. Gacek, A.C.; Pedrycz, W. *ECG Signal Processing, Classification and Interpretation: A Comprehensive Framework of Computational Intelligence*; Springer: London, UK, 2012.
32. Gupta, P.; Sharma, K.K.; Joshi, S.D. Baseline wander removal of electrocardiogram signals using multivariate empirical mode decomposition. *Healthc. Technol. Lett.* **2015**, *2*, 164–166. [CrossRef] [PubMed]
33. Wang, C.; Xiao, W. Second-Order IIR Notch Filter Design and Implementation of Digital Signal Processing System. *Appl. Mech. Mater.* **2013**, *347–350*, 729–732. [CrossRef]
34. Bolós, V.J.; Benítez, R. The Wavelet Scalogram in the Study of Time Series. In *Advances in Differential Equations and Applications*; Casas, F., Martínez, V., Eds.; Springer: New York, NY, USA, 2014; pp. 147–154.
35. Lilly, J.M.; Olhede, S.C. Higher-Order Properties of Analytic Wavelets. *IEEE Trans. Signal Process.* **2009**, *57*, 146–160. [CrossRef]
36. Mario, M.-O. Human Activity Recognition Based on Single Sensor Square HV Acceleration Images and Convolutional Neural Networks. *IEEE Sens. J.* **2019**, *19*, 1487–1498. [CrossRef]
37. Park, S.; Jeong, Y.; Kim, H.S. Multiresolution CNN for Reverberant Speech Recognition. In Proceedings of the 2017 Conference of the Oriental Chapter of International Committee for Coordination and Standardization of Speech Databases and Assessment Technique (O-COCOSDA), Seoul, Korea, 1–3 November 2017; pp. 1–4.
38. Li, Q.; Cai, W.; Wang, X.; Zhou, Y.; Feng, D.D.; Chen, M. Medical Image Classification with Convolutional Neural Network. In Proceedings of the 13th International Conference on Control, Automation, Robotics & Vision (ICARCV), Singapore, 10–12 December 2014; pp. 844–848.
39. Zhao, Z.; Zhou, N.; Zhang, L.; Yan, H.; Xu, Y.; Zhang, Z. Driver Fatigue Detection Based on Convolutional Neural Networks Using EM-CNN. *Comput. Intell. Neurosci.* **2020**, *2020*, 7251280. [CrossRef] [PubMed]
40. LeCun, Y.; Bengio, Y.; Hinton, G. Deep learning. *Nature* **2015**, *521*, 436–444. [CrossRef] [PubMed]
41. Ilievski, I.; Akhtar, T.; Feng, J.; Shoemaker, C. Efficient Hyperparameter Optimization for Deep Learning Algorithms Using Deterministic RBF Surrogates. In Proceedings of the Thirty-First AAAI Conference on Artificial Intelligence, San Francisco, CA, USA, 4–9 February 2017.
42. Shankar, K.; Zhang, Y.; Liu, Y.; Wu, L.; Chen, C.-H. Hyperparameter Tuning Deep Learning for Diabetic Retinopathy Fundus Image Classification. *IEEE Access* **2020**, *8*, 118164–118173. [CrossRef]
43. Neary, P. Automatic Hyperparameter Tuning in Deep Convolutional Neural Networks Using Asynchronous Reinforcement Learning. In Proceedings of the 2018 IEEE International Conference on Cognitive Computing (ICCC), San Francisco, CA, USA, 2–7 July 2018.
44. Peter, I. Frazier. Bayesian Optimization of Risk Measures. *arXiv* **2007**, arXiv:2007.05554v3.
45. Kawaguchi, K.; Kaelbling, L.P.; Lozano-Pérez, T. Bayesian Optimization with Exponential Convergence. *arXiv* **2018**, arXiv:1811.09558.
46. Snoek, J.; Larochelle, H.; Adams, R.P. Practical Bayesian Optimization of Machine Learning Algorithms. *arXiv* **2012**, arXiv:1206.2944.
47. Scholkmann, F.; Boss, J.; Wolf, M. An Efficient Algorithm for Automatic Peak Detection in Noisy Periodic and Quasi-Periodic Signals. *Algorithms* **2012**, *5*, 588–603. [CrossRef]
48. Task Force of the European Society of Cardiology the North American Society of Pacing Electrophysiology. Heart Rate Variability. *Circulation* **1996**, *93*, 1043–1065. [CrossRef]
49. Hoshi, R.A.; Pastre, C.M.; Vanderlei, L.C.M.; Godoy, M.F. Poincaré plot indexes of heart rate variability: Relationships with other nonlinear variables. *Auton. Neurosci.* **2013**, *177*, 271–274. [CrossRef]
50. Fitzgibbon, A.W.; Pilu, M.; Fisher, R.B. Direct Least Squares Fitting of Ellipses. In Proceedings of the 13th International Conference On Pattern Recognition, Vienna, Austria, 29 August 1996.
51. Morelli, D.; Rossi, A.; Cairo, M.; Clifton, D.A. Analysis of the Impact of Interpolation Methods of Missing RR-intervals Caused by Motion Artifacts on HRV Features Estimations. *Sensors* **2019**, *19*, 3163. [CrossRef]
52. Shaffer, F.; Ginsberg, J.P. An Overview of Heart Rate Variability Metrics and Norms. *Front. Public Health* **2017**, *5*, 258. [CrossRef] [PubMed]
53. Clifford, G.D.; Tarassenko, L. Quantifying errors in spectral estimates of HRV due to beat replacement and resampling. *IEEE Trans. Biomed. Eng.* **2005**, *52*, 630–638. [CrossRef]
54. Kim, K.K.; Kim, J.S.; Lim, Y.G.; Park, K.S. The effect of missing RR-interval data on heart rate variability analysis in the frequency domain. *Physiol. Meas.* **2009**, *30*, 1039–1050. [CrossRef]
55. Munoz, M.L.; van Roon, A.; Riese, H.; Thio, C.; Oostenbroek, E.; Westrik, I.; de Geus, E.J.; Gansevoort, R.; Lefrandt, J.; Nolte, I.M.; et al. Validity of (Ultra-)Short Recordings for Heart Rate Variability Measurements. *PLoS ONE* **2015**, *10*, e0138921. [CrossRef] [PubMed]
56. Ghawi, R.; Pfeffer, J. Efficient Hyperparameter Tuning with Grid Search for Text Categorization using kNN Approach with BM25 Similarity. *Open Comput. Sci.* **2019**, *9*, 160–180. [CrossRef]
57. Probst, P.; Wright, M.N.; Boulesteix, A.-L. Hyperparameters and Tuning Strategies for Random Forest. *WIREs Data Mining Knowl Discov.* **2019**, *9*, e1301. [CrossRef]

58. López, V.; Fernández, A.; García, S.; Palade, V.; Herrera, F. An insight into classification with imbalanced data: Empirical results and current trends on using data intrinsic characteristics. *Inf. Sci.* **2013**, *250*, 113–141. [CrossRef]
59. Seiffert, C.; Khoshgoftaar, T.M.; van Hulse, J.; Napolitano, A. RUSBoost: Improving Classification Performance When Training Data Is Skewed. In Proceedings of the 19th International Conference on Pattern Recognition (ICPR 2008), Tampa, FL, USA, 8–11 December 2008; pp. 1–4.
60. Krizhevsky, A.; Sutskever, I.; Hinton, G.E. ImageNet Classification with Deep Convolutional Neural Networks. In *Advances in Neural Information Processing Systems*; Pereira, F., Burges, C.J.C., Bottou, L., Weinberger, K.Q., Eds.; Curran Associates Inc.: Red Hook, NY, USA, 2012.
61. Simonyan, K.; Zisserman, A. Very Deep Convolutional Networks for Large-Scale Image Recognition. *arXiv* **2014**, arXiv:1409.1556.
62. He, K.; Zhang, X.; Ren, S.; Sun, J. Deep Residual Learning for Image Recognition. *arXiv* **2015**, arXiv:1512.03385.
63. Szegedy, C.; Liu, W.; Jia, Y.; Sermanet, P.; Reed, S.; Anguelov, D.; Erhan, D.; Vanhoucke, V.; Rabinovich, A. Going Deeper with Convolutions. *arXiv* **2014**, arXiv:1409.4842.
64. Melo, H.M.; Martins, T.C.; Nascimento, L.M.; Hoeller, A.A.; Walz, R.; Takase, E. Ultra-short heart rate variability recording reliability: The effect of controlled paced breathing. *Ann. Noninvasive Electrocardiol.* **2018**, *23*, e12565. [CrossRef] [PubMed]
65. Zhuang, F.; Qi, Z.; Duan, K.; Xi, D.; Zhu, Y.; Zhu, H.; Xiong, H.; He, Q. A Comprehensive Survey on Transfer Learning. *arXiv* **2020**, arXiv:1911.02685v3. [CrossRef]

Article

Enhancing Acceptance and Trust in Automated Driving trough Virtual Experience on a Driving Simulator

Philipp Clement [1,2,*,†], Omar Veledar [2,†], Clemens Könczöl [3,†] , Herbert Danzinger [2], Markus Posch [2], Arno Eichberger [1] and Georg Macher [4]

1 Faculty of Mechanical Engineering and Economic Sciences, Institute of Automotive Engineering, Graz University of Technology, 8010 Graz, Austria; arno.eichberger@tugraz.at
2 AVL List GmbH, Hans-List-Platz 1, 8020 Graz, Austria; omar.veledar@avl.com (O.V.); herbert.danzinger@avl.com (H.D.); markus.posch@avl.com (M.P.)
3 Faculty of Natural Sciences, Institute of Psychology, University of Graz, 8010 Graz, Austria; clemens.koenczoel@uni-graz.at
4 Faculty of Electrical and Information Engineering, Institute of Technical Informatics, Graz University of Technology, 8010 Graz, Austria; georg.macher@tugraz.at
* Correspondence: clement@student.tugraz.at or philipp.clement@avl.com; Tel.: +43-699-1987-0820
† These authors contributed equally to this work.

Citation: Clement, P.; Veledar, O.; Könczöl, C.; Danzinger, H.; Posch, M.; Eichberger, A.; Macher, G. Enhancing Acceptance and Trust in Automated Driving trough Virtual Experience on a Driving Simulator. *Energies* 2022, 15, 781. https://doi.org/10.3390/en15030781

Academic Editor: Wiseman Yair

Received: 1 December 2021
Accepted: 13 January 2022
Published: 21 January 2022

Publisher's Note: MDPI stays neutral with regard to jurisdictional claims in published maps and institutional affiliations.

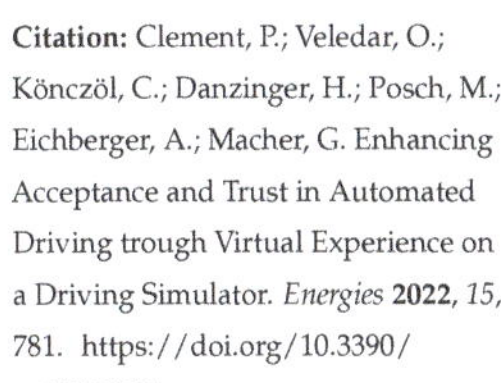

Abstract: As vehicle driving evolves from human-controlled to autonomous, human–machine interaction ensures intuitive usage as well as the feedback from vehicle occupants to the machine for optimising controls. The feedback also improves understanding of the user satisfaction with the system behaviour, which is crucial for determining user trust and, hence, the acceptance of the new functionalities that aim to improve mobility solutions and increase road safety. Trust and acceptance are potentially the crucial parameters for determining the success of autonomous driving deployment in wider society. Hence, there is a need to define appropriate and measurable parameters to be able to quantify trust and acceptance in a physically safe environment using dependable methods. This study seeks to support technical developments and data gathering with psychology to determine the degree to which humans trust automated driving functionalities. The primary aim is to define if the usage of an advanced driving simulator can improve consumer trust and acceptance of driving automation through tailor-made studies. We also seek to measure significant differences in responses from different demographic groups. The study employs tailor-made driving scenarios to gather feedback on trust, usability and user workload of 55 participants monitoring the vehicle behaviour and environment during the automated drive. Participants' subjective ratings are gathered before and after the simulator session. Results show a significant increase in trust ensuing the exposure to the driving automation functionalities. We quantify this increase resulting from the usage of the driving simulator. Those less experienced with driving automation show a higher increase in trust and, therefore, profit more from the exercise. This appears to be linked to the demanded participant workload, as we establish a link between workload and trust. The findings provide a noteworthy contribution to quantifying the method of evaluating and ensuring user acceptance of driving automation. It is only through the increase of trust and consequent improvement of user acceptance that the introduction of the driving automation into wider society will be a guaranteed success.

Keywords: automated driving (AD); driving simulator; expression of trust; acceptance; simulator case study; NASA TLX; advanced driver assistant systems (ADAS); system usability scale; driving school

1. Introduction

Advanced driver assistant systems (ADAS) and autonomous driving (AD) are stepwise turning into reality. According to the Society of Automotive Engineers (SAE) this process is defined in five levels. In Figure 1 the levels are depict from conventional manual driving—no driving automation (SAE Level 0)—through conditional automated driving

(SAE Level 3) with the fallback-ready user to the full driving automation (SAE Level 5), which covers all situations everywhere [1].

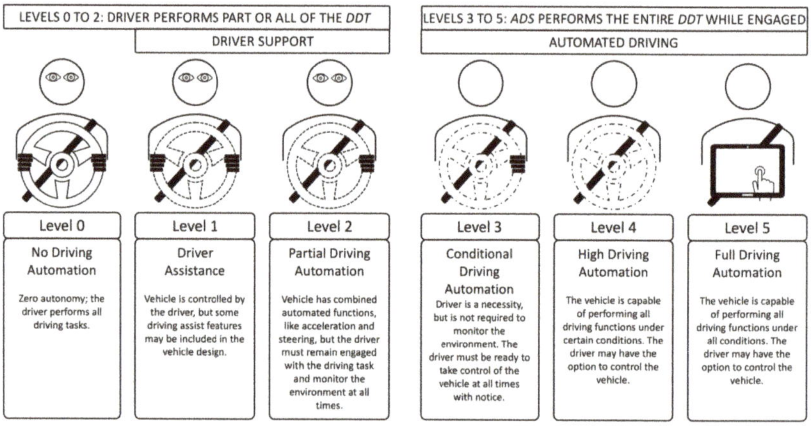

Figure 1. Society of Automotive Engineers (SAE) levels for driving automation following On-Road Automated Driving (ORAD) committee, SAE International [1] with Dynamic Driving Task (DDT).

In the intermediate steps, the human–machine interaction becomes an indispensable factor for the technological progress in the development [2,3], as it acts as a technological enabler that improves performance, safety, trustworthiness and comfort [4]. This is reflected by many studies concerning the interaction of humans with vehicle controls [5–7]. As there is a significant probability of harming the study participants, many studies are shifted from the real driving experience on the road into the virtual driving simulator environment, with particular attention on the selection of the appropriate simulator fidelity and validity [8]. This also turns out to be valid for future driver training programs with a potential expected to work even in countries with an already pretty low accident rate [9,10]. Moreover, simulators provide the opportunity to enhance training for safety-critical systems and situations [11–13]. The ongoing discussion of differences between real driving and the simulation has been overcome by the need for repeatability of the tests and the safety of the participants.

The future challenge on automated driving is that most of the time the task for the driver is to observe the environment and the vehicle performance without being actively involved. Then, occasionally in already near-critical situations the vehicle requests the driver to perform those critical situations immediately. These two contrary tasks challenge our society with the little knowledge about the technology and its interaction with human beings. Despite the potential benefits and influence of virtual training on the assessment of learning effects and trust in automation, there is a lack of sound research data on the effects of virtual training on those new systems. Besides study questionnaires, current research is ongoing to measure trust in automation in real-time scenarios, as in Azevedo Sá et al. [14].

1.1. The Technology Challenges

Multiple industrial sectors, including transportation and mobility, are experiencing a significant technological transformation. Considering the safety-critical nature of the technologies that drive this transformation, the automotive safety community has considerable interest in the development. However, the established safety engineering methods are somewhat failing to support these changes entirely. Thus, extensive effort is invested in developing safety engineering methods, safety design approaches and safety development processes that support this technological transformation [15].

AD is generally seen as one of the critical trends in the mobility sector and also one of the key components of the mentioned transformation [16]. To fulfil the anticipated AD evolution, there is a demand for progress in supporting technologies (e.g., sensors and actuators), particularly in artificial intelligence (AI) (at the edge) based approaches [17] and further to a combined driver evaluation out of non-obstrusive monitoring sensors and the AI technology [2]. AI technologies are increasingly used for critical applications because these approaches exceed the current state of the art (e.g., recognising patterns and inferring relationships), creating a high demand for AI in terms of realising automated and autonomous driving functions [18]. The outcomes are expected to be technically viable and compelling for humans, hence enticing user acceptance of AD. A lot of effort is put into completing the challenges related to trustworthy AI solutions to cater for human stakeholders [19]. Trust and safety are of utmost importance as to gain acceptance of AD functions.

The key challenge in this context is that the established safety engineering processes and practices (e.g., described by ISO 26262 [20] and ISO/PAS 21448 [21] for the automotive domain) have been only successfully applied in conventional model-based system development. In many industry contexts, none of the available safety standards have defined processes that explicitly consider the specifics of AI-based approaches. These factors include the requirements on dataset collections, the definition of performance evaluation metrics, the AI architectures, and the handling of uncertainty [18].

The ultimate goal of safety-critical systems is to maximise the evidence of a positive risk balance. Conventional safety engineering processes apply model-based system development to establish structured and human-understandable arguments of the risks inherent to the system under development. The risk-driven safety engineering processes implement a system function with the required service quality (i.e., safety integrity level). For AI-based systems, it is challenging to provide such service quality metrics. For example, AI-based concepts build upon probabilistic modelling, including random variables and probability distributions, to model situations and events. As such, an AI function model returns a probability distribution as output for specific input.

In general, AI-based approaches depend on the data used to parameterise its function and the process of parameterising (called training or learning). The quality of the used dataset and the choice of the AI architecture directly influence the quality of the function. In contrast, human-programmed model-based functions typically return a specific result and not a probability distribution to a particular input. In the context of this work, this AI-related topic also reappears in relation to the establishment of trust and acceptance of automated driving in general, and the selection of virtual experiences in particular. It contributes to the establishment of trust and acceptance of such systems, with the expectation to open the gate for implementation of new AI functionality in future.

1.2. Trust and Acceptance

Trust in automation is a key determinant for the adoption of automated systems and their appropriate use [22] and along with other factors, has a significant impact on interest in using an automated vehicle [23]. Studies suggest a strong connection between trust in technical solutions and user acceptance on the same level [24–26]. The key challenges in terms of acceptance of fully or partially autonomous vehicles are balanced between the needed interaction with the technology, the associated benefits and the hidden risks. In particular, it is the level of required driver's awareness in combination with the human trust towards automated decisions in all conditions that are associated with acceptance [24]. For example, the need for continuous monitoring limits user acceptance as it diminishes the benefit of securing the driver with more freedom for side tasks. The additional open questions that could limit user acceptance are presented by the need for the drivers to take over vehicle control at emergency scenarios or extreme conditions or when the driving automation fails [27], unless there is a dependable fail-operational strategy in place to take over the vehicle control [28]. Most of the user acceptance studies analyse the effects of cognitive factors, but with a limited impact from social aspects [25], since the decision-making process is heavily influenced by opinions that are most important to the people making these decisions [29] and their hedonistic motivations [30]. While that is empirically demonstrated [31], it is also expected that acceptance of the driving automation would be closely related to social influence but with a possibly different strength across different cultures, e.g., some cultures are likely to exert stronger on social influence than other ones [24]. Hence, there is a need for wider conclusive studies based on empirical data across demographic groups.

While acceptance is an important condition for the successful implementation of automated driving vehicles [32], recent research considers the relation between attitudes towards automated systems and their actual use [33]. Reduced acceptance level is also correlated to a lack of safety [34]. Furthermore, it is important to consider that currently AD has not yet been experienced by the majority of drivers [35]. Driving simulators are used to conquer the acceptance obstacle by determining the trust in AD functions and improvements of trust through experience while eliminating the need for complete vehicle prototypes and securing entire scenario repeatability [36]. It is the reproducible validation and demonstration of mature automated functions, their reliability, and safety that work towards securing societal acceptance [37]. Besides ensuring the durability of the technical solutions, increased user acceptance of the new automotive technologies crucially determines the sustainability of the consequent business implementation [38]. Sustainability is further supported through the trustworthiness of the solutions, which acts as an enabler for the development and implementation of an appropriate value creation strategy for maximisation of benefits amongst the engaging stakeholders [38].

In addition to focusing on the AD functions, the driving simulation potentially educates newly qualified drivers on unexpected critical situations on the road, which are generally unpredictable and not part of the driving license education. The potential of such training is evaluated by pre- and post-testing questionnaires for a broad spectrum of users, with attention to the low yearly driving experience group. There is a potential to increase acceptance and trust levels and situational experience through virtual training.

The consequent research questions are: can an advanced simulator (Section 2.4) experience improve trust in and the acceptance of automated driving, and are there any significant differences resulting from the driving simulations for different demographic groups (i.e., gender, age, experienced vs. inexperienced drivers, ADAS/AD experience)?

To answer these research questions we employ a unique combination of a state-of-the-art dynamic simulator in its characteristic measurement setup, the relevant scenarios targeting appropriate psychological assessment and a sizable study sample.

2. Materials and Methods

Simulators have the potential to raise the AD profile amongst the general population. To examine the impact of such exposure on a safe version of AD, a study comprising ten driving scenarios was designed to gather feedback on tailored questions from a defined sample group.

The expression of trust (EOT) questionnaire is the primary measurement tool in the present study to evaluate participant's subjective trust. The expression of trust is a modified version of the Trust into Automation questionnaire [39]. Each participant faces the questionnaire twice within the study to enable an initial estimate of trust and explore the change of trust in an AD system throughout the designed experiment. The questionnaire is used before the test as a baseline for each participant. Upon the test procedure, the participants face the same questionnaire again. This allows to measure the impact of such a simulator experience on the trust in the AD system.

The system usability scale (SUS) [40] is used to verify the suitability and usability of the simulator and the simulation within the testing environment, and the behaviour of the system itself. Furthermore, it indicates whether or not such a testing method is suitable for such a study.

A raw NASA TLX questionnaire [41,42] is used to research the workload of the participants. As the study does not request the participants to drive or react and control the vehicle on their own, the NASA TLX reflects the workload of the participants monitoring the vehicle behaviour and environment during the automated drive. This effort is expected to be low due to the eliminated need for active driving tasks from the participants.

2.1. Sample

A total of 60 out of 112 registered participants participated in the test procedure. The selection focused on obtaining a well-distributed study sample. Out of those 60 participants, 55 completed the test successfully. The results from 7 participants were removed due to incomplete datasets for the questionnaires. This yielded 48 complete datasets for the analysis of the questionnaires of expression of trust, raw NASA TLX, and the system usability scale.

The sample is grouped by the participants' demographic factors in the defined categories, such as age, gender, yearly driving range, driver assistance experience, and educational level. The distribution is as follows:

- gender: 27 male, 21 female, 0 neutral;
- age group: 21 (18–29 years), 19 (30–45 years), 8 (46–65 years);
- mileage km/year: 8 (<5000 km/yr), 12 (5001–10,000 km/yr), 28 (>10,001 km/yr);
- ADAS experience: 22 (without pre-experience), 26 (with pre-experience);
- education: 13 below Bachelor level, 35 above Bachelor level—according to ISCED [43].

2.2. Study Materials

Three different questionnaires were used to evaluate trust, system usability, and workload to obtain subjective ratings of these relevant aspects. They gather the participants' feedback on their global perception of trust in automation and the use of AD systems. The questionnaires are the EOT, the raw NASA TLX, and the SUS. The questions are presented using a tablet computer allowing the subjects to provide their rating by screen tapping. The ratings may be corrected up to the point when the participant confirms the questionnaire as complete. The answer is recorded in a decimal format. To support the participants to put in their intended answer, the graphical appearance of the answer input area is supported by a color code, smileys, and written explanation. An example of the visualization is shown in Figure 2.

Q1 How comfortable did you feel?

Figure 2. Example of a question presented to the participants [44].

The EOT is a modified version of Helldin et al. [45] and relies on the original questionnaire on trust in automation from Jian et al. [39]. The questions assess user trust in AD functionality on a seven-point Likert scale, spreading from do not agree at all to agree completely. The questionnaire is presented before the participants get into the cockpit and start the test execution, and after they leave the cockpit. This double testing allows measuring changes in the participants' trust subject to their experience in the test environment. The following questions are used in this questionnaire:

Q1 I understand how the automated driving system works—its goals, actions, and results.
Q2 I would use the automated driving system if they were available in my own car.
Q3 I think the actions of the automated driving system will have a positive impact on my own driving.
Q4 I put my faith in the system.
Q5 I think the automated driving system will ensure safety while driving.
Q6 I think the automated driving system is reliable.
Q7 I can trust the automated driving system.

Each question reflects upon a specific part of the human trust in automation. In the present study, all single answers of a participant are summarised, and a mean value is built according to Equation (1) to reflect an overall change of the response, from before to after the experiment, where $Q0_p$ is the new built overall score per participant and A_{p_i} is the single response to one of the questions one to seven of one participant.

$$Q0_p = \frac{1}{n} \sum_{i=1}^{n} A_{p_i} \tag{1}$$

The mean value is treated as a new dependent variable and processed together with the demographic interactions in the analysis of variances and correlations.

After the test execution, the usability of the system is assessed by the SUS [46]. The ten-question questionnaire is adapted to fit the boundaries of the study. It also adopts the advice from Grier et al. [47]. The questions used are:

Q1 I think I would like to use this assistance system frequently.
Q2 I thought the assistance system was unnecessarily complex.
Q3 I thought the assistance system was easy to use.
Q4 I think I would need the assistance of a technical person to use this system.
Q5 My impression was that the different functions in this assistance system were well integrated.
Q6 I felt there was too much inconsistency in this assistance system.
Q7 I can imagine that most people would learn to use this assistance system very quickly.
Q8 I found that the assistance system was very cumbersome to use.
Q9 I found the use of the assistance system to be convincing.
Q10 I had to learn a lot of things before I could start using this assistance system.

The possible answers range from absolutely disagree to absolutely agree. The numerical values behind the answers reach from 0 to 4, with four representing the highest value the participant's answer can be. Therefore, for questions 1, 3, 5, 7, and 9, the values range from 0 to 4 and for the negatively formulated questions 2, 4, 6, 8, and 10, the values range from 4 to 0 and their polarity is reversed prior to the statistical analysis. For the ten questions, the maximum summed up result is 40 points. The summed value is further scaled by a factor of 2.5 to obtain a scale from 0 to 100 for each participant. A mean value for the system is then generated out of all participants' answers [48,49].

The third post-testing questionnaire is the NASA TLX [50], a standardised questionnaire on participants' perceived workload. For the actual purpose, the questions are presented as single questions without the weighting of the question pairs, also called "raw TLX". This was chosen as an appropriate cut-off in terms of the participant's timing. Therefore, an overall workload calculation is excluded in the data analysis procedure as the official rules can not be followed [50].

Q1 Mental demand. How mentally demanding was the task?
Q2 Physical demand. How physically demanding was the task?
Q3 Temporal demand. How hurried or rushed was the pace of the task?
Q4 Performance. How successful were you in accomplishing what was required of you?
Q5 Effort. How hard did you have to work to achieve your level of performance?
Q6 Frustration. How insecure, discouraged, irritable, stressed, and angry were you?

The answer options to those questions range from very low to very high with exception of question 4, which offers an answer scale from perfect to failed on a seven-point scale.

2.3. Experimental Procedure

Each participant followed a standardised experimental procedure also described in Clement et al. [44]. Figure 3 offers a top-level visualization of this procedure, starting with the introduction phase and continuing on to the post-testing phase. These two phases are the measurement phases for investigating the change of attitude due to the participant's exposure to AD in a moving driving simulator.

In phase one, the "introduction phase", the participants are introduced to the testing procedure, informed about the relevant data protection rules (according to the GDPR [51]) and the appropriate measurements. This is done by a psychologist to ensure consent and to provide a trained professional to deal with unforeseen circumstances. Upon the initial talk, the participants' expression of trust in AD and their prior experience is evaluated by a questionnaire. Participants have the ability to withdraw from the experiment in any phase.

In phase two, the "get prepared phase", the participants are equipped with the necessary sensors and connected with the database for recording. They are seated in the simulator for further instructions, which is the first time they see it. The participants do not see or meet each other to avoid any influence in this regard. They get informed about the detailed testing procedure and the equipment used for it, especially the simulator vehicle cockpit they are seated in and which they need to handle. Especially the interaction with the human–machine interface (HMI) is explained, as the information about disabled functionality is displayed in the drivers dashboard and interaction with the presented Scenario Specific Questionnaire on Trust (SSQT) questions is done on the central infotainment.

In phase three, the "get used to it phase", the participants are able to drive in the virtual environment on their own, five minutes without the moving hexa-pot platform and then five minutes with the activated movement platform. Test questionnaires of the SSQT are presented in the central infotainment dashboard to avoid uncertainties within the experiment.

In phase four, the "test execution phase", the actual exposure to the high automated driving scenarios takes place. The participants are driving trough the scenarios with the task to observe the behaviour of the vehicle and its reactions to the environment and further answer the SSQT to provide the required feedback. All ten scenarios and their consecutive SSQTs are processed automatically one after another.

In phase five, the "get cleared phase", the participants leave the cockpit, once all scenarios are completed successfully and all sensors are removed, so they are able to recover before they move on to phase six.

In phase six, the "post-testing phase", the same questionnaire as in phase one is presented to assess the participants' expression of trust in the AD after their simulator experience to evaluate the differences. Additionally, the NASA TLX [50] questionnaire is presented to the participants for workload measurements without a pairwise weighting. Furthermore, the SUS [40] is used to evaluate the subjective usability of the testing system.

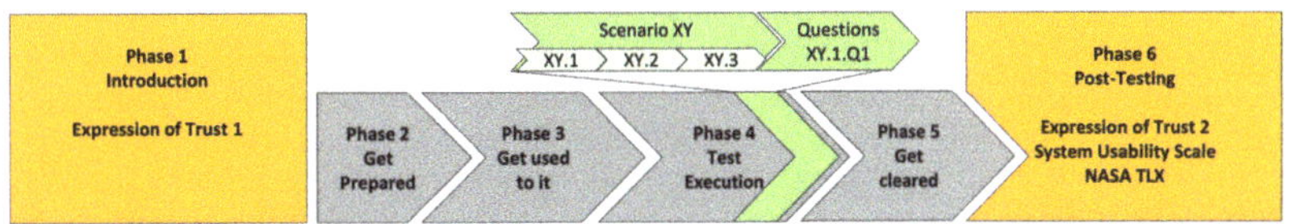

Figure 3. Experiment procedure for each individual participant. (XY represents the number of the scenario, XY.1 the first scene within the scenario XY, and XY.1.Q1 represents question one targeting the specific scene within the scenario XY).

2.4. Equipment and Techniques

For the virtual environment, an actual vehicle cockpit, resected of a real vehicle and modified with additional HMI displays and buttons, is mounted on a moving hexa-pod platform [52]. A 180° canvas displays the simulated scenery of a real road with three video projectors (4 K 100 Hz) for prevention against motion sickness. The platform is capable of performing movements in six degrees of freedom, three lateral directions with accelerating up to 6 m/s^2 in any direction and up to 1.5 m of movement, and rotating around the three axis as a result. The virtual environment is provided by Vires VTD [53], whereas the platform movement is calculated by a vehicle simulation model [54] for better accuracy. All components of the environment are synchronised by the co-simulation framework

Model.Connect [55]. The simulator is capable of performing all movements of normal driving conditions up to the aforementioned 6 m/s^2 accelerations [44].

The environment is regulated from a control (and preparation) room and is supported by a data server for data management. The automated data collection focuses on the participant's state, responses to the questionnaires, and the biometric data. The biometric sensor setup can be modified in accordance with the study design and the expected output. This study makes use of time-of-flight cameras, a chest belt for heart-rate measurements, a wristband for assessing skin conductivity and body temperature, and an eye-tracking device. The setup is characterised by the combination of the subjective psychological data assessment and the interactive subjective questionnaires.

2.5. Scenarios

Phase 4 of the testing procedure, shown in Figure 3, contains ten consecutive scenarios which are executed and standardised for each participant. The scenarios are designed to give the subject the impression of and experience with future AD. All scenarios contain potentially safety critical situations, which may influence participants' subjective perceived safety and therefore affect trust in automated driving systems. Even though participants do not get harmed in the simulated situations, they can neither predict the system's behaviour nor the final outcome. The simulated safety critical scenarios represent situations that may happen in real world driving with an estimated impact on trust and an influence on participants' subjective ratings. Each scenario is followed by an SSQT, supported by a reminding video of the action to be evaluated. The scenarios are scheduled automatically from the test automation service once the SSQT is completed. The ten scenarios are divided into the following 4 clusters (scenarios with similar use case [56]), which are also shown in Figure 4.

Simulated safety critical situations:

C1   Take-over manoeuvres with function decrease (2 scenarios);
C2   Emergency brake with bad weather conditions (2 scenarios);
C3   Narrow road with a child crossing (3 scenarios);
C4   Highway with construction site ahead (3 scenarios).

Each scenario is conducted in high driving automation mode (SAE level 4—no need for driver intervention, but still possible) [1], see also Figure 1, while the participants were tasked with observing the vehicle behaviour and the environment and providing their subjective feedback. The scenario clusters are designed in the same environment [56]. Their AD parameters differ so that two driving modes are available: sporty or comfortable. Compared to the comfortable configuration, the sporty configuration allows a shorter time to collision and higher de-/acceleration rates as well as smaller distances to follow or to stop. Beyond the description of the scenarios, they are published for all details in [57].

Cluster one describes a constant drive with 100 km/h approaching a vehicle with 70 km/h on a straight one-way two-lane road in an urban area. The driver is informed about the high driving automation mode in the HMI. As all parameters like speed difference, other traffic and range of vision for an automated take-over are met, the vehicle controller performs an automated take-over of the other vehicle. After the take-over, the ego vehicle automatically changes back to the initial lane and continues to drive with 100 km/h. Next, it is passed by another vehicle which splashes dirt on the ego vehicle's sensor setup. The driver gets informed about a decrease in take-over functionality. The next vehicle approached by the ego vehicle is driving with 70 km/h. As the sensor setup is not recovered yet, no automated take-over can be performed and the vehicle follows the slower vehicle in the lane in front of the ego vehicle. After a predefined time (10 s) of sensor cleaning, the take-over functionality is available again and displayed to the driver. As all mandatory conditions are met, another automated take-over is performed in the same manner as the first one. After changing back to the initial lane, the drive continues with 100 km/h and the scenario ends. This scenario is performed twice, once with each controller configuration.

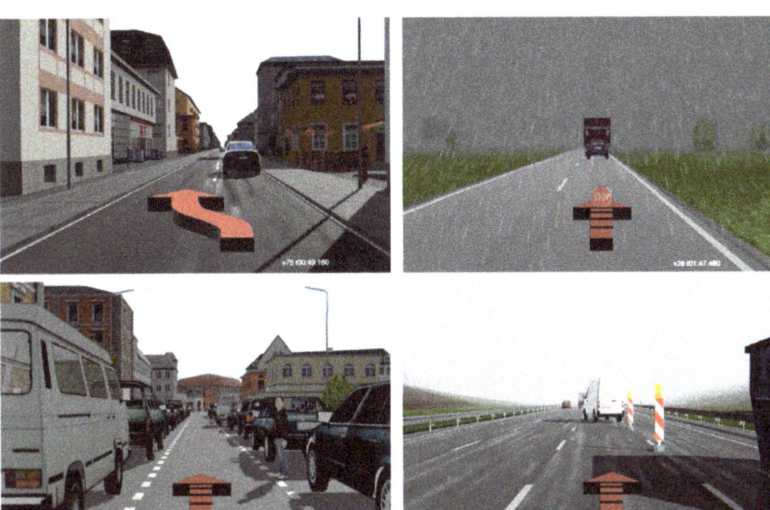

Figure 4. Pictures of critical events of the four scenario clusters: (**top left**) C1—at the beginning of the second take-over; (**top right**) C2—appearance of the stopped truck in the fog; (**bottom left**) C3—appearance of the child between the parked vehicles; (**bottom right**) C4—cut-in of the delivery van from the construction site with nearly zero speed.

Cluster two is described by the ego vehicle starting off and following another vehicle driving with 50 km/h in high driving automation mode on a straight two-way road. The speed remains constant and appropriate to the road conditions. After a defined time of 40 seconds following the vehicle ahead, the weather conditions deteriorate, as the appearing fog limits visibility. Due to the reduced visibility of the sensors, the vehicle's speed is automatically reduced by the automated driving controller to 30 km/h and the driver is informed about the issue via the HMI. The vehicle in front drives faster and disappears in the fog. As the drive continues, the vehicle ahead suddenly appears in the fog and seems to be stopping behind a truck that has already stopped in the fog. The automated driving system handles the emergency brake situation and automatically stops behind the vehicle in a suitable distance. Due to a higher speed, the *sporty* controller setting creates a more critical situation and a shorter stopping distance. No accident occurs. This scenario is performed twice, once with each controller configuration.

The three scenarios of cluster three are characterised by a high driving automation low speed drive in an urban residential area with parked vehicles. The road is straight. The ego vehicle approaches a narrow lane with vehicles parked on both sides of the one-way road with 30 km/h. The ego vehicle automatically changes lanes to the free middle lane and the speed is reduced further to 15 km/h due to reduced sensor range between the parked vehicles. After some driving time between the parked vehicles, suddenly a child appears between two parked vehicles on the right and crosses the road just ahead of the ego vehicle. The automated driving controller reacts to the situation and nearly stops <5 km/h the ego vehicle in front of the child. After the situation is cleared and the child leaves the scene, the controller automatically continues driving. This scenario is performed three times and the stopping distance and the deceleration values differ between the scenarios depending on controller configurations (1 & 3 comfortable and 2 sporty).

Cluster four implements three evolving scenarios. The consecutive scenario always extends the previous one by adding lousy weather in the second repetition and a delivery van cutting into the driving line in the third repetition, while the controller for the high driving automation is always set on comfortable mode. All scenarios are driven in high driving automation mode. The basic/first scenario is defined by driving on a three-lane motorway on the first lane with 100 km/h, which is then closed due to a construction site ahead with a speed limitation of 80 km/h. The automated driving controller automatically changes from the first to the second lane as all conditions for an automated lane change are met. The speed is reduced as required by the traffic signs and the construction site is passed successfully. After the construction site the speed is increased back to 100 km/h and the scenario ends. In the second scenario, the first scenario is repeated under poor weather conditions and reduces the visibility. The rest of the scenario remains unchanged, while in the third scenario, the weather conditions from the second scenario remain the same and an additional critical situation is provoked by a delivery van leaving the construction site right in front of the ego vehicle. The automated driving controller performs an emergency brake and reduces the speed from 80 km/h to 18 km/h to avoid a collision. The distance to the delivery van is reduced down to 2 m and creates a critical situation at higher speeds. The delivery van accelerates slowly as is typical for the vehicle type. After the construction site, the delivery van changes to the first lane. The ego vehicle lane is cleared and the automated driving controller accelerates to the initial speed of 100 km/h and the scenario ends.

3. Results

The results focus on the participants' subjective ratings on the questionnaires, measuring system usability, trust and workload. The results are divided into subchapters for each questionnaire. The subjective ratings are analysed using multifactorial analysis of variance (MANOVA) with repeated measurements with a significance level of $p_{unc} \leq 0.05$ of the null-hypothesis. The p-value index indicates the specific analysis. So p_{int} stands for the interaction of the relevant groups and the factor time, p_t for the factor time, and p_{gr} for the group factors. Moreover, Pearson correlations were calculated to examine the correlations between the different factors. The method used to process the gathered results is already made available in Clement et al. [44].

3.1. Expression of Trust

The evaluation of the expression of trust uses the mean value (Q0) to analyse the participants' trust within the sample throughout the experiment.

A tendency towards a significant interaction can be seen in combination with ADAS pre-experience ($F = 2.308$, $p_{int} = 0.136$, $\eta^2 = 0.048$). The group with ADAS pre-experience shows a lower increase in the overall response than the group without ADAS pre-experience. This can be seen in Figure 5 on the very left side at the Q0 values. The ADAS pre-experience group's results show that the ratings before and after the experiment are at a similar level but slightly increase for the post-test results. For the factor time, the rating changes highly significant from 4.58 to 4.98 point ($F = 10.160$, $p_t = 0.003$, $\eta^2 = 0.181$). For the factor ADAS pre-experience, a tendency towards a significant difference can be seen ($F = 2.940$, $p_t = 0.093$, $\eta^2 = 0.060$).

The mean values resulting from the Expression of Trust (EOT) questionnaire before and after the experiment are shown in Figure 5, including their standard deviation. The results are split into two key groups, with and without ADAS pre-experience. The figure shows this for the mean value (Q0) and for the questions Q1 to Q7. The details are in Table 1. The single questions show various noticeable and significant differences. In Q1 "I understand how the system works—its goals, actions and output", the interaction ($F = 10.619$, $p_{int} = 0.002$, $\eta^2 = 0.188$) and the factor ADAS pre-experience itself ($F = 6.462$, $p_{gr} = 0.014$, $\eta^2 = 0.123$) show a significant difference. Further, in Q3, the interaction reveals a significant difference between the two groups of the factor ($F = 6.384$, $p_{int} = 0.015$, $\eta^2 = 0.122$), as indicated in the empirical analysis above. In Q5, the factor pre-experience itself reveals two nearly significant different levels of the answers ($F = 3.962$, $p_{gr} = 0.052$, $\eta^2 = 0.079$), also indicated by the different empirical levels of the group answers. Ratings of participants without ADAS pre-experience are higher for all answers in the post-test questionnaire compared to the pre-test questionnaire. The ratings provided by the participants with ADAS pre-experience are less consistent with a higher variance before and after the experiment.

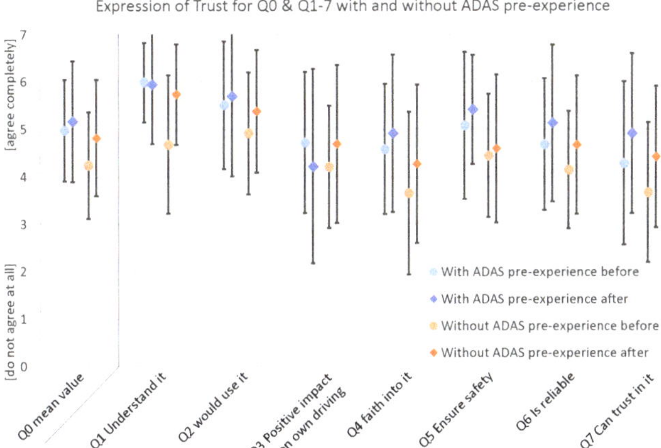

Figure 5. Expression of trust for question 0 (Q0—mean value) and detailed question 1–7 (Q1–7) with and without ADAS pre-experience, before and after the test.

Table 1. Mean values (m), standard deviation (sd) and ANOVA-results for ADAS pre-experience x Time.

| | with ADAS Pre-Experience | | | | without ADAS Pre-Experience | | | | ANOVA | |
	(m) before (sd)		(m) after (sd)		(m) before (sd)		(m) after (sd)		ADAS ($F/p_{gr}/\eta^2$)	Interaction ($F/p_{int}/\eta^2$)
Q0	4.98	(1.07)	5.17	(1.27)	4.24	(1.13)	4.82	(1.23)	*(2.940 / 0.093 / 0.060)*	*(2.308 / 0.135 / 0.048)*
Q1	5.99	(0.84)	5.94	(1.25)	4.68	(1.46)	5.74	(1.06)	*(6.462 / 0.014 / 0.123)*	*(10.619 / 0.002 / 0.188)*
Q2	5.51	(1.35)	5.70	(1.69)	4.92	(1.28)	5.38	(1.29)	*(4.535 / 0.217 / 0.033)*	*(0.593 / 0.445 / 0.013)*
Q3	4.72	(1.49)	4.22	(2.05)	4.21	(1.29)	4.69	(1.66)	*(0.001 / 0.973 / 0.000)*	*(6.384 / 0.015 / 0.122)*
Q4	4.58	(1.37)	4.91	(1.66)	3.65	(1.72)	4.27	(1.67)	*(3.382 / 0.072 / 0.068)*	*(0.548 / 0.463 / 0.012)*
Q5	5.08	(1.55)	5.41	(1.15)	4.44	(1.30)	4.59	(1.56)	*(3.962 / 0.052 / 0.079)*	*(0.260 / 0.612 / 0.006)*
Q6	4.68	(1.39)	5.13	(1.66)	4.14	(1.24)	4.67	(1.46)	*(1.749 / 0.192 / 0.037)*	*(0.068 / 0.795 / 0.001)*
Q7	4.28	(1.72)	4.91	(1.69)	3.67	(1.48)	4.42	(1.49)	*(1.885 / 0.176 / 0.039)*	*(0.077 / 0.783 / 0.002)*

Note: (F ... F-value, p_{gr} ... p-value for Group ADAS pre-experience, p_{int} ... p-value for interaction, η^2 ... effect size) of the expression of trust for Q0 to Q7 for the factors ADAS pre-experience (with/without) and time of measurement (before/after).

The factor gender is illustrated in Figure 6 regarding the mean value (Q0) and the single questions Q1 to Q7. Results show a tendency towards the same significant difference for time of measurement (before/after) for both groups, male and female. However, although they reflect the overall change of trust through the simulator session, they do not show a significant interaction regarding differences in Q0 ($F = 1.733$, $p_{int} = 0.195$, $\eta^2 = 0.036$). The global picture shows that the female group tended to rate the system lower than the male group before the experiment. After the experiment, the increase of the female group's rating was higher than the one of the male group but still did not surpass the level of the male group.

In Table 2, results for the analysis of variance for the expression of trust are given for the interaction between the factor time and the factor gender as well as the main effect for the factor gender. The main effect regarding the factor time is the same as mentioned in the analysis of ADAS pre-experience and not outlined separately. Significant differences can be seen for the interaction in Q4 ($F = 5.230$, $p_{int} = 0.026$, $\eta^2 = 0.102$) and for the factor gender in Q1 ($F = 5.510$, $p_{gr} = 0.023$, $\eta^2 = 0.107$).

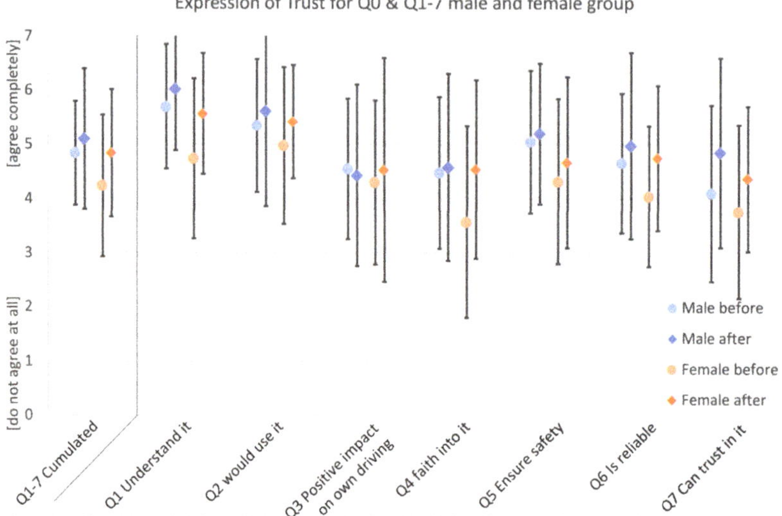

Figure 6. Expression of trust for question 0 (Q0—mean value) and detailed question 1–7 (Q1–7) for male and female groups, before and after the test.

Table 2. Mean values (m), standard deviation (sd) and ANOVA-results for Gender x Time.

	Male				Female				ANOVA	
	(m) before	(sd)	(m) after	(sd)	(m) before	(sd)	(m) after	(sd)	Gender ($F/p_{gr}/\eta^2$)	Interaction ($F/p_{int}/\eta^2$)
Q0	4.84	(0.96)	5.10	(1.30)	4.24	(1.31)	4.84	(1.18)	(1.801 / 0.186 / 0.038)	(1.733 / 0.194 / 0.036)
Q1	5.70	(1.15)	6.03	(1.14)	4.74	(1.48)	5.57	(1.12)	(5.510 / 0.023 / 0.107)	(1.836 / 0.182 / 0.038)
Q2	5.35	(1.23)	5.62	(1.76)	4.98	(1.45)	5.42	(1.05)	(0.597 / 0.443 / 0.013)	(0.223 / 0.639 / 0.005)
Q3	4.55	(1.30)	4.43	(1.68)	4.30	(1.52)	4.53	(2.07)	(0.032 / 0.859 / 0.001)	(0.701 / 0.406 / 0.015)
Q4	4.48	(1.40)	4.58	(1.73)	3.57	(1.77)	4.54	(1.65)	(1.193 / 0.280 / 0.025)	(5.230 / 0.026 / 0.102)
Q5	5.05	(1.32)	5.20	(1.30)	4.32	(1.53)	4.67	(1.58)	(2.867 / 0.097 / 0.059)	(0.344 / 0.560 / 0.007)
Q6	4.66	(1.29)	4.98	(1.72)	4.04	(1.30)	4.75	(1.34)	(1.256 / 0.268 / 0.027)	(1.377 / 0.246 / 0.029)
Q7	4.10	(1.63)	4.85	(1.75)	3.76	(1.60)	4.37	(1.34)	(1.001 / 0.322 / 0.021)	(0.089 / 0.767 / 0.002)

Note: (F ... F-value, p_{gr} ... p-value for group gender, p_{int} ... p-value for interaction, η^2 ... effect size) of the expression of trust for Q0 to Q7 for the factors gender (m/f) and time of measurement (before/after).

Analysis for the remaining demographic subgroups on the mean value (Q0) show no significant differences for the factor age ($F = 0.020$, $p_{gr} = 0.980$, $\eta^2 = 0.001$) and for the interaction with the factor time ($F = 0.183$, $p_{int} = 0.834$, $\eta^2 = 0.008$). For the factor driving experience, the values show no significant differences ($F = 1.075$, $p_{gr} = 0.350$, $\eta^2 = 0.046$), same as for the interaction with the factor time ($F = 1.045$, $p_{int} = 0.360$, $\eta^2 = 0.044$).

3.2. NASA TLX

The NASA TLX was performed as a raw test without the weighting of the questions in pairs. The evaluation of the questionnaire reveals that the factor age group ($F = 3.481$, $p_{gr} = 0.039$) had a significant influence on participants' ratings of Q5. The group of elderly participants and those with fewer kilometers per year report more effort to accomplish their performance. Moreover, the factor driving experience ($F = 4.278$, $p_{gr} = 0.019$) had a significant influence on participants' ratings of Q6. These factors as well as the factor ADAS are depicted in Figure 7. There were no significant interactions reported in the two-way analysis of variance as shown in Table 3. The differences are noteworthy for Q5. For the group with less than 5000 km/year, a descriptive analysis shows lower success for the questions Q4, Q5, and Q6.

Table 3. Results of the two-way ANOVA analysis for the NASA TLX.

	Q1	Q2	Q3	Q4	Q5	Q6
Unifactorial Analysis					Age Group *	Driving Exp *
(A) Age group vs. (B) Driving Exp.	(A vs. B)	(A vs. B)	(A * vs. B)	(A ** vs. B)	(A ** vs. B)	(A ** vs. B)
(A) ADAS Exp. vs. (B) Driving Exp.	(A vs. B)	(A vs. B)	(A vs. B)	(A vs. B)	(A vs. B)	(A vs. B *)
(A) Age Group vs. (B) ADAS Exp.	(A* vs. B)	(A vs. B)	(A vs. B)	(A vs. B)	(A vs. B)	(A vs. B)
(A) Gender vs. (B) ADAS Exp.	(A vs. B)	(A vs. B)	(A vs. B)	(A vs. B)	(A vs. B)	(A vs. B)
(A) Gender vs. (B) Age Group	(A vs. B)	(A vs. B)	(A vs. B)	(A vs. B)	(A vs. B *)	(A vs. B)
(A) Gender vs. (B) Driving Exp.	(A vs. B)	(A vs. B)	(A vs. B)	(A vs. B)	(A vs. B)	(A vs. B *)

Note: ** indicates $p \leq 0.01$, and * indicates $p \leq 0.05$ X. * indicates a significant difference for the factor, and (X vs. X)* indicates a significant interaction.

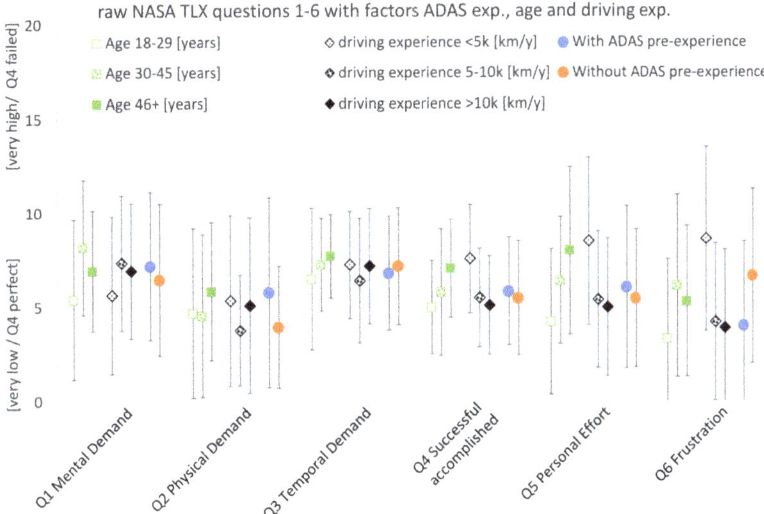

Figure 7. NASA TLX for questions 1–6 (Q1–6) for the groups with and without ADAS pre-experience, the groups of driving experience per year, and the age groups.

3.3. System Usability Scale

The system usability evaluated for each demographic sample group shows results around the overall value of 77.49 out of 100. The value is generated from the sum of all single questions for each subject multiplied by 2.5 to correspond to the target maximum scale of 100 according to Rauer [49]. Figure 8 shows the results for each demographic subgroup within the participants. The groups are all in a similar range and distribution.

A two-way analysis of variance revealed only one significant interaction of the age group with the driving experience ($F = 4.347$, $p_{gr} = 0.005$). This may be explained by a small number of participants in the age group of 46+ years, which results in a biased p-value sensitive to a distribution based on small sample size, even though according to Bangor et al. [46], the system used reflects an overall score in the 3rd quartile, which is acceptable and in terms of rating between good and excellent for the whole sample.

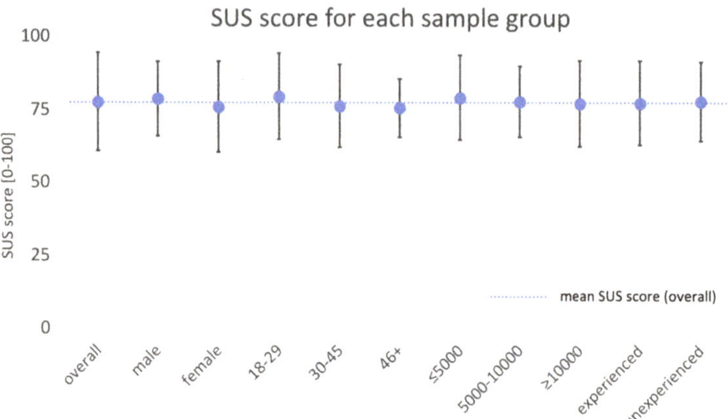

Figure 8. SUS score for the demographic sample groups.

The results for single questions are shown in Figure 9. Since the questions are positively and negatively polarised, the negative ones were inverted prior to the analysis. Results suggest an easy-to-use and quickly adjustable system with no unnecessary complexities. Its usage is not cumbersome and it demands no assistance to be used. The results also reveal that the system has some inconsistency (2.67 out of 4) and is not fully convincing (2.64 out of 4) to the overall sample, but still far better than the 50% quantile.

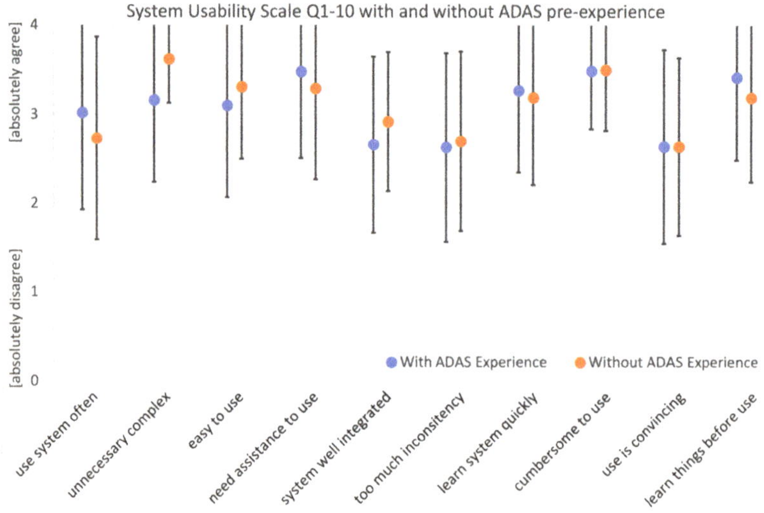

Figure 9. SUS scores for the single questions for the groups with and without ADAS pre-experience, inverted for negative questions.

3.4. Correlations between Workload, Trust and Usability

To identify the correlation between the participants' ratings, Pearson correlations between all questionnaires and single items were calculated. In Table 4, an overview of the major calculations is given using the mean values (Q0) for the SUS and the EOT, as they have higher reliability due to being built on the sub-questions of the related questionnaires and therefore get more information into the variance. The values show the correlation r and their significance.

Table 4. Pearson correlations between system usability scale (SUS), expression of trust (EOT) and NASA TLX items.

		1	2	3	4	5	6	7	8	9
1	**SUS_Q0**	-	0.37 *	0.63 ***	−0.47 **	−0.05	−0.40 **	−0.40 **	−0.17	−0.46 ***
2	**EOT_BEFORE_Q0**		-	0.75 ***	−0.14	0.21	−0.07	−0.23	−0.21	−0.31*
3	**EOT_AFTER_Q0**			-	−0.29	0.03	−0.16	−0.31 *	−0.06	−0.37 *
4	**NASA_Q1 (mental demand)**				-	0.48 **	0.24	0.22	0.41 **	0.28
5	**NASA_Q2 (physical demand)**					-	0.06	0.03	0.16	0.07
6	**NASA_Q3 (temporal demand)**						-	0.14	0.22	0.40 **
7	**NASA_Q4 (performance)**							-	0.38 **	0.42 **
8	**NASA_Q5 (effort)**								-	0.35 **
9	**NASA_Q6 (frustration)**									-

Note: *** indicates $p \leq 0.001$, ** indicates $p \leq 0.01$, and * indicates $p \leq 0.05$.

The results reveal the following significant findings. SUS (Q0) and EOT before (Q0) were found to be highly significant and moderately positively correlated ($r = 0.37$, $p = 0.01$). SUS (Q0) and EOT after (Q0) were found to be highly significant and highly positively correlated ($r = 0.63$, $p = 0.00$). EOT before (Q0) and EOT after (Q0) were found to be highly significant and highly positively correlated ($r = 0.75$, $p = 0.00$). Scatter plots of these correlations can be found in Figure 10. These positive correlations imply that participants who rated the system usability low also rated the expression of trust (before and after) lower, and those who rated the system usability higher also have a higher expression of trust. The correlation between SUS and EOT after is higher than SUS and EOT before. Moreover, significant correlations between several NASA TLX single items and the SUS (Q0), the EOT (Q0) before, and the EOT (Q0) after were found as illustrated in Table 4.

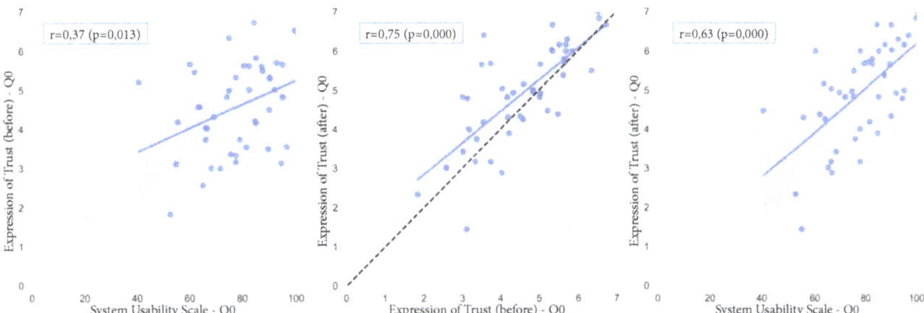

Figure 10. Pearson correlations for the combination of the accumulated questions Q0 with best fitted line. (r ... Pearson correlation coefficient, p ... significance of the correlation).

Results that reveal significant correlations between NASA TLX and SUS as well as NASA TLX and EOT after, they are consistently negatively correlated. This means the higher participants rate the workload, the lower they rate the usability and trust. These results reflect a high internal validity of the experiment. Furthermore, these correlations suggest a high validity of the applied questionnaires, as results point to a consistent and contextual evaluation within the measurement tools.

To check for spurious correlations and improve understanding of the correlations, especially regarding correlations' linearity and data outliers, the correlations are visually analysed using a scatter plot and the correlation line. The major correlations are depicted in Figure 10. The scatter plots do not show a tendency for a non-linearity and are well scattered around the correlation line. A visual comparison of the expression of trust (Q0) before and after the experiment reveals a general increase of trust as there are more data points above the 45° dotted line. A point on the dotted line would mean an equal rating in Q0 before and after the experiment and indicates no change throughout the experiment. A data-point below the dotted line indicates a decrease of trust, and a data-point above it means an increase of trust. This approach supports the findings of the analysis of variance reported in Section 3.1. Apart from the inter-questionnaire correlations, the highest correlation and significance levels relate to the system usability scale with questions SUS Q1 (would use the system frequently) and SUS Q9 (found the system to be convincing) in correlation to the post-testing measurement of the expression of trust with questions Q2, Q4, Q6, and Q7.

4. Discussion

We quantified the increase of trust in the "AD system under test" by exposing participants to a driving simulator and evaluating their subjective perception before and after that experiment session. As expressed by the mean value (Q0) of the Expression of Trust questionnaire, the increase of trust was significant for all groups between the two points of measurement. This study shows an appropriate way to increase and evaluate trust in a simulator study.

Participants with ADAS pre-experience entered the study with higher confidence in such systems compared to the group without ADAS pre-experience, as depicted in Figure 5. Despite the high starting confidence level, the simulator session increases their trust in the AD system, indicating a high level of validity between the driving simulation and the real-world experience. There is a significant increase in trust for both groups. The group without ADAS pre-experience shows a much higher growth than the already pre-experienced. Besides the single questions Q1 and Q3 of the EOT, no significant difference can be seen between the groups. Q1 hints that the pre-experienced group could not gain further understanding, whereas the inexperienced group could significantly increase their knowledge of such a system. We found that pre-experienced participants might see a negative impact on their driving style the more they learn about the possibilities of automated driving systems and their driving behaviour in safety-critical situations, as a decrease in the Q3 was found. In contrast, the group without ADAS pre-experience sees a positive impact on their driving style, as they express a willingness to use the system in the future.

The analysis of demographic differences and similarities between genders shows the same tendency towards gaining trust (Q0) in the system through the simulator experience. The subgroups express a similar behaviour, besides Q1, with only one significant difference in the interaction of the groups and the time of measurement, which reflects the faith in the system shown in Figure 6. Female participants are more sceptical about their faith in the system before the simulator session compared to the male group. The simulator session affected their attitude towards the applied automated driving systems, i.e., their trust increased, reaching a similar level as the male participants. This effect can also be observed as a tendency in all other questions, even though there is no significant difference. This trend can also be seen within a descriptive analysis. It may reflect that the female group was more sceptical before the experiment but reached a level similar to one of the male participants after the experiment.

The NASA TLX analysis reveals an overall expected low workload of the participants. As the questionnaire is provided after the complete test sequence, it reflects the workload of the entire simulator session and does not differentiate between the single scenarios. There are significant negative correlations between the NASA TLX and both the system usability scale and the expression of trust. Since both questionnaires measure similar aspects with different purposes, the effect can be assumed as valid. The negative correlations between the trust in the system and the workload throughout the simulator session are significant, meaning that a higher workload correlates with less trust within the participants. The participants' ratings were below the 50% line for the overall workload evaluation, which suggests a low workload within the simulator session. This was expected, as full AD minimised the physical and mental demands. Nevertheless, some subgroups showed more strain within the experiment. As shown in Figure 7, the groups of higher ages and lower yearly mileage had a lower success rate, a higher personal effort and a higher frustration after the simulator session. Hence, there are only tendentially significant differences between the groups.

The participants' ratings within the System Usability Scale (SUS) show a usability that can be seen as good to excellent. As already discussed in Bangor et al. [46] and Brooke [48], this method is suitable for evaluating the overall system usability. We hypothesise that the participant knowledge of the early developmental state of the system may have influenced their evaluations. With a mean of around 77.5 points out of 100, the response indicates a very good system although it also implies that following a normal distribution, around 50% of the participants rated the system lower than that and therefore only good instead of very good. The assessment of the mean and standard deviation of all demographic subgroups reveals that they are all on a similar level regarding their usability evaluation. This points to a neutral, not too complicated and realistic experimental design, which may be seen as an essential prerequisite for reliable results.

Regarding the Pearson correlation with the combination of both the pre- and post-questionnaires for the expression of trust, one can derive that smaller trust in the initial level results in a higher increase of trust after the simulator session. This is expressed in the middle of Figure 10 by the fitted line being above the 45° line on the left side while touching the 45° line at the top end, which may also reflect a ceiling effect as the scale is limited to 0–7 and may not be extended in the post-session questionnaire. An increase from a higher level may not be measured that accurately. A comparison of the pre- and post-correlations with the system usability shows a more homogeneous rating in the post-evaluation. The correlation with the pre-questionnaire is smaller ($r = 0.37$, $p = 0.013$) reflecting that the system's behaviour is unknown prior to the simulator session. Nevertheless, the lower part of the system usability scale shows a slight decrease in the expression of trust. In contrast, the upper part did increase noticeable in a clockwise turn of the fitted line in the left and right part of Figure 10.

Both correlations suggest different clusters, one below and one above the fitted line of Figure 10. An analysis of the correlations with a separation into the different demographic groups reveals noticeable findings which can be seen in Figure 11. It shows that the group without ADAS pre-experience seems to be located closer to the cluster below the fitted line than the group with ADAS pre-experience. It also reveals that the trust of the group with a low annual driving experience increases its mean ratings in trust throughout the simulator session and has the highest intern group delta according to their system usability rating.

The findings of the study respond to the demands of standardisation in a human-centric approach to manage handover and takeover between the vehicle and the human for SAE level 3 automation [1]. Besides the required compliance with the existing standards, the monitoring of the human state during the interaction with the AD functions also provides precious feedback on the machine's performance which can be used for improving the intelligent machine itself [19] and provides the foundations for further human-centred development with the flair of a humanistic AI control. Considering that ensuring dependability of such systems relies on AI approaches, it is still an open issue that lacks standard industrialisation solutions [15]. Ensuring dependability, despite the complexity and changing nature of the systems due to adaptation and learning, is a precondition for public trust and acceptance [58]. The findings of the study help to quantify human behaviour and define measurable parameters that are also working towards standardisation of the concept of safety-critical autonomous or AI-enhanced applications.

In addition, the presented work can support the establishment of trust and acceptance measures of AD in general and AI-based approaches in particular. Such measures may provide the foundation of necessary acceptance and standardisation related to human perception and trust in AD systems.

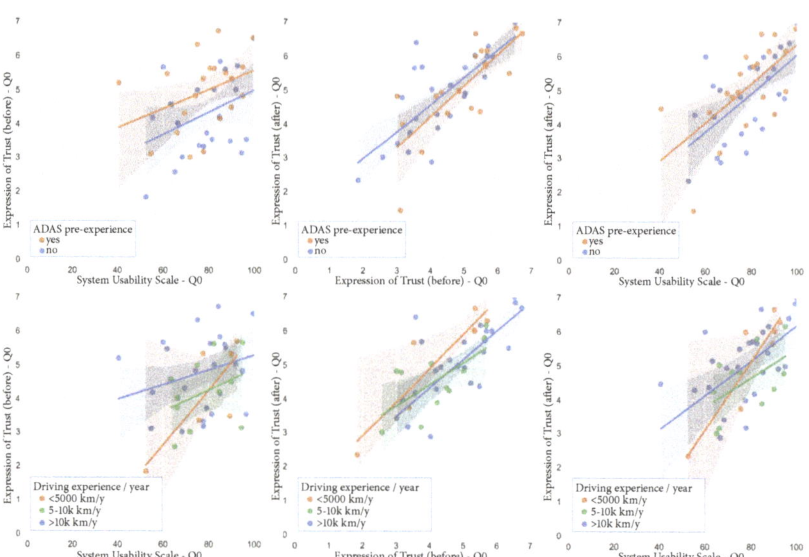

Figure 11. Pearson correlations with descriptive analysis of potential clusters regarding ADAS pre-experience and driving experience.

Limitations

A limiting factor in this study may be seen in its virtual reality nature within a simulator setting. This may provide an impression and sensation of safety which would not be experienced in reality and result in a more comfortable and less critical driver state in the simulator. We suggest that the direction of the effect is the same though the effect size may be smaller.

We would expect the NASA TLX questionnaire to lead to a higher level of workload when facing the same situation in a vehicle in a real-world scenario like on the road or a test track. This influence is expected to be lower for the Expression of Trust questionnaire due to the research focus on the difference between pre- and post-evaluation. In both cases, the participants are aware of the virtual character in the simulator. Therefore, the participants' rating focuses on the complete simulator session rather than on a single scenario.

Moreover, we wish to point out that the number of participants was not completely balanced between all groups analysed in the calculations. This results in a small number of participants within some specific groups and may lead to a limited generalisation for specific results while not harming the big picture of the findings.

Measuring the effects of AD on human perception was concluded on a highly advanced system, which may lead to a limited generalisation regarding the behaviour of the system within a high situational variance in the real world.

5. Conclusions

While the driving simulator offers a chance to immerse in AD functionality, it comes with the limitation posed by the virtual environment. The missing risk of harm can support the participant to gain sufficient impressions of the AD behaviour and therefore significantly increase the participant's trust in the system. When comparing the analysis of participants' impressions before and after the simulator session, the conducted experiment shows that this increase of trust holds for all demographic subgroups on a similar level. For example, the mean increase in trust resulting from the described experience in the driving simulator is measured 3.8% for study participants with and 13.7% without previous ADAS experience. Concerning the gender diversion an increase of trust of 5.4% for the male and 14.5% for the female participants was measured. Group analyses do not reveal significant differences. Hence, we suggest a reduced need for precise and equally balanced demographic groups. Single significant differences within the subgroups are only recognised when using single inquiry lines (specific aspects within the questionnaires).

According to the participants' subjective perception, relying on the system usability scale, the system was able to provide sufficient performance (77.49%) for the conducted evaluation in a reliable virtual environment, substantiated by a homogeneous distribution of the demographic group mean ratings on the questionnaire on a high level.

For SAE level 4 AD in a simulator, participants' subjective workload appears to be low on all singular aspects of the NASA TLX. This reflects the definition of the automated driving mode with no objective task for the driver in the operational design domain. The factors age and driving experience have a significant influence on the participant's workload. Further, the measurement provided evidence that there is a significant relationship between the trust in the system and the participant's workload. In essence, the collected experimental data consistently indicates that the desired increase of trust (as a proxy to improve user acceptance) could be supported by reducing the need for user workload in vehicles.

At present, the integration of the highly automated and autonomous operation of safety-critical systems is still in an early stage. Therefore, an industry-agreed state of the art for measuring trust in autonomous or AI-based systems does not yet exist. Standards addressing the specific aspects of autonomous and/or AI-based technology for safety-critical systems are under development, providing essential further steps in such directions. Although this paper is not directly focusing on standardisation, the presented measurements addressing the establishment of trust and definition of measurable parameters create a foundation for further support for designers and engineers in the conception of safety-critical autonomous or AI-enhanced applications.

Future investigations into the complex combination of human drivers in an AD personal transportation system are required, using statistical analysis in combination with AI to understand and improve the future user trust in the AD technology.

Author Contributions: Conceptualization, P.C.; methodology, P.C.; software, H.D. and M.P.; validation, P.C. and C.K.; investigation, P.C., H.D. and M.P.; resources, P.C. and O.V.; writing—original draft preparation, P.C., C.K. and O.V.; writing—review and editing, P.C., C.K., O.V., H.D. and G.M.; visualization, P.C.; editing, A.E.; supervision, A.E.; project administration, P.C.; funding acquisition, O.V. All authors have read and agreed to the published version of the manuscript.

Funding: This research was funded by the European Union's Horizon 2020 research and innovation program under grant agreements No 723324 (TrustVehicle) and No 871385 (TEACHING). The publication is funded by the Open Access Funding by the Graz University of Technology.

Institutional Review Board Statement: The study was conducted according to the guidelines of the Declaration of Helsinki, and approved by the Ethics Committee of the University of Surrey protocol code 353003-352994-38756154 and date of approval 27 August 2018.

Informed Consent Statement: Informed consent was obtained from all participants involved in the study.

Data Availability Statement: The scenarios conducted during the study are available at [57], the datasets of the results are not available in public as they are restricted to the informed consent and the GDPR rules provided to and signed with all participants. They are stored in a restricted data area of AVL List GmbH as stated to the subjects during the study for no longer than is necessary for the purposes of the study.

Conflicts of Interest: The authors declare no conflict of interest. The funders had no role in the design of the study; in the collection, analyses, or interpretation of data; in the writing of the manuscript, or in the decision to publish the results.

Abbreviations

The following abbreviations are used in this manuscript:

AD	Automated Driving
ADAS	Advanced Driver Assistant Systems
AI	Artificial Intelligence
ANOVA	Analysis of Variance
DDT	Dynamic Driving Task
EOT	Expression of Trust
GDPR	General Data Protection Regulation
GSR	General Safety Regulation
HMI	Human–machine Interface
HR	Heart rate
MANOVA	Multifactorial Analysis of Variance
MDPI	Multidisciplinary Digital Publishing Institute
NASA TLX	NASA Task Load Index
SAE	Society of Automotive Engineers
SSQT	Sequence Specific Questionnaire on Trust
SUS	System Usability Scale

References

1. On-Road Automated Driving (ORAD) Committee. *Taxonomy and Definitions for Terms Related to Driving Automation Systems for on-Road Motor Vehicles*; SAE International: Warrendale, PA, USA, 2021. [CrossRef]
2. Davoli, L.; Martalò, M.; Cilfone, A.; Belli, L.; Ferrari, G.; Presta, R.; Montanari, R.; Mengoni, M.; Giraldi, L.; Amparore, E.G.; et al. On Driver Behavior Recognition for Increased Safety: A Roadmap. *Safety* **2020**, *6*, 55. [CrossRef]
3. Gasser, T.M.; Seeck, A.; Smith, B.W. Rahmenbedingungen für die Fahrerassistenzentwicklung. In *Handbuch Fahrerassistenzsysteme*; Springer: Berlin/Heidelberg, Germany, 2015. [CrossRef]
4. Marcano, M.; Tango, F.; Sarabia, J.; Castellano, A.; Pérez, J.; Irigoyen, E.; Díaz, S. From the Concept of Being "the Boss" to the Idea of Being "a Team": The Adaptive Co-Pilot as the Enabler for a New Cooperative Framework. *Appl. Sci.* **2021**, *11*, 6950. [CrossRef]
5. Jeon, M.; Walker, B.N.; Yim, J.B. Effects of specific emotions on subjective judgment, driving performance, and perceived workload. *Transp. Res. Part F Traffic Psychol. Behav.* **2014**, *24*, 197–209. [CrossRef]
6. Egerstedt, M.; Hu, X.; Stotsky, A. Control of a car-like robot using a virtual vehicle approach. In Proceedings of the 37th IEEE Conference on Decision and Control (Cat. No.98CH36171), Tampa, FL, USA, 16–18 December 1998; Volume 2, pp. 1502–1507. [CrossRef]
7. Deter, D.; Wang, C.; Cook, A.; Perry, N.K. Simulating the Autonomous Future: A Look at Virtual Vehicle Environments and How to Validate Simulation Using Public Data Sets. *IEEE Signal Process. Mag.* **2021**, *38*, 111–121. [CrossRef]
8. Wynne, R.A.; Beanland, V.; Salmon, P.M. Systematic review of driving simulator validation studies. *Saf. Sci.* **2019**, *117*, 138–151. [CrossRef]
9. Ogitsu, T.; Mizoguchi, H. A study on driver training on advanced driver assistance systems by using a driving simulator. In Proceedings of the 2015 International Conference on Connected Vehicles and Expo (ICCVE), Shenzhen, China, 19–23 October 2015; pp. 352–353. [CrossRef]
10. Saetren, G.B.; Pedersen, P.A.; Robertsen, R.; Haukeberg, P.; Rasmussen, M.; Lindheim, C. Simulator training in driver education-potential gains and challenges. In *Safety and Reliability-Safe Societies in a Changing World*; Taylor & Francis Group: Abingdon, UK, 2018.
11. FFG. Integrating Mental Training Techniques into Simulator Training. Available online: https://projekte.ffg.at/projekt/2758303 (accessed on 30 November 2021).
12. Schwarz, M.; Harfmann, J. Essential Tools for Safety Culture Development in Air Traffic Management. In *Aviation Psychology. Applied Methods and Techniques*; Koglbauer, I.V., Biede-Straussberger, S., Eds.; Hogrefe Publishing: Goettingen, Germany, 2021; pp. 33–50.
13. Talker, C.M.; Kallus, K. Anticipatorily Controlled Top-Down Processes Influence the Impact of Coriolis Effects. In Proceedings of the 18th International Symposium on Aviation Psychology, Dayton, OH, USA, 4–7 May 2015.
14. Azevedo Sá, H.; Jayaraman, S.; Esterwood, C.; Yang, X.J.; Robert, L.; Tilbury, D. Real-Time Estimation of Drivers' Trust in Automated Driving Systems. *SSRN Electron. J.* **2020**. [CrossRef]
15. Macher, G.; Akarmazyan, S.; Armengaud, E.; Bacciu, D.; Calandra, C.; Danzinger, H.; Dazzi, P.; Davalas, C.; Gennaro, M.C.D.; Dimitriou, A.; et al. Dependable Integration Concepts for Human-Centric AI-Based Systems. In *International Conference on Computer Safety, Reliability, and Security*; Springer: Berlin/Heidelberg, Germany, 2021; pp. 11–23.
16. Veledar, O. New Business Models to Realise Benefits of the IoT Technology within the Automotive Industry. Ph.D. Thesis, Vienna University of Economics and Business—Executive Academy, Vienna, Austria, 2019. [CrossRef]

17. Bierzynski, K.; Calvo Alonso, D.; Gandhi, K.; Lehment, N.; Mayer, D.; Nackaerts, A.; Neul, R.; Peischl, B.; Rix, N.; Röhm, H. AI at the Edge, 2021 EPoSS White Paper. 2021. Available online: https://cora.ucc.ie/handle/10468/11495 (accessed on 29 December 2021).

18. Aptiv; Audi; Baidu; BMW; Continental; Daimler; FCA; HERE; Infineon; Intel; et al. Safety First for Automated Driving. 2019. Available online: https://www.aptiv.com/en/newsroom/article/automotive-and-mobility-industry-leaders-publish-first-of-its-kind-framework-for-safe-automated-driving-systems (accessed on 29 November 2021).

19. Bacciu, D.; Akarmazyan, S.; Armengaud, E.; Bacco, M.; Bravos, G.; Calandra, C.; Carlini, E.; Carta, A.; Cassarà, P.; Coppola, M.; et al. TEACHING-Trustworthy autonomous cyber-physical applications through human-centred intelligence. In Proceedings of the 2021 IEEE International Conference on Omni-Layer Intelligent Systems (COINS), Barcelona, Spain, 23–25 August 2021; pp. 1–6. [CrossRef]

20. ISO-International Standardization Organisation. ISO 26262 Road Vehicles-Functional Safety. 2018. Available online: https://www.iso.org/standard/68383.html (accessed on 30 November 2021).

21. ISO-International Organization for Standardization. ISO/WD PAS 21448 Road Vehicles-Safety of the Intended Functionality, Work-in-progress. Available online: https://www.iso.org/standard/70939.html (accessed on 20 August 2021).

22. Körber, M.; Baseler, E.; Bengler, K. Introduction matters: Manipulating trust in automation and reliance in automated driving. *Appl. Ergon.* **2018**, *66*, 18–31. [CrossRef]

23. Ward, C.; Raue, M.; Lee, C.; D'Ambrosio, L.; Coughlin, J.F. Acceptance of automated driving across generations: The role of risk and benefit perception, knowledge, and trust. In *International Conference on Human-Computer Interaction*; Springer: Berlin/Heidelberg, Germany, 2017; pp. 254–266.

24. Dimitrakopoulos, G.; Uden, L.; Varlamis, I. Chapter 7-User acceptance and ethics of ITS. In *The Future of Intelligent Transport Systems*; Elsevier: Amsterdam, The Netherlands, 2020; pp. 85–91. [CrossRef]

25. Zhang, T.; Tao, D.; Qu, X.; Zhang, X.; Zeng, J.; Zhu, H.; Zhu, H. Automated vehicle acceptance in China: Social influence and initial trust are key determinants. *Transp. Res. Part C Emerg. Technol.* **2020**, *112*, 220–233. [CrossRef]

26. Kaur, K.; Rampersad, G. Trust in driverless cars: Investigating key factors influencing the adoption of driverless cars. *J. Eng. Technol. Manag.* **2018**, *48*, 87–96. [CrossRef]

27. Kyriakidis, M.; de Winter, J.C.; Stanton, N.; Bellet, T.; van Arem, B.; Brookhuis, K.; Martens, M.H.; Bengler, K.; Andersson, J.; Merat, N.; et al. A human factors perspective on automated driving. *Theor. Issues Ergon. Sci.* **2019**, *20*, 223–249. [CrossRef]

28. Druml, N.; Ryabokon, A.; Schorn, R.; Koszescha, J.; Ozols, K.; Levinskis, A.; Novickis, R.; Nigussie, E.; Isoaho, J.; Solmaz, S.; et al. Programmable Systems for Intelligence in Automobiles (PRYSTINE): Final results after Year 3. In Proceedings of the 2021 24th Euromicro Conference on Digital System Design (DSD), Palermo, Spain, 1–3 September 2021; pp. 268–277. [CrossRef]

29. Klöckner, C.A. The dynamics of purchasing an electric vehicle–A prospective longitudinal study of the decision-making process. *Transp. Res. Part F Traffic Psychol. Behav.* **2014**, *24*, 103–116. [CrossRef]

30. Ribeiro, M.A.; Gursoy, D.; Chi, O.H. Customer Acceptance of Autonomous Vehicles in Travel and Tourism. *J. Travel Res.* **2021**. [CrossRef]

31. Barth, M.; Jugert, P.; Fritsche, I. Still underdetected–Social norms and collective efficacy predict the acceptance of electric vehicles in Germany. *Transp. Res. Part F Traffic Psychol. Behav.* **2016**, *37*, 64–77. [CrossRef]

32. Nordhoff, S.; Van Arem, B.; Happee, R. Conceptual model to explain, predict, and improve user acceptance of driverless podlike vehicles. *Transp. Res. Rec.* **2016**, *2602*, 60–67. [CrossRef]

33. Beggiato, M.; Krems, J.F. The evolution of mental model, trust and acceptance of adaptive cruise control in relation to initial information. *Transp. Res. Part F Traffic Psychol. Behav.* **2013**, *18*, 47–57. [CrossRef]

34. Macher, G.; Druml, N.; Veledar, O.; Reckenzaun, J. Safety and Security Aspects of Fail-Operational Urban Surround perceptION (FUSION). In *Model-Based Safety and Assessment*; Papadopoulos, Y., Aslansefat, K., Katsaros, P., Bozzano, M., Eds.; Springer International Publishing: Berlin/Heidelberg, Germany, 2019; pp. 286–300. [CrossRef]

35. Hartwich, F.; Beggiato, M.; Krems, J.F. Driving comfort, enjoyment and acceptance of automated driving–effects of drivers' age and driving style familiarity. *Ergonomics* **2018**, *61*, 1017–1032. [CrossRef]

36. Hanzl, G.; Haberl, M.; Eichberger, A.; Fellendorf, M. Human Driver's Acceptance of Automated Driving Systems Based on a Driving Simulator Study. In *International Forum on Advanced Microsystems for Automotive Applications*; Springer: Berlin/Heidelberg, Germany, 2020; pp. 186–195.

37. Watzenig, D.; Horn, M. Introduction to Automated Driving. In *Automated Driving: Safer and More Efficient Future Driving*; Springer International Publishing: Berlin/Heidelberg, Germany, 2017; pp. 3–16. [CrossRef]

38. Veledar, O.; Armengaud, E.; Happ Botler, L.; Damjanovic-Behrendt, V.; Jaksic, S. Steering Drivers of Change: Maximising Benefits of Trustworthy IoT. In *Systems, Software and Services Process Improvement. EuroSPI 2021. Communications in Computer and Information Science*; Springer: Berlin/Heidelberg, Germany, 2021; Volume 1442, pp. 663–674.

39. Jian, J.Y.; Bisantz, A.M.; Drury, C.G. Foundations for an Empirically Determined Scale of Trust in Automated Systems. *Int. J. Cogn. Ergon.* **2000**, *4*, 53–71. [CrossRef]

40. U.S. General Services Administration—Technology Transformation Services. System Usability Scale (SUS). 2013. Available online: https://www.usability.gov/how-to-and-tools/methods/system-usability-scale.html (accessed on 12 January 2022)

41. Said, S.; Gozdzik, M.; Roche, T.R.; Braun, J.; Rössler, J.; Kaserer, A.; Spahn, D.R.; Nöthiger, C.B.; Tscholl, D.W. Validation of the Raw National Aeronautics and Space Administration Task Load Index (NASA-TLX) Questionnaire to Assess Perceived Workload in Patient Monitoring Tasks: Pooled Analysis Study Using Mixed Models. *J. Med. Internet Res.* **2020**, *22*, e19472. [CrossRef] [PubMed]

42. Grier, R.A. How High is High? A Meta-Analysis of NASA-TLX Global Workload Scores. *Proc. Hum. Factors Ergon. Soc. Annu. Meet.* **2015**, *59*, 1727–1731. [CrossRef]

43. UNESCO Institute for Statistics. International standard classification of education: ISCED 2011. *Comp. Soc. Res.* **2012**, *30*. [CrossRef]

44. Clement, P.; Danzinger, H.; Veledar, O.; Koenczoel, C.; Macher, G.; Eichberger, A. Measuring trust in automated driving using a multi-level approach to human factors. In Proceedings of the 24th Euromicro Conference on Digital System Design (DSD), Palermo, Italy, 1–3 September 2021; Leporati, S.F., Vitabile, A.S., Eds.; IEEE Computer Society: Palermo, IT, USA, 2021; pp. 410–417. [CrossRef]

45. Helldin, T.; Falkman, G.; Riveiro, M.; Davidsson, S. Presenting system uncertainty in automotive UIs for supporting trust calibration in autonomous driving. In *Proceedings of the 5th International Conference on Automotive User Interfaces and Interactive Vehicular Applications, Leeds, UK, 28–30 October 2013*; Terken, J., Ed.; ACM: New York, NY, USA, 2013; pp. 210–217. [CrossRef]

46. Bangor, A.; Kortum, P.; Miller, J. Determining What Individual SUS Scores Mean: Adding an Adjective Rating Scale. *J. Usability Stud.* **2009**, *4*, 114–123.

47. Grier, R.A.; Bangor, A.; Kortum, P.; Peres, S.C. The System Usability Scale. *Proc. Hum. Factors Ergon. Soc. Annu. Meet.* **2013**, *57*, 187–191. [CrossRef]

48. Brooke, J. SUS: A quickdirty usability scale. In *Usability Evaluation in Industry*; Taylor and Francis: London, UK, 1996; p. 252.

49. Rauer, M. Quantitative Usablility-Analysen mit der System Usability Scale (SUS)-Nachrichten, Tipps & Anleitungen für Agile, Entwicklung, Atlassian-Software (JIRA, Confluence, Bitbucket, ...) und Google Cloud. 2011. Available online: https://blog.seibert-media.net/blog/2011/04/11/usablility-analysen-system-usability-scale-sus/ (accessed on 20 April 2021).

50. Index, The NASA TLX Tool: Task Load. TLX @ NASA Ames-Home. Available online: https://human-factors.arc.nasa.gov/groups/TLX/ (accessed on 19 April 2021).

51. EU. General Data Protection Regulation EU3016/0679: GDPR. Available online: https://eur-lex.europa.eu/legal-content/EN/ALL/?uri=CELEX:32016R0679 (accessed on 19 April 2021).

52. CRUDEN. Automotive Driving Simulators-Cruden. Available online: https://www.cruden.com/automotive-driving-simulators/ (accessed on 29 December 2021).

53. VIRES. VTD Virtual Test Drive. Available online: https://www.mscsoftware.com/de/virtual-test-drive (accessed on 19 April 2021).

54. AVL List GmbH. AVL VSM™ Vehicle Simulation. Available online: https://www.avl.com/-/avl-vsm-4- (accessed on 19 April 2021).

55. AVL List GmbH. AVL Model.Connect™. Available online: https://www.avl.com/de/-/model-connect- (accessed on 19 April 2021).

56. Clement, P.; Danzinger, H.; Quinz, P.; Hillbrand, B.; Hartavi, A.E.; Kasikci, K.Z. Assessment Concept for TrustVehicles. In *Enhanced Trustworthiness and End User Acceptance of Conditionally Automated Vehicles in the Transition Period*; Watzenig, D., Schicker, L.M., Eds.; Lecture Notes in Intelligent Transportation and Infrastructure; Springer International Publishing: Berlin/Heidelberg, Germany, 2020; pp. 107–128. [CrossRef]

57. TrustVehicle H2020 Project. Experiment Scenarios Video Description of Scnearios. Available online: https://www.youtube.com/channel/UCdsw-Md5OxfsgFyQY7DtXpA (accessed on 4 November 2021).

58. Macher, G.; Diwold, K.; Veledar, O.; Armengaud, E.; Römer, K. The Quest for Infrastructures and Engineering Methods Enabling Highly Dynamic Autonomous Systems. In *Systems, Software and Services Process Improvement*; Walker, A., O'Connor, R.V., Messnarz, R., Eds.; Springer International Publishing: Berlin/Heidelberg, Germany, 2019; pp. 15–27.

Article

Digitalize the Twin: A Method for Calibration of Reference Data for Transfer Real-World Test Drives into Simulation

Martin Holder [1,†,‡], Lukas Elster [2,‡] and Hermann Winner [2,*]

[1] Independent Researcher, 64287 Darmstadt, Germany; Martin.Holder@wihi.tu-darmstadt.de
[2] Institute of Automotive Engineering, Technical University of Darmstadt, 64287 Darmstadt, Germany; lukas.elster@tu-darmstadt.de
* Correspondence: hermann.winner@tu-darmstadt.de
† Martin Holder was formerly with TU Darmstadt.
‡ These authors contributed equally to this work.

Abstract: In the course of the development of automated driving, there has been increasing interest in obtaining ground truth information from sensor recordings and transferring road traffic scenarios to simulations. The quality of the "ground truth" annotation is dictated by its accuracy. This paper presents a method for calibrating the accuracy of ground truth in practical applications in the automotive context. With an exemplary measurement device, we show that the proclaimed accuracy of the device is not always reached. However, test repetitions show deviations, resulting in non-uniform reliability and limited trustworthiness of the reference measurement. A similar result can be observed when reproducing the trajectory in the simulation environment: the exact reproduction of the driven trajectory does not always succeed in the simulation environment shown as an example because deviations occur. This is particularly relevant for making sensor-specific features such as material reflectivities for lidar and radar quantifiable in dynamic cases.

Keywords: virtual validation; automated driving; ground truth; reference measurement; calibration method; simulation

Citation: Holder, M.F.; Elster, L.; Winner, H. Digitalize the Twin: A Method for Calibration of Reference Data for Transfer Real-World Test Drives into Simulation. *Energies* 2022, 15, 989. https://doi.org/10.3390/en15030989

Academic Editor: Arno Eichberger

Received: 29 November 2021
Accepted: 21 January 2022
Published: 28 January 2022

Publisher's Note: MDPI stays neutral with regard to jurisdictional claims in published maps and institutional affiliations.

1. Introduction

"Ground truth (GT) data was obtained using an real time kinematic (RTK)-based global navigation satellite system (GNSS) device and provides accuracy of up to $\pm 3\,\mathrm{cm}$". Such statements are often found in research articles to justify the quality of reference data accompanying data acquisition for various tasks [1,2]. In the automotive context, RTK-aided GNSS is widely used for obtaining positions. There is no doubt that RTK-based GNSS methods can achieve accuracies in the cm range. However, this applies only to the position determination of the antenna and under favorable operating conditions of the GNSS receiver. If one is, however, interested in the position information of another reference point, e.g., the center of the vehicle's rear axle, the translational offsets between the antenna and the respective point must be determined very precisely. In complex geometries such as vehicles, further aids are needed for this. Uncertainties in the determination of these offsets can be hardly avoided. For this reason, it is unclear whether the specified precision of the device can also be achieved in its installed state.

In this work, we address the issue of the trustworthiness of reference data obtained with GNSS devices. We aim to refine the notion of GT in the context of environmental perception with different sensor modalities. It must be ensured that the reference measurement shows higher credibility against other sensors used, e.g., lidar or radar sensors. To determine this, reference measurements are required to determine the credibility of the reference, called the "super-reference". Figure 1 contextualizes the aforementioned term "super-reference" in comparison to GT and a reference sensor.

The GT can only be measured with finite accuracy, which we call the super-reference. Thereby, GT is only approximated by the super-reference, leaving a minor deviation to GT. A super-reference is typically only available in limited and controllable circumstances. A reference sensor, however, is optimized for the practical application at the cost of potentially higher GT deviation. To achieve higher trustworthiness in the accuracy of the reference, a super-reference is used for its calibration. Ultimately, when validating sensors, it is of interest to determine measurement uncertainties, which result from the difference between sensor and reference measurement.

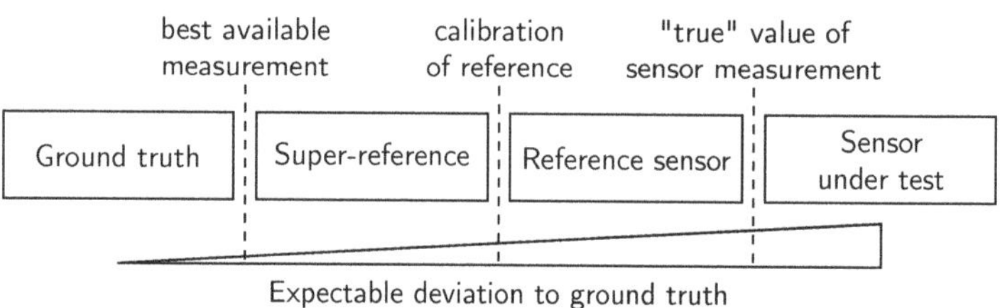

Figure 1. Relationship between GT, super-reference, reference sensor, and sensor under test.

The main interest of this article lies in increasing the trustworthiness in reference data, which enables the reenacting of real-world test drives in virtual environments. This is of particular importance in the development and validation of sensor models for the virtual validation of automated driving (AD), as reference data are required. The basic idea of transferring test drives to simulation is admittedly not new. However, our paper specifically deals with the calibration process of positioning measuring devices and discusses the achievable accuracy.

This paper is structured as follows. First, we discuss the need for the careful calibration of measurement devices that are used for collecting reference data. Next, an overview of previous research on obtaining GT data and sensor principles employed for this purpose is presented. We present stationary and dynamic calibration experiments, which serve as a reference and are thereby eligible for the calibration of measurement devices. In the practical application of our experiments, we show that the proclaimed accuracy of the positioning devices is not always met. Finally, we show the achievable precision when reenacting a real-world test drive in two simulation environments. The source code for creating scenarios with real driven trajectories based on GNSS measurements is made available.

2. A Motivational Example: Can We Trust Our Reference?

After data collection with sensors in real-world scenarios, faithful reenacting of the driven scenario in simulation is tedious, but of high interest for virtual validation aspects. There are various types of measurement phenomena that are inherent in the sensor measurement principle and can manifest as measurement artefacts. These can cause deviations between the obtained measurement result and GT. A simple yet illustrative example is the limited resolution of the (discrete) distance measurement with radar and lidar sensors, which causes quantization errors in the determination of the (continuous) distance to an object. For this reason, reference sensors are needed that are capable of measuring the movement and position of vehicles with high accuracy, precision, and reliability.

Even for simple scenarios, such as a follow-up drive with an Adaptive Cruise Control (ACC) system, one can observe non-stationary behavior when inspecting the movements of the vehicles in close detail, although the vehicle movements were subjectively perceived by the occupants as stationary. If the movement of the vehicle in front is now recorded by a sensor, further sensor-specific uncertainties are superimposed on its perception.

Figure 2 shows an example of a measurement record of an ACC drive run at 40 km/h with a medium time gap to the front vehicle. The measured variables used for the ACC function are read out from the radar sensor, which supplies the object information for the ACC system, via the vehicle Controller Area Network (CAN).

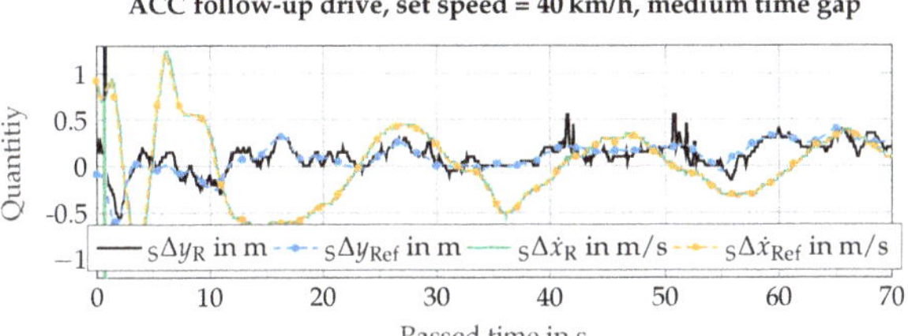

Figure 2. Variation of lateral distance and longitudinal speed between ego car and object of interest (OOI) in an ACC scenario. The left index "S" denotes the sensor coordinate system, "R" the radar sensor and "Ref" the GNSS-based reference system.

Several aspects emerge from the measurement record shown in the figure: the relative longitudinal velocity $_S\Delta\dot{x}$ shows a variation bandwidth of ±0.5 m/s, which is around one order of magnitude above the velocity resolution of automotive radar sensors. In the lateral direction, i.e., $_S\Delta y$, the object fluctuates within its lane at around ±0.2 m, which can hardly be noticed with visual inspection by a human driver. The object reported by the used radar sensor shows the pronounced discretization of the lateral measurement. The direct transfer of the radars' measured variables into the simulation would lead to sudden, physically implausible jumps in an object's trajectory. Although the radial velocity of objects is measured by the radar with high precision, the discretization by the sensor used in this example makes post-processing necessary in order to obtain a feasible motion profile.

There is reasonable hope that high-accuracy motion analyzers, which combine GNSS position measurements as well as accelerations and angular rates captured by inertial measurement unit (IMU) sensors, can be used to capture the motion of agents with high precision. An exemplary device is the GeneSys Automotive Dynamic Motion Analyzer (ADMA) or the RT device series by OXTS. The corresponding measurements of such a device are also shown in Figure 2 and denoted by "Ref". However, the question remains as to the actual accuracy of measuring motion and transferring the motion to the simulation. The central question of this paper is therefore as follows: How much accuracy does a reference measurement system really provide and how does one perform its calibration?

3. Related Work

Real-world traffic is a suitable data source for developing and testing automated driving functions because it is highly diverse and has random characteristics. Moreover, regarding simulation aspects, real-world data offer the highest possible quality for the validation of simulation models. Consequently, there are several previously reported approaches to transfer a real-world test drive into the simulation.

Roughly speaking, two categories can be found. These are, on the one hand, object list-based approaches. Here, the object list from sensors or a fused sensor cluster is taken as the starting point for scenario reconstruction. The goal of this method is to prepare real data for a scenario-based testing approach in the simulation. Regarding the study of absolute accuracy, previous work in the field of reference sensing deals with obtaining the accuracy that is achievable with contemporary automotive grade perception sensors.

3.1. Object List-Based Approach

In the literature, there are approaches known in which the object lists of the sensors are used to transfer the recorded scenario into a simulation. These methods aim to extract a concrete scenario from the measurement data in the sense of scenario-based testing. Logical scenarios can be abstracted from this. A typical pipeline consumes sensor data (e.g., object lists, point clouds, etc.) and compiles a standardized scenario description using the

OpenScenario/OpenDrive language format. A typical example is the framework proposed by Wagner et al. that relies on lidar sensors [3]. It is capable of sensing road objects as well as road semantics (e.g., road geometry, lane markings, etc.). There are also a number of commercial suppliers in this field that compile scenario data from sensor readings, such as [4,5].

These methods require a comprehensive object list as well as additional sensor data to facilitate the inference of the road properties or street layout. This prerequisite is not fulfilled in many everyday traffic situations, such as the occurrence of occlusions, or the cutting in/out of objects. Object detection is generally not reliable in such moments. This constraints can be gradually resolved by manual annotations of sensor data. There are no uniform quality standards, to the best of our knowledge, for the required accuracy of such methods. Based on the published information about these procedures, there is an impression that a visual inspection is performed by experts or the test engineers.

Services that provide so-called GT information for the annotation of sensor data (lidar point cloud, camera images), such as [6,7], are not within the scope of this paper because no reference data are used for this purpose. Instead, recorded sensor data, which may be calibrated extrinsically to each other when multiple sensor modalities are used, are annotated in a manual or (semi-)automated fashion.

3.2. Reference Sensors

With the help of reference sensors (e.g., high-precision GNSS measurement technology or laser scanners), the position of traffic participants can be obtained within the respective measurement accuracy. Data sets such as KITTI, nuScenes, WaymoOpen, etc., therefore provide GT information of traffic participants obtained from an automotive-grade laser scanner mounted on the roof of the ego vehicle. This approach provides useful results for annotating bounding boxes such as those used for labeling in machine learning methods. Minor inaccuracies in the labeling, so-called label noise, can even increase the robustness of the learning algorithm under certain circumstances. In order to use bounding boxes that are labeled in this way as a reference when transferring the scenario to the simulation, a specification of the accuracy over several time steps is required. This is not given in most data sets. The suitability of automotive-grade lidar sensors was investigated in a paper by Schalling et al. [8]. However, the limitations of lidar sensors with respect to the factors influencing their measurement result prevent their justification as GT sensors.

Thorough research on referencing the reference system ("super-referencing") has been presented by Brahmi [9]. His focus is on the evaluation of object-based advanced driver assistance system (ADAS) systems. The basic ideas presented in his thesis can essentially be applied to the problem of this paper, namely the transfer of a real test drive to a simulation.

In a paper by Steinhard, the suitability of a lidar sensor system for GT determination is investigated [10]. As with Brahmi, a high-precision laser scanner with sub-mm resolution serves as a super-reference.

3.3. Gaps in State of the Art

The determination of GT is mostly done via RTK-based GNSS or high-precision lidar sensors with mm-scale resolution e.g., Leica D5. In this context, however, there is no verification that the proclaimed accuracy is actually met under all circumstances. Previous experiments, such as the work from Brahmi [9], have indeed identified the need for a calibration procedure with reference sensors. What remains unresolved so far is to study the fidelity of "GT" in dynamic cases, as well as the stationary analysis of the yaw angle between two reference systems, which is of the utmost interest in reflectivity studies and signal drift.

The digitalizing of a test run relies on the position accuracy of the RTK-based GNSS device. However, it lacks the discussion of whether the proclaimed accuracy is maintained during dynamic situations. Modern lidar and high-resolution radar sensors have distance resolutions in the cm range. If sensor models are to be validated, high demands are therefore made on the accuracy of the trajectory reproduction in the simulation.

4. Calibration Aspects: The Need for a Super-Reference

At this point, a discussion of the term "GT" in the context of automotive simulation is needed to obtain a common understanding of it. It is often used to describe the true state of an object, and potentially also the future state, e.g., in terms of planned actions. Thus, there is a state that can be estimated or measured. Its true value is called GT. It is initially irrelevant how GT is determined. The only relevant aspect is that the GT value serves as a reference against other methods for determining a certain value (measurement, estimation). Especially in the field of virtual environments, which consider 3D representations of objects, the term can be used in a broader sense: it covers material assignments, reflectively properties, as well as geometry detailing, and others.

When a "GT" is obtained with a prospective device, the resulting deviations can be conceptualized in terms of "accuracy" and "precision". The term "accuracy" is defined as *"the degree to which the result of a measurement or calculation matches the correct value or a standard"* [11]. Moreover, the term "precision" is defined as *"the quality of being exact, accurate and careful"* [12].

GT can hardly claim to be completely accurate. It represents rather a value that can be faithfully measured to the best of one's knowledge and belief, as well as up to the accuracy of the measurement equipment used. Prominent examples are object states, such as its longitudinal and lateral positions, as well as the object's orientation. Measurement errors of all kinds, as they are present in all measuring instruments, mean that GT can basically only be obtained with finite accuracy. Nevertheless, the measurement data obtained using the highest-precision device are considered to be a GT measurement. Consequently, a GT to the "GT" is needed. Thus, for verification of the reference sensor, a more accurate reference is needed, the so-called "super-reference". We define the term "super-reference" as follows:

"Comparing the result ξ obtained by device $\mathcal{A}$ to that of device $\mathcal{B}$. The underlying measurement principle of $\mathcal{B}$ is fundamentally different to $\mathcal{A}$, i.e., $\mathcal{B}$ is invariant to error sources of $\mathcal{A}$. Measuring ξ by means of $\mathcal{B}$ is characterized by high fidelity, accuracy, repeatability, and intuition. $\mathcal{B}$ is thereby seen as a super-reference for obtaining ξ".

In order to distinguish the term "super-reference" from the calibration of a measuring device, the definition of calibration is considered. Calibration is defined as *"to mark units of measurement on an instrument so that it can be used for measuring something accurately"* [13]. Therefore, the usability of a measuring device for determining the "GT" is qualified by a calibration procedure.

The "super-reference" principle is demonstrated using position measurements with GNSS. A GNSS device is chosen to serve as a reference measurement technique. To determine the shortest distance between two GNSS points, their Euclidean distance according to the obtained GNSS positions can be used. The result is subject to all errors affecting the GNSS measurements and can only be seen as correct within ± 2 cm. A super-reference for calibrating this method is given by a length-measuring device such as a tape measure or meter stick, which usually have an accuracy level in the sub-mm range according to EC Regulation 2004/22/EC [14]. Thereby, the demand for accuracy during the setup of the measurement to obtain these values has to be absolutely exact regarding experimental conduct.

4.1. Super-Referencing in Automotive Use Cases

The current state of an object is given by its translational and rotational degrees of freedom and the respective rates of change and accelerations, which are defined according to ISO 8855 [15]. In a Cartesian frame, these would be $\mathbf{x} = [x, y, z, \phi, \theta, \psi]$ along with $\dot{\mathbf{x}} = [\dot{x}, \dot{y}, \dot{z}, \dot{\phi}, \dot{\theta}, \dot{\psi}]$ and $\ddot{\mathbf{x}} = [\ddot{x}, \ddot{y}, \ddot{z}, \ddot{\phi}, \ddot{\theta}, \ddot{\psi}]$, as well as $\dddot{\mathbf{x}}$ when also considering jerk.

For the calibration of these 24 quantities, only the longitudinal acceleration values offer a natural reference value: standard acceleration due to gravity (approx. 9.81 m/s^2 [16]) can be calculated for different locations and altitudes [17]. An acceleration sensor measuring along the axis pointing to the center of the earth can be referenced via this value. Aids are required for calibrating the other measured variables. For example, translational distances can be referenced via auxiliary means, such as the aforementioned meter stick. With respect to manufacturing tolerances, high-precision Computerized Numerical Control (CNC) machinery would provide sufficient accuracy for calibration rotation angles [18].

Finding a super-reference is more difficult for velocities. Although the speed of sound defines a reference, it is beyond relevant velocities in the automotive domain. Furthermore, the specified velocity resolution of precision-measuring instruments such as ADMA or OXTS is in the range of less than 0.01 m/s. This is an order of magnitude above the velocity resolution of automotive radar sensors via the Doppler effect [19] (p. 272).

Technically, velocity can be determined by the change in location within a time interval. However, this requires very high sampling rates in the automotive context, as the following calculation example illustrates: let an object's longitudinal velocity $\dot{x} = 10$ m/s and the lowest possible distance between two measurement points $\Delta x = 5$ cm be the parameters of the measurement setup; the necessary sampling frequency f_s is calculated by the time difference between $\dot{x}$ and the sum of $\dot{x}$ and the velocity accuracy $\Delta\dot{x} = 0.01$ m/s. Then, the following consideration is valid under the assumption of constant velocity.

$$f_s = \frac{1}{\frac{\Delta x}{\dot{x}} - \frac{\Delta x}{\dot{x}+\Delta\dot{x}}} \approx 200\,\text{kHz} \tag{1}$$

This sampling frequency exposes high demands on typical measurement devices and is therefore beyond the scope of our considerations.

4.2. Materials and Methods for Practical Super-Referencing

The following section is organized as follows: first, the ADMA is described. Next, the different experimental setups for super-referencing the lateral y_{SRef} and longitudinal x_{SRef} position in stationary and dynamic cases with the corresponding materials, as well as determination of the yaw angle, are described. Thereby, the index "SRef" denotes the super-reference measurement. In the automotive sensor modeling and validation context, these values are of the utmost interest.

The ADMA-G-PRO+ by Genesys Offenburg GmbH is available as a reference measurement technique in this study. Because of the high accuracy of up to ± 2 cm [20], high sampling frequencies of up to 1000 Hz and the possibility to use the device as standalone, as well as the combination of two systems, the methods and results can be generalized for comparable devices. Next to the position, the yaw angle accuracy is specified by $\pm 0.05°$ [21] and the velocity is measured with an accuracy of less than ± 0.01 m/s.

The ADMA is mounted via a rack on the vehicle. To configure the device, the mounting offset between its measuring center and the GNSS antenna is required. The ADMA is capable of outputting the poses and their derivatives in a defined point of interest (POI), provided that their positions with regard to its measuring center are known. In our case, we define and measure two POIs: the center of the rear axle and the connection point of a tow bar in the front/back of the vehicle. We use cross line lasers, a measurement tape, and meter rods to determine the described aforementioned offsets with an accuracy of ± 2 mm. Additional supporting points are obtained by photogrammetry measurement of the vehicle.

4.2.1. Calibration of Lateral and Longitudinal Position in Stationary Conditions

To determine the correct measurement procedure during the setup of the ADMA and antenna in the vehicle, a stationary calibration experiment has to be conducted to ensure lateral and longitudinal positioning correctness. The accuracy of the measurement device can be determined by two reference points. These points must be known with regard to their geodetic or Cartesian position. One of these reference points marks the origin of a local coordinate system, of which one axis spans through the second reference point. For a positioning device, a given lateral/longitudinal displacement between the POI and the measurement origin is to be indicated. When one of them is brought to zero, the other quantity can be determined directly. The method is applicable for a single or dual car setup.

For the single-car calibration, the vehicle equipped with the positioning measurement device is placed along one of the axes of the reference coordinate system; see Figure 3a. The measured longitudinal component should now indicate zero, while the lateral component can be determined with a reliable distance measurement device such as a meter stick. The remaining errors indicate the calibration offsets of the positioning device, such as in the aforementioned mounting offsets.

For determining the position of two cars with regard to each other, the setup is fundamentally similar. The rear axles of two vehicles are placed parallel to each other, resulting in zero displacement in the longitudinal direction. The lateral distance can now be obtained in the same way with a meter stick. To ensure the correct positioning of the vehicle's POI at the position $_{\mathrm{L}}x = 0$ in a local coordinate system "L", a cross line laser is used. The super-reference measurement of $\Delta_\mathrm{L} y_\mathrm{SRef}$ is done by means of two cross line lasers focusing on the middle axis of the vehicles, as visualized in Figure 3b. The measured values are then compared to the output of the GNSS device.

(a)   (b)

Figure 3. Dual-car calibration setup. Super-reference is provided by perpendicular cross line laser lines. (**a**) Zero longitudinal offset (i.e., $\Delta_\mathrm{L}x_\mathrm{SRef} = 0$) between the vehicles is verified by cross line laser through center of rear axles. (**b**) Lateral offset is obtained by measuring the distance between cross line laser lines focusing along the vehicle's middle axis.

4.2.2. Yaw Angle

The yaw angle and, in turn, the relative orientation between vehicles is among the relevant quantities in evaluating movement patterns in road traffic. Given the sensitivity of the reflectivity of vehicles with regard to the aspect angle for radar and lidar sensors, its accurate determination is highly desirable.

When using IMU-based systems for angle measurement, drift of the displayed angle may occur. This error is caused by the integration of the measured rotation rate and the angular acceleration by the IMU. An offset error can hardly be avoided, which results in a higher drift after a longer operating time, without correction by additional efforts. This so-called drift stability is usually provided in the sensor specification.

As a super-reference for the yaw angle, the cosine theorem is used: it determines the enclosed angles from the given side length of a triangle, i.e., $\cos(\psi_\mathrm{SRef}) = \frac{x_1^2 + x_2^2 - d^2}{2x_1 x_2}$. The measurement setup for the stationary yaw angle super-reference is shown in Figure 4. This experiment is suitable as a super-reference, because the underlying measurement principle is completely different in comparison to the device under test.

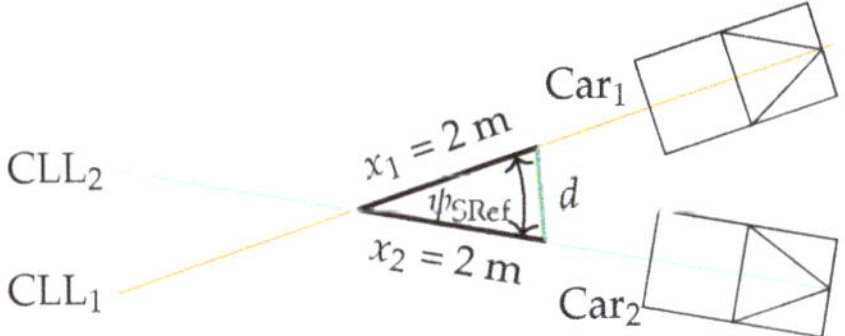

Figure 4. Measurement setup for yaw angle super-reference.

We use two cross line lasers, which are positioned in the same directions as the x-axes of the two cars, to obtain the origin of the straight x_1 and x_2. Cross line lasers are aligned so that they point exactly through the centerline of the vehicles. The manufacturer's logo on the trunk and the shark radio antenna on the roof serve as support points when aligning the lasers. Starting from the intersection of the laser lines, the side lengths of the triangle can now be determined. The edges x_1 and x_2 are determined using a 2 m long meter stick. Meter rods offer accuracy classes in the sub-mm range, which is considered adequate for the intended use here. This simplifies angle determination by means of the law of cosine because two side lengths are already fixed. The length of d is measured by a measurement tape and ψ is calculated by the three given lengths and the cosine theorem. Five measurements are made within 36 min. The accuracy of this measurement method can be calculated based on the Gaussian error propagation. The values for the error propagation are $x_1 = x_2 = 2 \pm 0.005$ m, $d_{min} = 0.902 \pm 0.005$ m and $d_{max} = 1.529 \pm 0.005$ m.

$$\Delta\psi_{max,SRef} = \left|\frac{\partial\psi}{\partial x_1}\right|\Delta x_1 + \left|\frac{\partial\psi}{\partial x_2}\right|\Delta x_2 + \left|\frac{\partial\psi}{\partial d}\right|\Delta d = \pm 0.085 \text{deg} \tag{2}$$

4.2.3. Absolute Positioning in Dynamic Case

To investigate the absolute accuracy of the reference measurement technique in the dynamic case, the following experiment is proposed: a vehicle passes through three light barriers designated as Lb_1, Lb_2, and Lb_3. These are aligned perpendicular to the roadway. The timesteps $t_{Lb_{1...3}}$ at which a light barrier is crossed mark the point in time with zero longitudinal offset between the light barrier and the front point of the vehicle in a light barrier-centered coordinate system. In addition, a foam line is drawn perpendicular to the road. The measurement principle of the super-reference is again completely different to the ADMA and therefore this experiment is suitable as a super-reference. The full measurement setup is illustrated in Figure 5.

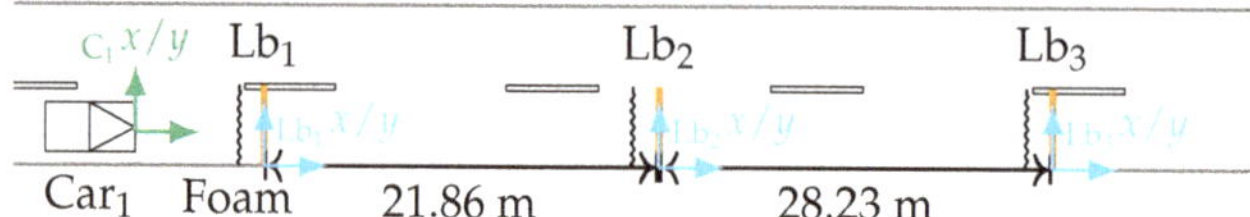

Figure 5. Measurement setup for super-referencing absolute positioning in the dynamic case.

The error of the reference system is found at each light barrier as

$$\epsilon_{lat}(t_{Lb_{1...3}}) = C_1 x_{Ref}(t_{Lb_{1...3}}) - Lb_{1...3} x_{SRef}(t_{Lb_{1...3}}). \tag{3}$$

and when crossing the foam line, the lateral error can be determined based on the tire marks that remain on the foam. The lateral offset can only be determined at the wheels. The imprint of the tires is determined with a measurement tape and gives the lateral distance between the light barrier and the wheels. To account for the offset between the front of the vehicle and the front axle, the foam line is applied in front of the light barrier with an offset by this amount to minimize errors due to yaw angles. In other words, the longitudinal offset is known at the time at which the light barrier is crossed and should be zero. The longitudinal error is obtained at $t_{Lb_{1...3}}$ for each light barrier and reads:

$$\epsilon_{long}(t_{Lb_{1...3}}) = C_1 y_{Ref}(t_{Lb_{1...3}}) - Lb_{1...3} y_{SRef}(t_{Lb_{1...3}}). \tag{4}$$

The experiment is conducted with the vehicle passing the light barriers at a constant velocity of 30 km/h and with an initial set speed of $\dot{x}_{C_1} = 100$ km/h at Lb_1 and braking. When the vehicle is decelerated while passing through the light barriers, the accuracy of the positioning in the dynamic case can be studied. Crossing the barriers with constant velocity indicates the potential sensitivity of positioning errors to velocity.

Three SICK WL 12-2 light barriers that have a specified delay time of $330\,\mu s$ are chosen for use in the experiment. The light barriers are connected to a second ADMA that is placed stationary next to Lb$_2$. Time synchronization between both devices is given by timestamps conveyed in the GNSS signal. Both ADMAs operate with $f_s = 1000\,\text{Hz}$ to minimize the positioning error due to sampling discretization.

The position of the light barriers in GNSS coordinates is measured with the RTK-aided Piksi Multi GNSS Module by Swift Navigation. The position of the point is averaged by measurement over $60\,\text{s}$. The verification of these GNSS coordinates is given as it matches the distance between the light barriers, which is determined by a measuring tape with mm accuracy. The spherical GNSS coordinates are converted into an East-North-Up (ENU) coordinate system based on the WGS84 ellipsoid, which is a metric Cartesian system.

4.2.4. Relative Positioning between Vehicles in Dynamic Case

To determine the accuracy of the ADMA in the dual measurement setup under dynamic conditions, a constant distance between the two vehicles can be used. A tow bar mounted between two vehicles fulfills the requirement between the respective mounting points, also while driving. The position of the towing lugs on the vehicles relative to the ADMA is defined as a POI. By using the positioning information obtained, the calibration goal is to obtain the length of the tow bar, denoted $l_{tb,\text{Ref}}$, which is assumed constant when neglecting strain effects of materials. Then, the resulting error, i.e., $\epsilon_{tb} = l_{tb,\text{Ref}} - l_{tb,\text{SRef}}$, is obtained, which should give zero for an ideal measurement. Measured length $l_{tb,\text{SRef}}$ by a measuring tape of the tow bar is defined as the Euclidean distance of the measured mounting points in Cartesian world coordinates, i.e.,

$$
l_{tb,\text{Ref}} = \left\| \begin{pmatrix} {}_L x_2 - {}_L x_1 \\ {}_L y_2 - {}_L y_1 \end{pmatrix} \right\|_2
\tag{5}
$$

Car$_1$ accelerates from standstill to a given set speed. After a period of constant velocity, the front vehicle brakes the convoy to standstill. The velocity is controlled by Car$_1$'s speed limiter, while Car$_2$ rolls behind in towing mode, i.e., neutral gear position. Three velocity profiles were studied, each with multiple repetitions.

1. $0 \rightarrow 30\,\text{km/h} \rightarrow$ maintaining $\rightarrow 60\,\text{km/h} \rightarrow$ maintaining $\rightarrow 30\,\text{km/h} \rightarrow$ maintaining $\rightarrow 0$
2. $0 \rightarrow 30\,\text{km/h} \rightarrow$ maintaining $\rightarrow 0$
3. $0 \rightarrow 80\,\text{km/h} \rightarrow$ maintaining $\rightarrow 0$

The profiles differ in the duration and intensity of acceleration or deceleration, as well as the duration of cruising at "constant" speed. In this way, the influence of these motion phases on the error can be studied. It is to be noted that the set speed of the speed limiter is the speedometer value, which is above the actual GT speed. The general scenario setup is shown in Figure 6.

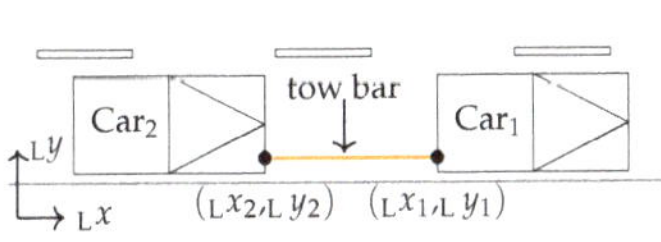

Figure 6. Measurement setup for dynamic dual super-reference with a tow bar in a local coordinate system L.

5. Super-Referencing Results Obtained in Practical Experiments

The proposed super-reference methods were performed at the August Euler airfield near Darmstadt, Germany, between April and September 2021. The ADMA devices used were mounted, measured, and initialized according to the manufacturer's instructions. A 2015 VW Golk Mk7 and a 2018 Mercedes S Class V222 were available as test vehicles.

5.1. Yaw Angle

Figure 7 shows the results of our yaw angle referencing experiment. It compares the heading angle as calculated from the law of cosine to the measured value from the ADMA, i.e., $\epsilon_\psi = |\psi_{\text{Ref}} - \psi_{\text{SRef}}|$. The experiment was conducted five times at various positions and data were collected for around 60 s each. The vehicles were moved only for the purpose of changing position and were otherwise stationary, especially during the determination of the super-reference, which took a couple of minutes. Stationary operating conditions particularly favor the occurrence of yaw angle drift. The drift objectively shows little effect and the deviations are less than 1 deg even after 36 min. It should be noted that the IMU and GNSS fusion system utilizes the dynamic movements of the device. Such a stationary experiment over a long time is challenging for the system. Drift is therefore an expected side effect.

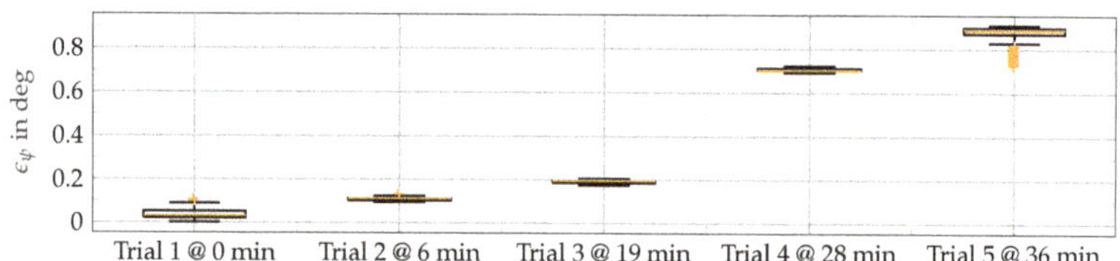

Figure 7. Statistical analysis of heading angle error ϵ_ψ.

5.2. Absolute Positioning in Dynamic Case

The results of the super-referencing absolute positioning in the dynamic case by using light barriers (see Section 4.2.3) are shown in Figure 8. The lateral and longitudinal errors are denoted by ϵ_{lat} and ϵ_{long}, respectively.

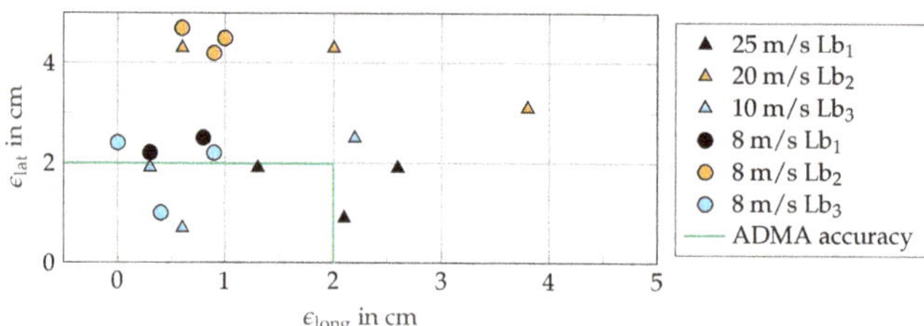

Figure 8. Lateral ϵ_{lat} vs. longitudinal error ϵ_{long} obtained by light barriers and foam.

In general, high longitudinal and lateral precision in the three trials of every experiment and light barrier position is achieved. It is to be noted that ϵ_{long} is larger with a higher speed of the vehicle as it crosses the light barrier. Therefore, low velocities should be used as target velocities to achieve sufficient accuracy or devices with higher sampling frequencies. This is explained with measurement errors due to the light barrier's time delay $\Delta t_{\text{Lb}} = 330\,\mu s$ [22]. This explains the decreasing deviation in the longitudinal direction with decreasing speed, visible by the triangle markers. The delay results in a worst-case error at $25\,\text{m/s}$ of

$$\epsilon_{\text{long,max}} = \dot{x}_{\text{C}_1,\text{max}}\Delta t_{\text{Lb}} = 8\,\text{mm}\,@\,25\,\text{m/s}. \tag{6}$$

The remaining deviation is the error of the ADMA and the positioning error of the experimental setup. The lateral error ϵ_{lat} of our calibration method shows deviations higher than the proclaimed accuracy of the ADMA consistently present at the second light barrier. It shows deviations of around 2.5 cm from the proclaimed accuracy and indicates the experimental setup error. The ADMA's error in the absolute dynamic case with a low velocity is always positive and differs between $0\,\mathrm{cm}$ and $3.8\,\mathrm{cm}$ in the longitudinal and $0.5\,\mathrm{cm}$ and $4.5\,\mathrm{cm}$ in the lateral direction.

5.3. Relative Positioning in Dynamic Case

The relative positioning error in the dynamic case is obtained by estimating the length of a tow bar mounted between two vehicles while driving; see Section 4.2.4. Figure 9 shows exemplary results obtained during one trial of the experiment. It is structured as follows: the error, which is obtained when estimating the tow bar length, i.e., ϵ_{tb}, varies within $\pm 3\,\mathrm{cm}$. Because of the dual measurement setup, the worst-case error based on (5) and $\Delta x_{1/2} = \Delta y_{1/2} = \Delta x = \pm 2\,\mathrm{cm}$ is:

$$\Delta \epsilon_{tb} = \sqrt{(\Delta x_1 + \Delta x_2)^2 + (\Delta y_1 + \Delta y_2)^2} = 2\sqrt{2}\Delta x = 5.7\,\mathrm{cm} \tag{7}$$

Therefore, the deviation of the devices is in accordance with their specification. Longitudinal acceleration in Car_1 or Car_2 with the fixed coordinate system shows little difference due to the mechanical coupling by the tow bar, which causes crabbing at the rear car. Moreover, the velocity profile is shown and does not indicate a strong correlation between error dynamics and longitudinal acceleration.

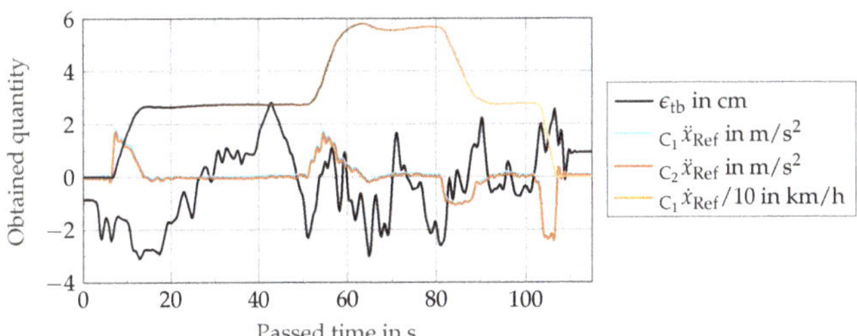

Figure 9. Exemplary measurement reading from one out of five trials. The distance error in cm, longitudinal acceleration, and velocity of the front vehicle are shown. Note that velocity is scaled for better readability. The color gradient with the velocity indicates the running time.

In Figure 10, the influence of velocity and acceleration on ϵ_{tb} is shown. The time course of the velocity or acceleration profile is coded in the color gradient from black to light brown and all trials of the experiment are shown. Studying the sensitivity of ϵ_{tb} to velocity reveals three consistent characteristics for all tests; see the left column in Figure 10.

1. The error shows a fluctuation range of approximately 2 cm during quasi-stationary driving and matches the specification.
2. During acceleration and braking phases, the error remains at a tolerable constant value within the fluctuation range.
3. When reaching standstill, the error settles at a certain value, which lies inside the specification of the dual measurement setup.

No consistent correlations follow from the acceleration profile, as shown in the right column of Figure 10. However, it can be seen that the error also changes during the acceleration phases in the range of a few cm. It is worth noting, however, that the error profile shows some consistency when the acceleration profile is similar, as shown in the portion highlighted by a light blue ellipse in the right column and the first row of Figure 10.

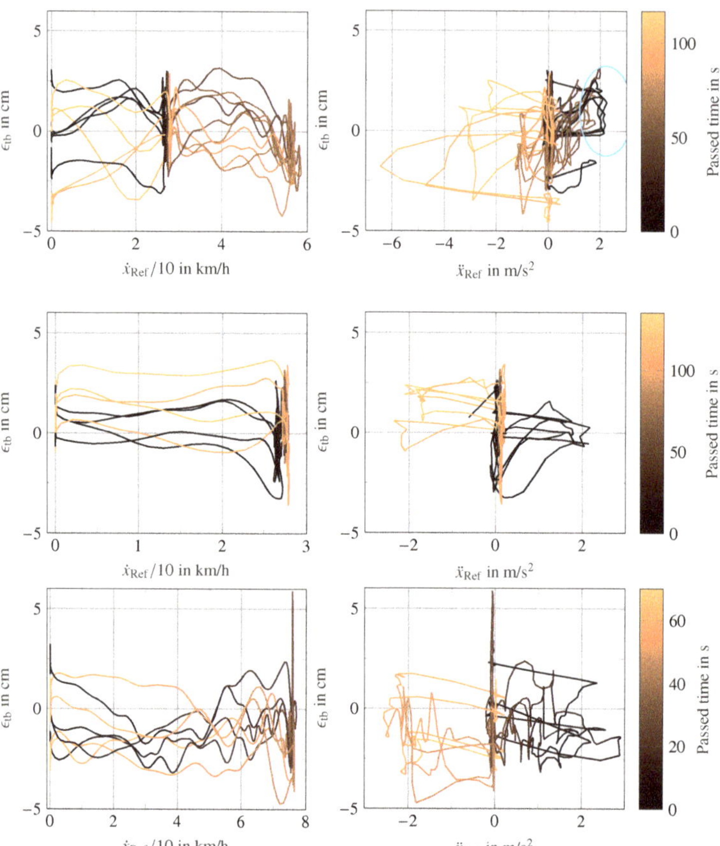

Figure 10. In the left column, ϵ_{tb} vs. velocity is shown for the different velocity profiles mentioned in Section 4.2.4. During areas of "constant" speed, the distance error settles within the accuracy of the measurement devices. In the right column, ϵ_{tb} vs. acceleration is shown. The distance error dynamics show only low sensitivity to acceleration. All trials are depicted and running time is denoted by the line's color gradient.

6. Feasibility of Transferring Real-World Test Drives to Simulation

The main interest in using reference sensors in the context of virtual validation is ultimately to transfer real-world test drives to virtual environments. Under the so-called "Measurement2Sim" method, modern simulation tools such as IPG CarMaker, Vires VTD, or CARLA are able to control an actor's position based on a given trajectory. The tow bar experiments are suitable to represent a simulation's capability to render recorded measurements in the movement of objects. For this purpose, these experiments were transferred to two different simulation environments: Sim1 and Sim2.

The results are given in Figure 11 and are organized as follows. The left column shows the error ϵ_{Sim1} of the first simulation and the right column shows the error ϵ_{Sim2} of the second simulation. The topmost figures show five trials of the experiment, where the two vehicles undergo two phases of acceleration and deceleration with semi-stationary drive in between, i.e., from 0 to 30 km/h, 30 km/h to 60 km/h, back to 30 km/h, and finally to 0. The middle figures show the experiment with 30 km/h and the bottom figures with 80 km/h. The figures visualize the error between the reference measurement, as discussed in Section 5.3, and the simulation environment. Zero error would indicate that the measurement of the distance between vehicles, obtained either in simulation or via the reference measurement, exactly corresponds to the length of the tow bar.

$$\epsilon_{\text{Sim}1/2} = (l_{\text{tb},\text{Sim}1/2} - l_{\text{tb},\text{Ref}}) \tag{8}$$

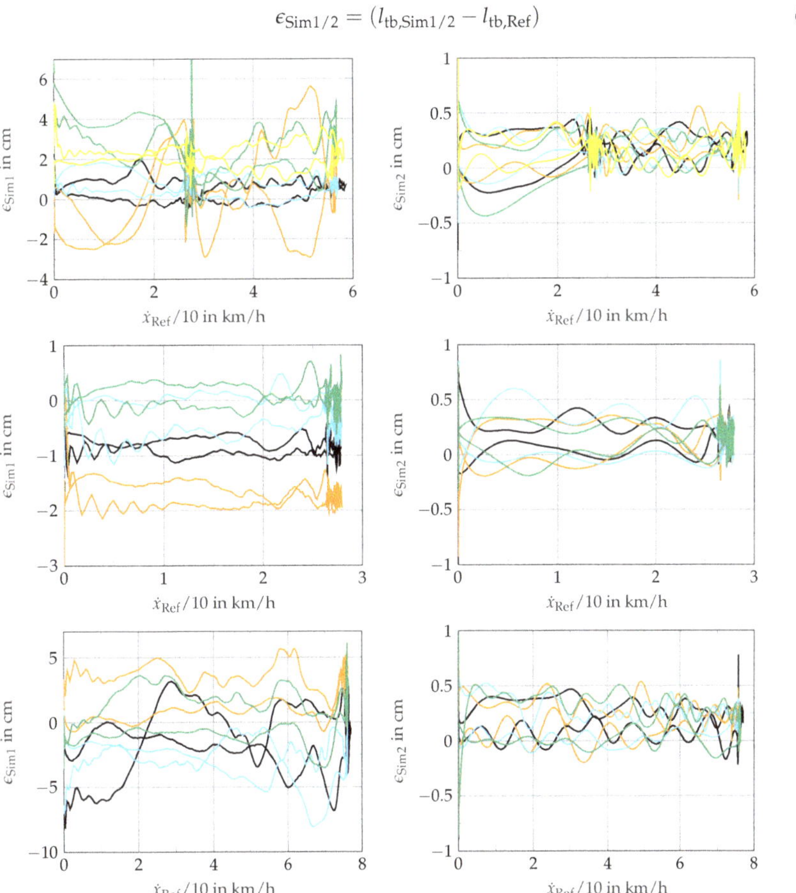

Figure 11. In the left column, $\epsilon_{\text{Sim}1}$ vs. velocity is shown for the different velocity profiles in the first simulation. In the right column, $\epsilon_{\text{Sim}2}$ vs. the same profiles is shown in the second simulation. Each trial is visualized with a different color.

The experiments show that the resulting deviations vary between trials through all trials of the experiments. The error in both simulation tools thereby shows sensitivity to velocity: from the results shown, it can be concluded that the error becomes less with lower speeds, while showing the largest error during phases of acceleration or deceleration. In the first simulation during phases of semi-stationary velocities, the error occasionally extends the proclaimed accuracy of the ADMA. In the second simulation, in turn, the errors are always in the accuracy range.

Our results show that the reenacting of test drives performs best with the first simulation tool when the velocities of the vehicles are kept fairly constant and the accelerations are low, i.e., less than $2\,\text{m}/\text{s}^2$. The absolute deviation between measurement and simulation is in orders of magnitude exceeding the distance resolution of lidar sensors or high-resolution radar sensors. This makes the comparison of simulation to measurement considerably more difficult, since the basis of comparison shows deviations.

7. Discussion

In this paper, we present four different experimental setups to obtain super-reference measurements. With the proposed methods, the confidence in GNSS- and IMU-based reference data for lateral and longitudinal positions can be strengthened in stationary and dynamic cases, as well as the drift analysis of the stationary yaw angle. We note that the highest precision is required when setting up the measurement equipment in order to achieve useful results in terms of super-referencing. During our experimental setups, we encountered the necessity of excellent measuring conditions regarding GNSS measurement devices, because the accuracy of the device's data is highly dependent on the surrounding conditions. Effects such as multipath propagation, shading by other objects, and loss of differential GNSS and RTK connection result in deviation that is an order of magnitude above ideal conditions.

Our experiments reveal the strengths and weaknesses of the reference system under study, the ADMA. The stated measurement accuracy is almost consistently met. The yaw angle measurement quantifies the expected drift of the device. The reference system confirms the proclaimed accuracy during the light barrier experiment. The experiment shows the difficulty in verifying the position accuracy by means of the super-reference, showing less deviation than the system under test. In the dynamic dual measurement setup with the tow bar, the deviation always lies within the specification.

Our comparison of the simulation and real test drive shows a new possibility of verifying the fidelity of so-called "Measurement2Sim" methods. Not only the transfer of the trajectory into the simulation is a source of deviations between measurements and simulation, but also the simulation tool itself provides errors due to the trajectory discretization. The results between the two simulation tools differ clearly. The sources of the deviations cannot be directly identified. When the "Measurement2Sim" method is used in the context of validation of sensor models, it has to be noted that the deviation must not exceed the accuracy of the sensor itself. In the case of lidar, typically, accuracy lies within in the centimeter range. The second simulation tool is better suited to reproducing sensor effects in sensor simulation models with the "Measurement2Sim" method. This simulation tool converts the trajectories very well on the basis of an OpenScenario xosc file based on the measurements of $\mathbf{x} = [x, y, z, \phi, \theta, \psi]$. For future verification and validation experiments in combination with "Measurement2Sim" methods, we highly recommend the analysis of the transfer error of the measurement into the simulation.

Regarding virtual validation by means of digital twins, our results indicate that sample validation using "reference measurement sensors" can hardly be achieved. This is of particular importance when considering the accuracy of perception sensors, which is close to the stochastic deviation margin of the reference measurement system. Rather, our findings strengthen the argumentation for stochastic validation approaches that explicitly take the measurement uncertainties of the reference system into account.

Author Contributions: M.H. and L.E.: conceptualization, methodology, formal analysis, investigation, software, visualization, writing—original draft preparation. H.W.: conceptualization, methodology, supervision, writing—review and editing. All authors have read and agreed to the published version of the manuscript.

Funding: This work received funding from *VIVID* grant number 16ME0173, promoted by the German Federal Ministry for Education and Research (BMBF), based on a decision of the Deutsche Bundestag. We also acknowledge support by the Open Access Publishing Fund of the Technical University of Darmstadt.

Institutional Review Board Statement: Not applicable.

Informed Consent Statement: Not applicable.

Data Availability Statement: The source code for the conversion of the ADMA data into an xosc file is available on *TUdatalib* under http://https://tudatalib.ulb.tu-darmstadt.de/handle/tudatalib/2993, accessed on 9 November 2021.

Acknowledgments: The authors would like to thank Dominik Huber from GeneSys Elektronik GmbH for his support during the commissioning of the ADMA measurement system in our vehicles, Jonathan Knerr and Anthony Ngo for their contributions to earlier versions of our method and Timm Ruppert for preparing the Piksi Multi GNSS Module System, and Felix Glatzki for supporting the measurement campaign.

Conflicts of Interest: The authors have no conflict of interest. The funders had no role in the writing of the manuscript.

Abbreviations

The following abbreviations are used in this manuscript:

ACC	Adaptive cruise control
AD	Automated driving
ADAS	Advanced driver assistance system
ADMA	Automotive dynamic motion analyzer
CAN	Controller area network
GNSS	Global navigation satellite system
GT	Ground truth
IMU	Inertial measurement unit
POI	Point of interest
RTK	Real-time kinematics

References

1. Ramezani, M.; Wang, Y.; Camurri, M.; Wisth, D.; Mattamala, M.; Fallon, M. The Newer College Dataset: Handheld LiDAR, Inertial and Vision with Ground Truth. In Proceedings of the 2020 IEEE/RSJ International Conference on Intelligent Robots and Systems (IROS), IEEE, Las Vegas, NV, USA, 24 October 2020–24 January 2021; pp. 4353–4360. [CrossRef]
2. Skoglund, M.; Petig, T.; Vedder, B.; Eriksson, H.; Schiller, E.M. Static and dynamic performance evaluation of low-cost RTK GPS receivers. In Proceedings of the 2016 IEEE Intelligent Vehicles Symposium (IV), Gothenburg, Sweden, 19–22 June 2016; pp. 16–19. [CrossRef]
3. Wagener, A.; Katz, R. Automated scenario generation for testing advanced driver assistance systems based on post-processed reference laser scanner data. In *Fahrerassistenzsysteme 2016*; Springer Fachmedien Wiesbaden: Wiesbaden, Germnay, 2018; pp. 175–190.
4. Automotive Artificial Intelligence (AAI) GmbH. Scenario Cloning and Extraction—Closing the Loop between Reality and Simulation. 2016. Available online: https://www.automotive-ai.com/technologies/scenario-cloning (accessed on 15 April 2021).
5. Atlatec Gmbh. Scenarios for Simulation. 2021. Available online: https://atlatec.de/scenarios-for-simulation/ (accessed on 15 April 2021).
6. Annotel. Annotation & Ground Truth. 2021. Available online: https://www.annotell.com/annotation (accessed on 15 April 2021).
7. Scale AI, Inc. 3D Sensor Fusion for Simulation. 2021. Available online: https://scale.com/3d-sensor-fusion (accessed on 15 April 2021).
8. Schalling, F.; Ljungberg, S.; Mohan, N. Benchmarking LiDAR Sensors for Development and Evaluation of Automotive Perception. In Proceedings of the 2019 4th International Conference and Workshops on Recent Advances and Innovations in Engineering (ICRAIE), Kedah, Malaysia, 27–29 November 2019.
9. Brahmi, M. Bewertung der Objektbasierten Umfeldwahrnehmung für Fahrerassistenzsysteme Mithilfe von Referenzsystemen. Ph.D. Thesis, Technische Universität Braunschweig, Braunschweig, Germany, 2020.
10. Steinhard, N.; Kaufmann, S.; Sven, R.; Lages, U.; Goerick, C.; Noutangnin, Y. Lidar Based Object Tracking Evaluation Forautomotive Applications. 2016. Available online: https://www.honda-ri.de/pubs/pdf/3216.pdf (accessed on 15 April 2021).
11. Oxford University Press "Accuracy". Oxford Leaner's Dictionary of Academic English. 2021. Available online: https://www.oxfordlearnersdictionaries.com/definition/academic/accuracy (accessed on 18 September 2021).
12. Oxford University Press "Precision". Oxford Leaner's Dictionary of Academic English. 2021. Available online: https://www.oxfordlearnersdictionaries.com/definition/english/precision (accessed on 18 September 2021).
13. Oxford University Press "Calibrate". Oxford Leaner's Dictionary of Academic English. 2021. Available online: https://www.oxfordlearnersdictionaries.com/definition/english/calibrate (accessed on 18 September 2021).
14. European Parliament and Council. Regulation 2004/22/EG. 2004. Available online: https://eur-lex.europa.eu/legal-content/EN/ALL/?uri=CELEX%3A32004L0022 (accessed on 19 September 2021).
15. International Organization for Standardization. ISO 8855:2011 Road Vehicles—Vehicle Dynamics and Road Holding Ability Vocabulary. 2011. Available online: https://www.iso.org/standard/51180.html (accessed on 19 September 2021).

16. Bureau International des Poids et Mesures. Resolution 2 of the 3rd CGPM. Declaration on the Unit of Mass and on the Definition of Weight; Conventional Value of gn. 1901. Available online: https://www.bipm.org/en/committees/cg/cgpm/3-1901/resolution-2 (accessed on 19 September 2021).
17. Physikalisch-Technische Bundesanstalt. g-Extractor. 2007. Available online: http://www.ptb.de/de/org/1/11/115/g-extractor.htm (accessed on 27 September 2021).
18. Bosetti, P.; Bruschi, S. Enhancing positioning accuracy of CNC machine tools by means of direct measurement of deformation. *Int. J. Adv. Manuf. Technol.* **2012**, *58*, 651–662. [CrossRef]
19. Winner, H. Radarsensorik. In *Handbuch Fahrerassistenzsysteme: Grundlagen, Komponenten und Systeme für Aktive Sicherheit und Komfort*; Winner, H., Hakuli, S., Lotz, F., Singer, C., Eds.; Springer Fachmedien Wiesbaden: Wiesbaden, Germany, 2015; pp. 259–316. [CrossRef]
20. Hessische Verwaltung für Bodenmanagement und Geoinformation. 2021. SAPOS Hessen HEPS. Available online: https://sapos.hvbg.hessen.de/service.php#EPS (accessed on 27 September 2021).
21. GeneSys Elektronik GmbH. *ADMA-G-Pro+ Technical Documentation*; GeneSys Elektronik GmbH: Offenburg, Germany, 2020.
22. Sick AG. *Technical Documentation SICK SENSICK WL12-2*; Sick AG: Düsseldorf, Germany, 2007.

Article

Effects of Automated Vehicle Models at the Mixed Traffic Situation on a Motorway Scenario

Xuan Fang [1,*], Hexuan Li [2], Tamás Tettamanti [1], Arno Eichberger [2] and Martin Fellendorf [3]

1   Department of Control for Transportation and Vehicle Systems, Faculty of Transportation Engineering and Vehicle Engineering, Budapest University of Technology and Economics, Műegyetem rkp. 3., H-1111 Budapest, Hungary; tettamanti.tamas@kjk.bme.hu
2   Institute of Automotive Engineering, TU Graz, 8010 Graz, Austria; hexuan.li@tugraz.at (H.L.); arno.eichberger@tugraz.at (A.E.)
3   Institute of Highway Engineering and Transport Planning, TU Graz, 8010 Graz, Austria; martin.fellendorf@tugraz.at (M.F.)
*   Correspondence: fangxuan@edu.bme.hu

Abstract: There is consensus in industry and academia that Highly Automated Vehicles (HAV) and Connected Automated Vehicles (CAV) will be launched into the market in the near future due to emerging autonomous driving technology. In this paper, a mixed traffic simulation framework that integrates vehicle models with different automated driving systems in the microscopic traffic simulation was proposed. Currently, some of the more mature Automated Driving Systems (ADS) functions (e.g., Adaptive Cruise Control (ACC), Lane Keeping Assistant (LKA), etc.) are already equipped in vehicles, the very next step towards a higher automated driving is represented by Level 3 vehicles and CAV which show great promise in helping to avoid crashes, ease traffic congestion, and improve the environment. Therefore, to better predict and simulate the driving behavior of automated vehicles on the motorway scenario, a virtual test framework is proposed which includes the Highway Chauffeur (HWC) and Vehicle-to-Vehicle (V2V) communication function. These functions are implemented as an external driver model in PTV Vissim. The framework uses a detailed digital twin based on the M86 road network located in southwestern Hungary, which was constructed for autonomous driving tests. With this framework, the effect of the proposed vehicle models is evaluated with the microscopic traffic simulator PTV Vissim. A case study of the different penetration rates of HAV and CAV was performed on the M86 motorway. Preliminary results presented in this paper demonstrated that introducing HAV and CAV to the current network individually will cause negative effects on traffic performance. However, a certain ratio of mixed traffic, 60% CAV and 40% Human Driver Vehicles (HDV), could reduce this negative impact. The simulation results also show that high penetration CAV has fine driving stability and less travel delay.

Keywords: traffic evaluation; simulation and modeling; connected and automated vehicle

Citation: Fang, X.; Li, H.; Tettamanti, T.; Eichberger, A.; Fellendorf, M. Effects of Automated Vehicle Models on the Mixed Traffic Situation on a Motorway Scenario. *Energies* **2022**, *15*, 2008. https://doi.org/10.3390/en15062008

Academic Editor: Mario Marchesoni

Received: 30 January 2022
Accepted: 7 March 2022
Published: 9 March 2022

Publisher's Note: MDPI stays neutral with regard to jurisdictional claims in published maps and institutional affiliations.

1. Introduction

With the rapid development of autonomous driving technology, Autonomous Vehicles (AV) have entered the operational stage in the road transport system. It is foreseeable that, in the near future, the proportion of AV will gradually increase. However, extensive autonomous driving is still out of reach. Considering the enormous possession of conventional vehicles, the first possibility of autonomous driving to implement on the road is the mixed traffic flow. This possibility will first appear in the motorway scenario, which is much simpler than urban roads. The mixing of AV and conventional vehicles will definitely have a significant impact on the performance of motorway traffic.

AV refer to the vehicles that can achieve the environment perception, route planning, decision making, and vehicle control functions in a highly intelligent and safe manner through the advanced on-board sensors, controllers, and actuators. The Society of Automotive Engineers (SAE) divides autonomous driving into six levels from Level 0 to Level 5 according to the need of the amount of driver intervention [1]. Level 0–Level 2 are defined as Advance Driver Assistant System (ADAS) while Level 3–Level 5 are defined as high-level automatic driving system. As high-level autonomous vehicles carry massive electronic devices, they are usually based on electric vehicles [2,3]. Connected and Automated Vehicles (CAV) refer to autonomous vehicles with integrated communication systems and network technologies to realize intelligent information transfer, exchange, and sharing between vehicles and everything (other vehicles, transport infrastructure, passersby, clouds, etc.). CAV have the capability of complex environment perception, intelligent decision making, and collaborative control, which can realize safe, efficient, comfortable, and energy-saving driving.

In this paper, we mainly focus on the Highly Autonomous Vehicles (HAV) and CAV simulation where HAV are defined as autonomous vehicles with Level 3 automation technology introduced in [4]. Vehicles with increasing levels of automation will fuse information from on-board multi-sensors and systems, allowing the vehicle to perceive the surrounding traffic and to locate itself precisely. Meanwhile, systems can enable the piloting of the vehicle with little or no human intervention during highly automated driving. Furthermore, the CAV model in this paper refers to vehicles with dedicated short-range communication technologies based on highly automated driving function, which allows vehicles to communicate with their surroundings, including infrastructure and other vehicles. In addition, it can provide drivers with real-time information about road and traffic conditions, as well as a wide range of connectivity services.

According to the market forecast of [5], the share of HAV and CAV in new car sales will increase from about 10% in 2025 to about 50% in 2035. Therefore, it is particularly important to evaluate the existing traffic scenario, driven by the huge market prospect.

Vehicle automation and communication technologies are considered promising approaches to improve the efficiency, safety, and environmental protection of traffic systems. Numerous studies have investigated the impacts of autonomous vehicles on traffic with simulation technology. However, the current Traffic Analysis, Modeling, and Simulation (TAMS) tools are not adequate for evaluating CAV or HAV driving behavior. Changes of the driving behavior parameter even had the opposite effect in different microscopic traffic simulation tools [6]. The reasons for this are as follows. First, for the CAV model, most TAMS tools cannot simulate vehicle inter-connectivity, i.e., V2V communication information sharing. Additionally, the majority of driving models are unrealistic, and many existing models require parameter calibration. Refs. [7–9] introduce the approaches to use empirical data to calibrate Wiedemann 99 model in Vissim in order to replicate CAV and HAV driving behavior. This method requires much time for collecting road data, which reduce the cost of modeling but require a lot of effort in training the samples as well as data statistics. In addition, most of them did not systematically evaluate the lateral/longitudinal control model, and [10,11] apply a linearized ACC model to perform speed control while considering only the following distance. To real driving conditions, the driver's desired speed should also be considered as a significant input to the system. Finally, Ref. [12] introduces HAV and CAV simulation models where control strategy is simplified. Although this approach reduces the difficulty of modeling, it does not reflect the actual vehicle driving behavior. In order to realistically reflect the driving behavior of HAV and CAV on the highway, we propose a driving model based on the Highway Chauffeur (HWC) function, which is introduced and defined in [13].

2. Methodology

For the HAV model, the proposed functionality (HWC) is defined as conditional automated driving function (SAE level 3—Conditional Automation) for standard driving, that is based on requirements and conditions defined by PEGASUS in [13]. PEGASUS is a research project which aims for a definition of a standardized procedure for the testing and experimenting of automated vehicle systems in simulation and real environments. Regarding the conditional automation on the guidance level, HWC function shall be capable of controlling the vehicle in longitudinal and lateral direction if the current vehicle state allows it. Additionally, the CAV model is defined by a lane change warning system on the basis of HWC, which ascertains the surrounding vehicle motion states based on V2V communication. The safe distance between ego vehicle and rear vehicle in the target lane is analyzed according to the goal of both collision avoidance and vehicle following safety.

2.1. Simulation Platform

Traffic analysis, modeling, and simulation is a mature field; several simulators are available. Each simulator has its own advantages in simulating real-world traffic based on a different car-following model. Typical TAMS tools applied by traffic engineers are PTV VISSIM [14], Simulation of Urban MObility (SUMO) [15], CORSIM [16], and Paramics [17]. Vissim is a microscopic road traffic simulator developed by PTV Group. Due to the comprehensive simulation diversity (motor vehicle module, bicycle module, pedestrian module, public transportation module, traffic timing module, etc.), as well as multi-dimension and efficiency of traffic simulation parameters, it is widely used in consulting firms, academia, and the public sector in the field of road traffic simulation.

Vissim provides a user-friendly Graphical User Interface (GUI), which means the user does not need to write programs manually to call different simulation modules and set up simulations. In addition to visual applications, Vissim also offers script-based modeling, which is very useful when users aim to dynamically access and control Vissim objects during simulation. This can be achieved through the COM (Component Object Model) interface, a technology that realizes inter-process communication between software with various programming language (e.g., C++, Phyton, Visual Basic, Java, Matlab, etc.).

The Vissim COM interface defines a hierarchical model with a head called IVissim, which represents the Vissim object. Under IVissim, there are different objects in which the functions and parameters of the simulator originally provided by the GUI can be controlled by programming. The Vissim-COM programming is introduced through Matlab Script for the co-simulation framework. For this Vissim-COM interface, [18] introduced a detailed development of the simulation environment. It is capable of performing all simulation sequences with the flexibility to allow the user to calibrate parameters and finally generate statistical plots automatically.

2.2. External Driving Model

The models are implemented in Vissim described in [18,19] using the External Driving Model interface (DLL). This interface provides the possibility to implement driver models with defined driving behavior. Similar functions are available through a Python interface called Traffic Control Interface (TraCI) in SUMO [20]. The whole DLL driving model is illustrated in Figure 1, which consists of three models. Considering that the traffic participants in Vissim cannot individually set up a sensor model to perceive the surrounding obstacles, DLL provides a possibility to obtain specific parameters of the surrounding vehicles (e.g., relative distance and velocity, heading angle, etc.) passed from Vissim. As a consequence, this perception information is gathered in the sensor model then sent to the driving model (HAV and CAV) as input. In parallel, driving models receive sensor fusion data and current vehicle dynamics data from Vissim to calculate lateral and longitudinal control commands, which are fed back to Vissim and the movement of the traffic vehicle is completed in the loop. The three models are described in more detail in the following subsections.

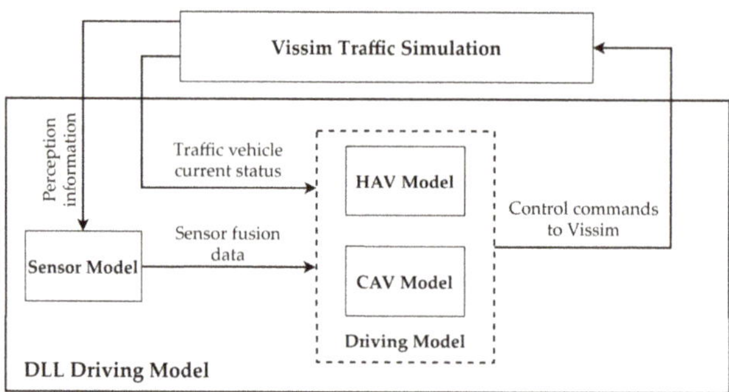

Figure 1. Overview of the DLL driving model.

2.3. Sensor Model

There are several different sensors built in a modern car supported to assist the driver or even drive autonomously. Therefore, the sensor model plays an important role in the adaptive control of the ego car and objects perception. Due to the limitations of Vissim, it is not possible to provide sensor models, a simplified sensor model is therefore presented here. In [21], an advanced driver assistance system is introduced from Toyota, which has been commercially realized in Japan in 2021. This system has multi-modal sensors covering the complete periphery of 360 degrees. Hence, the sensor model should have the ability to detect the surrounding objects, especially traffic participants upstream and downstream of the adjacent lane. Additionally, the most important parameter is to set the effective range of the sensor detection. Namely, only traffic participants within the detection range are considered. With the development of sensor technology, long-range radar is a range capability up to 150–200 m, presented in [22]. Therefore, a maximum detection range of 200 m is defined in the sensor model. The main function of the sensor model is to receive the specific parameters of the traffic participants from Vissim and transmit them to the driving function model by combining sensory data and fusion algorithms. Hence, Time to Collision (TTC), the lane change decision, and other perception signals are introduced in the subsequence.

2.3.1. Time to Collision Calculation

TTC is used to determine the time difference between the current time to a future moment when a potential crash will happen. It is a snapshot of the currently prevailing conditions and is only valid if the conditions stay stable. Nevertheless, it is useful for the prediction of potential crashes and for classifying the time-based safety distance to other traffic participants. The calculation of the TTC starts with the position of a car in dependency of the driving velocity, and the acceleration is described by Equation (1), where s and v are relative distance and velocity between ego and target car acquired from Vissim, respectively, and a is collected based on current ego car driving dynamics. The calculation of TTC t is therefore, the solution (Equation (2)) to Equation (1).

$$s = \frac{a}{2} \cdot t^2 + v \cdot t \tag{1}$$

$$t_{1,2} = \frac{-v \pm \sqrt{v^2 - 2 \cdot a \cdot s}}{a} \tag{2}$$

$$D = v^2 - 2 \cdot a \cdot s \tag{3}$$

Decisive for the solution of the equation for TTC $= t_{1,2}$ is the term under the square root, also called determinant D in Equation (3). There are three possible cases shown in Figure 2.

- $D < 0$: there is no collision expected. For example, if a car is traveling behind another car with the same speed and both start braking at the same moment, but the ego car decelerates more than the target car so that a collision does not occur.
- $D > 0$: in comparison, if the ego vehicle decelerates much less than the target vehicle, a crash will theoretically happen at TTC_1 and TTC_2. Due to the quadratic function, there are two solutions, however, only one of them is valid as the other one is a theoretical moment in the past in most cases.
- $D = 0$: there is only one result for the TTC calculation.

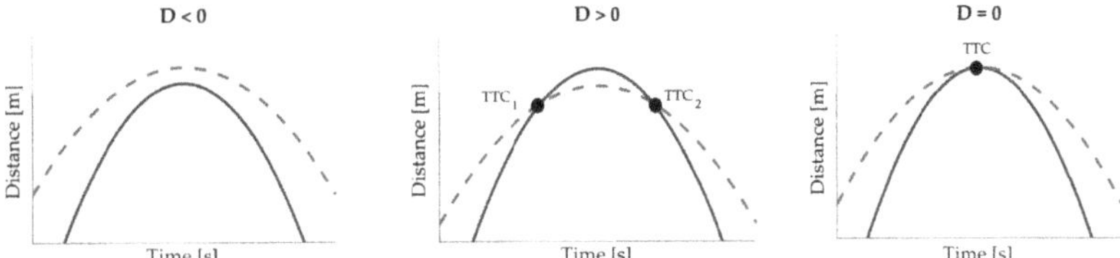

Figure 2. TTC outcome possibility.

When the TTC calculation is positive, it means that the velocity of the ego car is faster than that of the target car. Namely, the ego car is accelerating relative to the target car. Conversely, the TTC is negative, which means the ego car is decelerating relative to the target vehicle. In addition, the TTC is assumed to be the maximum value when there is no vehicle within the front detection distance or maintaining the same speed as the target car. Finally, When the TTC is towards zero, this is a very hazardous situation, meaning that the distance between vehicles is decreasing and a potential collision may occur. Therefore, TTC, as an important output of the sensor model, will be used as an essential condition to determine the occurrence of the lane change.

2.3.2. Decision Making

Decision making is another important function in the sensor model. In the previous section, TTC has been determined. The decision to change lanes based on the left-hand overtaking rule of the road is therefore initialized according to the TTC, which are predefined by the decision-making algorithm. Figure 3 illustrates a decision-making process to decide lane change direction. These commands will be as an internal signal transmitted to the driving function model. Before the vehicle reaches cruising speed, if the TTC detected in the same lane is highest, it means the target vehicle is far enough away to be safe. Therefore, the ego car can continue to drive on the current lane unless the adjacent lane TTC is higher and the lane change condition is satisfied; in this case, the sensor model will send a lane change decision signal to the driving model. Even during the lane change execution, the sensor model continues the TTC calculation between the surrounding cars in simulation iteration and the ego car in order to change the lane change decision at any time.

2.3.3. Other Relevant Signals

Other signals related to sensor sensing can be read directly from the Vissim simulation environment via the DLL interface, without additional computation. These signals will be used in the driving function model as well. These signals include:

- The relative speed, acceleration, and distance of the surrounding vehicles are used to adjust longitudinal and lateral control.
- Adjacent lane detection free space is used to determine lane change conditions.
- The position information of adjacent lanes for the multiple-lane road; according to the highway overtaking rules, overtaking on the right hand should be prohibited.
- For CAV, the sensor model is responsible for receiving the broadcasted V2V information and transmitting it to the control module.

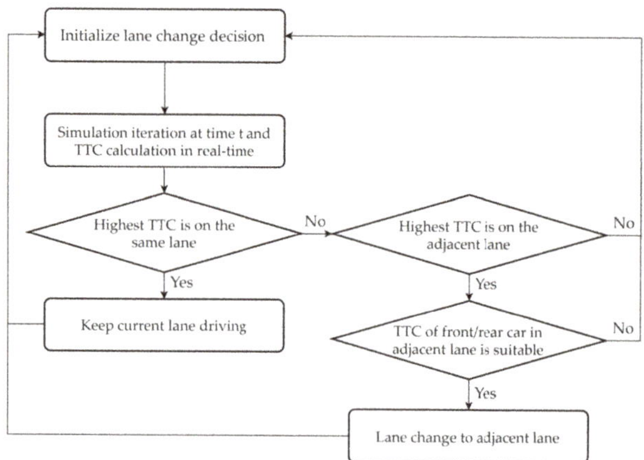

Figure 3. Process of decision making based on TTC calculation.

2.4. Highly Automated Vehicle Model

The HAV model is implemented based on HWC function defined in [13], which is the most advanced vehicle automation technology, operating on motorways only. The HAV model should adapt to all types of traffic conditions. The physical environment and driving function is virtually reconstructed and simulated in Vissim. In addition, the vehicles can carry out maneuvers in a fully autonomous and safe manner. The HAV model is primarily responsible for the longitudinal and lateral control of the ego car, which are introduced respectively in the subsequence.

2.4.1. Longitudinal Control

As one of the already serialized and common longitudinal controllers, the Adaptive Cruise Control (ACC) has the task to maintain a desired longitudinal speed or distance to a preceding vehicle. The norm International Organization for Standardization (ISO) 22179:2009 [23] defines the Full Speed Range Adaptive Cruise Control (FSRA), which allows control not only while free-flowing but also for congested traffic conditions. The system regulates the velocity of the ego vehicle depending on the vehicles in front and other traffic objects. Furthermore, if the FSRA-type system is used, the controller attempts to stop behind an already tracked vehicle within limited deceleration capabilities. The presented FSRA algorithm is developed based on a longitudinal vehicle model, the speed and distance controller introduced in [24]. The overall control scheme of the FSRA implementation process is depicted in Figure 4. The input signal is separated into three types, sensor inputs, ego vehicle states, and desired vehicle states. For the sensor input, they are relative speed δv and distance δs to a target vehicle provided, respectively, by the sensor model and transmitted to the distance control. As shown in Figure 1, the current traffic vehicle states can be read directly from Vissim. Therefore, the longitudinal ego vehicle velocity v_x is transmitted to the distance controller and speed controller. For the desired states, the desired velocity v_d and safe distance s_d are predefined by the user. Additionally, an acceleration controller is used for developing longitudinal control algorithms, which means that the distance and speed controller generates the desired acceleration $a_{s,d}$ and $a_{c,d}$, respectively. The control logic calculates a final desired acceleration, a_d, which is forwarded to control a Vissim traffic vehicle through a DLL interface. All the values of model parameters are set according to [25], which depends on reasonable literature.

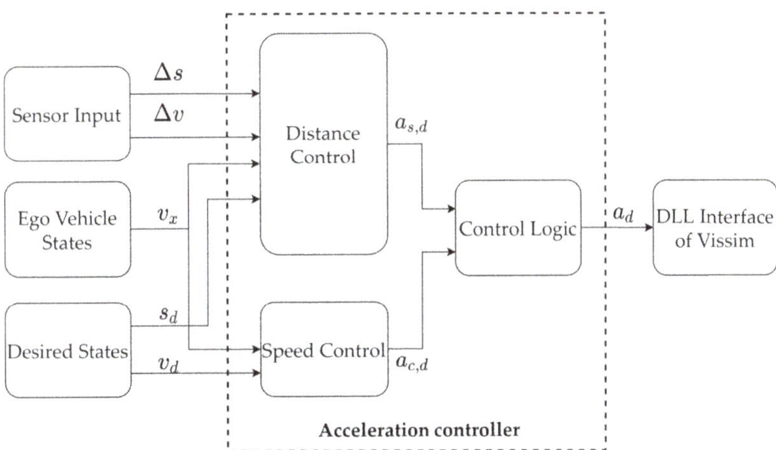

Figure 4. Longitudinal control schema in the driving model.

2.4.2. Lateral Control

The lateral control of the HAV mainly simulates the lane change function for a vehicle. For an autonomous lane change on SAE level 3, there is still no ISO norm defined. However, the ISO 21202:2020 [26] norm deals with Partially Automated Lane Change Systems (PALS). It describes basic control strategies, basic driver interface elements, minimum requirements for reaction to failure, minimum functionality requirements, and performance test procedures for a PALS. However, this will only be possible on a road where no pedestrian or other non-motorized vehicle is taking part in the traffic. For autonomous lane change of HAV, it has to observe the position of the car within the lane as well as the adjacent lanes and obstacles in the vicinity. Meanwhile, the ego vehicle can initiate a lane change on its own, as defined by PEGASUS [13]. Figure 5 presents a complete lateral control logic. The sensor model determines lane change decision based on the target car in front of the ego car on the same lane and the surrounding driving situation in the adjacent lane. It thus sends a fused calculation of the TTC and lane change indication f_{lc} from sensor model to Trajectory Planning Block (TPB). Meanwhile, TPB receives the time consumed for the whole process of lane change t_{lc}, vehicle speed v_m at the moment the lane change is triggered, and the lateral displacement h in real-time from Vissim. However, consider a corner case where an accelerating car may suddenly drive from behind during a lane change. In this case, TPB should abort the lane change action and re-plan back to the initial lane based on the rear TTC Δt_r provided by the sensor model. In the end, TPB converts the inputs to the heading angle of the vehicle, lane change active command and lane change direction. These signals are transmitted to the DLL interface to control the Vissim traffic vehicles until lane change action is finished. Therefore, lane change trajectory and the back-planning trajectory generation in TPB are described as two important functions in the following:

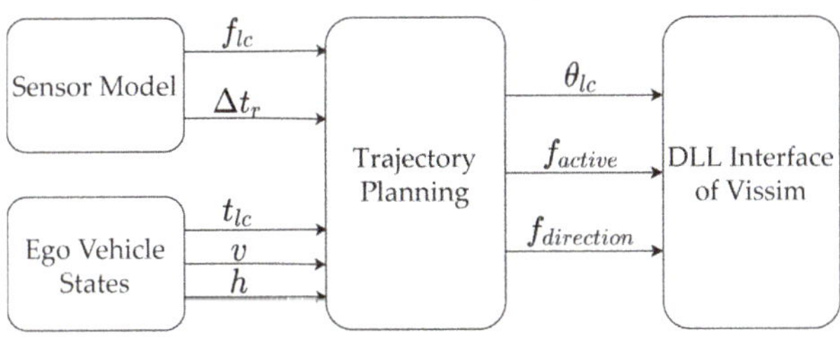

Figure 5. Lateral control schema in the driving model.

Regarding the lane change trajectory generation, the algorithm for the lane change behavior and the simulation implementation are presented in [27,28]. Based on these investigations, it is then known that the lane change is a time-based behavior. Therefore, the displacement of the vehicle from the center of the road can be derived in the trajectory equation for the lane change. The lateral $y_{lat}(t_{lc})$ and longitudinal $y_{long}(t_{lc})$ trajectories are calculated by the polynomial Equations which are determined in Equations (4) and (5), respectively. With this method, a smooth trajectory is composed of only a few points. Meanwhile, the acceleration profile is calculated by Equation (6) in order to better and realistically match the lateral motion, where t_m is maneuver time and calibrated to $6s$, which corresponds to an average time according to [27]. Referring to the description of the longitudinal behavior in [28], the maximum acceleration value can be set to $a_{max} = 1.2$ m/s^2. For a complete lane change action $h_l = 3.5$ m, which represents the displacement from one center line to the other (lane width). The entire process of lane change has an acceleration at the beginning phase and a gradual decrease in speed after the maneuver is completed. As shown in Figure 6a, the ego car detects a slower car ahead and that the adjacent lane is available for a lane change. Thus, a trajectory is generated.

$$y_{lat}(t_{lc}) = \left(\frac{-6h_l}{t_m^5}\right) \cdot t_{lc}^5 + \left(\frac{15h_l}{t_m^4}\right) \cdot t_{lc}^4 + \left(\frac{-10h_l}{t_m^3}\right) \cdot t_{lc}^3 \tag{4}$$

$$y_{long}(t_{lc}) = v_m t_{lc} \tag{5}$$

$$a(t_{lc}) = a_{max} \cdot sin\left(\frac{2\pi}{t_m} \cdot t_{lc}\right) \tag{6}$$

In back-planning trajectory generation, Δt_r is continuously checked as a safety standard until the lane change is completed. Lane change is aborted if the safety criterion fails to be met. Figure 6b presents a typical scenario. In this scenario, the ego car plans to change lane to pass the slower Car 1 in front of it, but a fast approaching Car 2 forces the ego car to abandon the lane change and move back to the initial lane. At this moment, h in Equation (4) is adjusted according to the lateral displacement between the current vehicle position and the initial point. Therefore, the lane change abortion path is generated by Equations (4)–(6). The HAV drives back to its original lane following the abortion path. With the lane change abortion mechanism, lane change safety of HAV can be guaranteed.

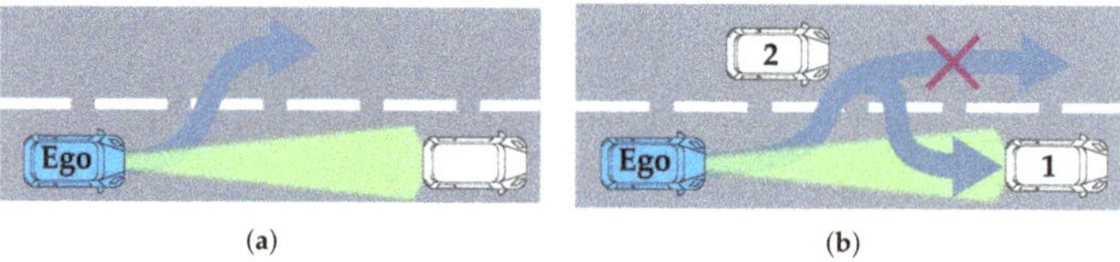

(a)   (b)

Figure 6. Lane change trajectory generation (**a**) a complete lane change trajectory planning; (**b**) an abortion trajectory generation.

2.5. Connected and Automated Vehicle Model

Based on the HAV model introduced in the previous section, the CAV model is proposed to simulate a realistic environment with a V2V cooperative lane change function. Wireless technologies are rapidly evolving. This evolution provides opportunities to use these technologies in support of advanced vehicle safety applications and crash avoidance countermeasures [29]. Compared to the HAV model, CAV share their positions, speeds, accelerations, and states with each other, which has a greater impact on the upstream vehicles in the adjacent lane. Therefore, in this section, the CAV model focuses more on the cooperation and communication between vehicles. The preliminary application communication scenario requirements are defined in [30]. Lane change warning function should support a maximum distance of up to 150 m.

As the CAV has the basic functions of the HAV illustrated in Figure 7, it has the same lateral and longitudinal control mechanism and commands signals *cmd* sent to DLL interface. Meanwhile, the V2V block receives state signals from HAV model and Vissim (i.e., current ego car speed v and acceleration a, rear TTC Δt_{rear} and lateral displacement h). Additionally, in order to simulate communication between vehicles, the ego car will broadcast an encapsulated lane change warning messages $COM_{message}$ once lane change flag f_{active} is triggered, with a radius of 150 m around the current ego car. Therefore, all CAV within the signal coverage area will receive this signal, and the relevant vehicle will be able to adjust its speed based on received encapsulated information. A typical scenario is shown in Figure 8; Car 3 is a CAV with V2V communication and follows Car 2. When the ego car detects Car 1 ahead, and the adjacent lane meets the lane change conditions, a lane change warning messages are broadcast before the lane change is triggered. After Car 3 receives the warning messages from the ego car, it will change the target car from Car 2 to ego car in longitudinal control. The ego car V2V broadcast communication messages to Car 3 in real-time during the lane change. Car 3 changes longitudinal movement according to V2V signals to reserve safe space for the ego car during the lane change.

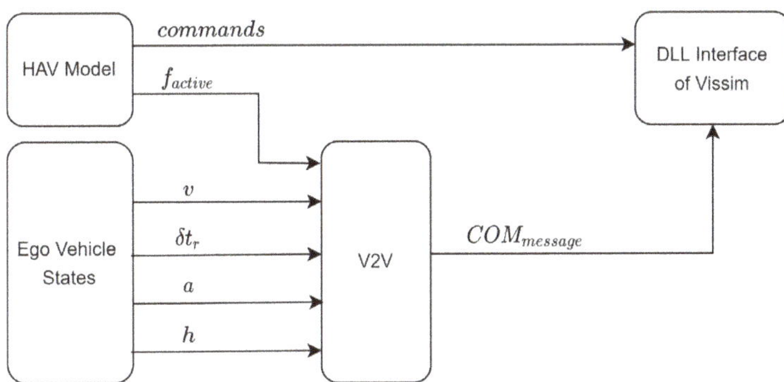

Figure 7. CAV model in the driving model.

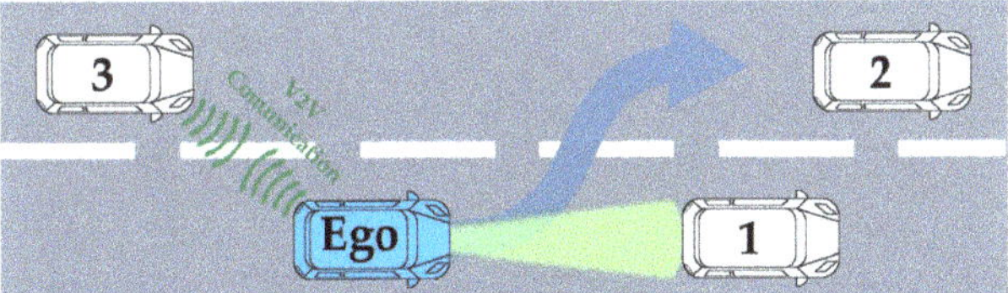

Figure 8. Cooperative lane change for CAV.

3. Simulation

The emerging AV will definitely change the travel demand; however, whether this change is positive or negative is still under research. To simulate the current realistic traffic conditions, traffic flow was generated based on the data measured by KIRA (Transportation Information System Database of Hungary). Based on historical information provided by this database, the volume on the main road was set to 1440 vehicles/h, and the volume on the ramp was set to 312 vehicles/h, with eight percent of them Heavy Goods Vehicles (HGV). As mentioned above, the HAV and CAV will be introduced to the road system gradually. Based on this view, 31 scenarios representing different vehicle model combinations were simulated. Scenarios 1–11 contain CAV and Human Drive Vehicles (HDV), the penetration rates of CAV ranging from 0% to 100% with 10% step. HDV are represented by the calibrated Wiedemann 99 model in Vissim. Similarly, scenarios 12–21 contain the HAV model and HDV, the penetration rates of HAV ranging from 0% to 100% with 10% steps.

Scenario 22–31 are a mix of three vehicle models with 20% steps. Each simulation involves a 1 h period, from 480 s to 4080 s with the first 480 s as warm-up time.

The test scenario is the upstream part of the M86 motorway. This road is close to the town of Csorna in northwestern Hungary (Győr-Moson-Sopron County, West Transdanubia region), connecting Szombathely with Győr, towards Budapest. The M86 is part of the TEN-T network [31] and also part of Hungarian State Public Road Network. Currently, the M86 is only in service between Szombathely and Csorna, with plans to extend north and south. This road is constructed to support the development and testing of autonomous vehicles. Figure 9 illustrates the overall 3.4 km profile of the M86 where the four sections of the road are marked:

- Section 1: two 3.50 m wide lanes are available for vehicles to travel. However, this section is connected to other roads. Thus, there is a ramp, which allows traffic from another motorway to merge into the main M86 road. Additionally, there are additional acceleration lanes connected to the ramp. Each vehicle can adjust vehicle speed in order to safely merge into the traffic. Therefore, this section of road has a great impact on the traffic speed in the simulation due to the complex traffic environment, which also proposes a challenge to the driving model.
- Section 2: This is a common two-lane section of approximately 300 m long, with two 3.50 m wide lanes. There will be some traffic merging into the main road from the acceleration lane coming out of the ramp extension.
- Section 3: At the end of the extended acceleration section, the dual carriageway will merge into a single carriageway, so there will be a lot of lane changes generated in this section.
- Section 4: The last section is a single lane with 3.5 m width up to the roundabout. This section of the road is relatively simple and has no lane changing behavior.

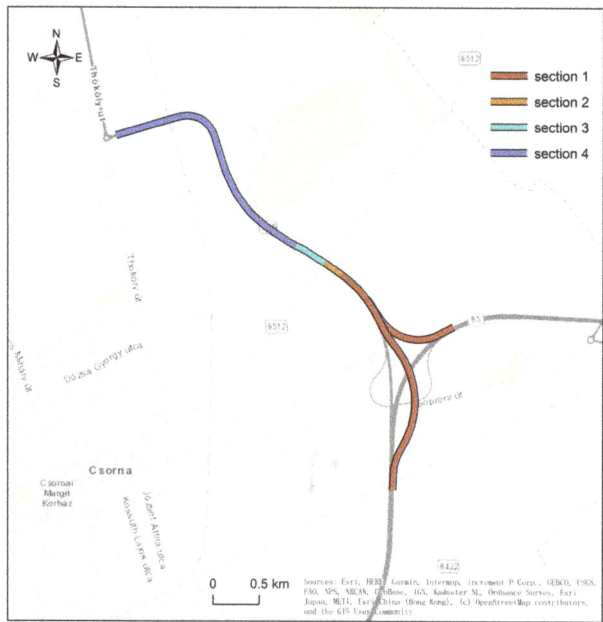

Figure 9. The upstream network of the M86 motorway.

In order to make the simulation scenario reproduce the real road conditions and environment, a digital twin-based M86 motorway is generated, including every detail of test environments at high accuracy. Ref. [32] introduced high precision mapping to build an ultra-high definition map based on the road geometry. Meanwhile, the M86 road network has been used in [33] as virtual test scenarios, and its accuracy can reach ±2 cm, this accuracy is enough to ensure the effectiveness of the test and verification. As consequence, our simulation results are more realistic.

4. Result and Discussion

Through the presented comprehensive simulation system, the operation process of different proportions of autonomous vehicle models were simulated. The simulated data over the whole network are collected every 60 s interval. As the most indicative and intuitive parameters for the network status, the average speed of the network, the total travel time, and the average delay are used to evaluate traffic efficiency. The average speed is calculated by dividing total distance vehicles traveled by total travel time. The average delay is calculated by dividing total delay by the number of vehicles in the network plus the number of vehicles that have arrived. This delay is obtained by subtracting the actual distance traveled in the time step and desired speed from the duration of the time step.

Table 1 shows the simulation results of the mix of three vehicle models. The negative impact of the introduction of SAE level 3+ AV on traffic efficiency is evident observed from the simulated data. Compared to CAV, this negative effect is worse when HAV isintroduced to the network. This can be explained by an over perfect Wiedemann 99 model and as, compared to the human drivers that may take aggressive driving behavior, SAE level 3+ AV will not take any risky behavior when changing lane. Furthermore, SAE level 3+ AV require much larger gaps to perform a lane change than human drivers, which causes congestion at the merging area.

Table 1. Traffic performance evaluation of the simulated network.

Vehicle Composition			Average Speed over the Network (km/h)	Total Travel Time (s)	Average Delay (s)
HAV	**CAV**	**HDV**			
0%	0%	100%	97.33	3462.04	7.32
0%	20%	80%	90.69	3721.97	7.33
0%	40%	60%	90.00	3699.27	5.36
0%	60%	40%	93.20	3577.72	3.38
0%	80%	20%	89.25	3738.47	2.13
0%	100%	0%	81.02	4133.17	0.51
20%	0%	80%	89.70	3776.68	7.12
40%	0%	60%	87.98	3793.02	5.18
60%	0%	40%	89.35	3756.64	3.46
80%	0%	20%	84.44	3997.94	2.21
100%	0%	0%	68.13	5085.81	1.03
20%	20%	60%	88.82	3774.40	5.32
20%	40%	40%	84.07	3967.40	4.43
20%	60%	20%	80.65	4125.98	2.70
20%	80%	0%	70.21	4125.98	0.78
40%	20%	40%	84.41	3941.23	4.03
40%	40%	20%	78.98	4223.81	2.87
40%	60%	0%	64.31	5221.23	1.04
60%	20%	20%	78.19	4253.08	2.62
60%	40%	0%	58.58	5834.29	1.35
80%	20%	0%	62.22	5496.21	1.08

To intuitively present the relationship between the penetration rate of the three vehicle models and the average speed, Figure 10a was drawn in ternary plots. It graphically depicts the penetration rate of the three vehicle models from 0% to 100% as the three sides in an equilateral triangle. The color inside the triangle indicates the average speeds over the network. At every point within the triangle, the ratio of each combination is inversely proportional to the distance from the corner. Combining Table 1 and Figure 10a, it can be known that in the mixed flow, 40% CAV and 60% HDV show the best traffic efficiency, with the highest travel speed and the shortest travel time. Although the introduction of SAE level 3+ AV has a negative impact on total travel time and average speed, average delays in mixed traffic flow are significantly reduced. Especially in 100% CAV scenarios, the average delay dropped from 7.32 s to 0.51 s. The standard deviation plot of average speed in Figure 10b demonstrated the huge advantage of mixed flow consisting of CAV and HDV on the traffic stability. A smaller standard deviation means that the speed measurements are closer to the mean speed, which represents that the vehicles on the network can travel at a relatively uniform speed.

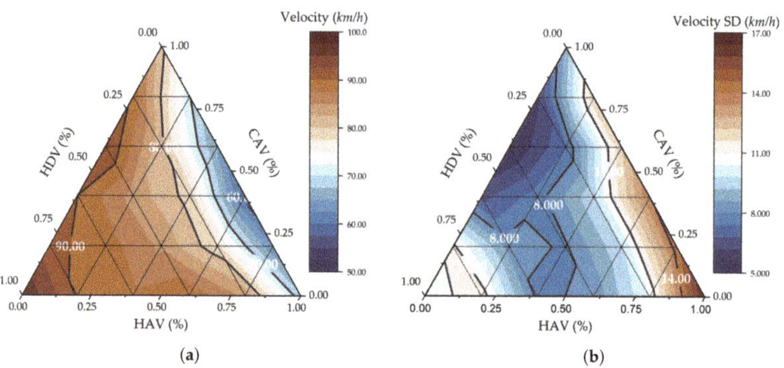

Figure 10. (**a**) Average speed over the network with mixed traffic; (**b**) Standard deviation of average speed over the network with mixed traffic.

Figures 11 and 12 present the changes of average speed with simulation time for various penetration rates of HAV and CAV, respectively. Overall, the introduction of HAV or CAV individually will cause the speed drop. For mixed traffic flow of HAV and HDV, 30% HAV with 70% HDV can generally keep the average speed at 100 km/h. Congestion can be observed at the end of simulation on the 100% HAV scenario; the average speed drops down to 40 km/h. It is foreseeable that, as the simulation time increases, the network will be fully blocked. This phenomenon can be explained by the much larger gap required by the HAV than human drivers when changing lanes. In addition, due to comfort considerations, the maximum acceleration of HAV is smaller than that of HDV, which results in the fact that when the network is full of HAV, the traffic downstream of the bottleneck decreases, and the upstream situation deteriorates.

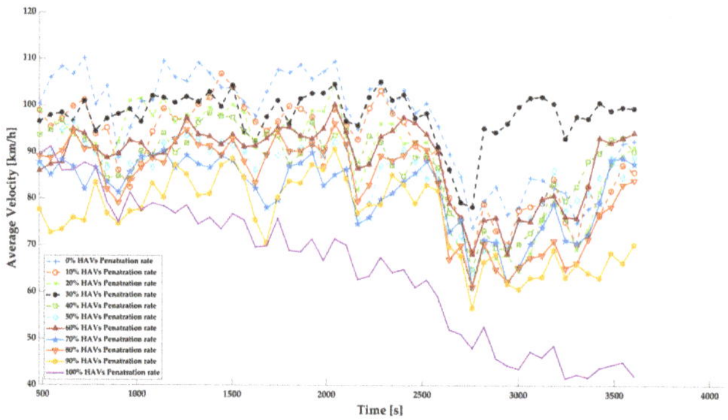

Figure 11. Average speed over the network for the various HAV penetration rates.

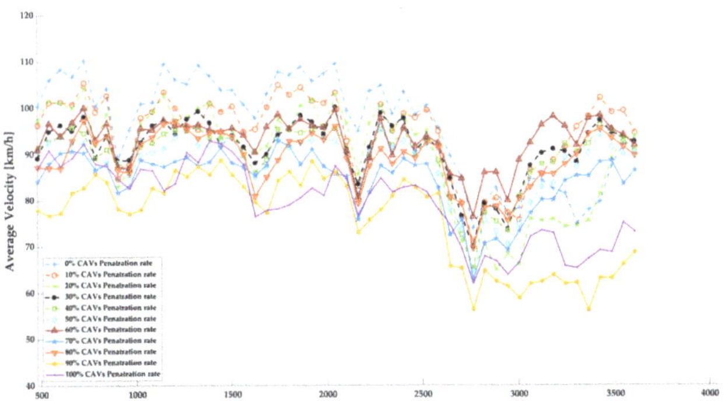

Figure 12. Average speed over the network for the various CAV penetration rates.

Although the introduction of CAV individually also has a negative impact on the road network performance, due to the connection and cooperation functionality of CAV, the distribution of average speed is more concentrated than that of HAV. In the later stages of the simulation, as the penetration rate of CAV in the network increases, they can make full use of their connectivity to travel with small gaps, change lane faster, and absorb shock wave. Focusing on speed change, we see that under a high CAV penetration rate, the speed of the network shows a continuous growth tendency.

5. Conclusions

This paper demonstrates the potential effects of the introduction of HAV and CAV on a real-world network. A microscopic traffic simulation framework that integrates vehicle models with different automated driving functions was constructed. These functions were implemented as an external driver model in the microscopic traffic simulator PTV Vissim. The framework was tested in a detailed digital twin based on the M86 motorway located in the southwest of Hungary. A case study consisting of different scenarios was performed to declare the effects of various combinations of HDV, HAV, and CAV. The traffic demand was obtained from real traffic counts. The possible combinations in 10% and 20% steps of the variable penetration rates per vehicle model formed 31 simulations. Each simulation was performed within a 1 h time period. Simulation results indicate the introduction of HAV and CAV deteriorating network performance. HDV outperformed HAV and CAV because HDV may take aggressive driving behaviors and is able to function over the speed limit. This characteristic is magnified by the presence of the ramp in the network. Among multitude scenarios with mixed traffic flow, the combination of 60% CAV and 40% HDV possess the optimal traffic performance in terms of average speed, total travel time, and average delay.

Due to the connectivity between CAV, the uniformity of speed was better in scenarios with high CAV penetration rates, which led to the excellent driving stability and the inhibition of the formation of traffic oscillations. In addition, the high CAV penetration rates in the network result in a significant reduction in traffic delays.

Author Contributions: Conceptualization, X.F. and H.L.; methodology, H.L.; simulation, X.F.; validation, X.F. and H.L.; writing—original draft preparation, X.F. and H.L.; writing—review and editing, X.F., H.L., T.T. and A.E.; supervision, T.T. and A.E; project administration, T.T., A.E. and M.F. All authors have read and agreed to the published version of the manuscript.

Funding: The research was supported by the Hungarian Government and co-financed by the European Social Fund through the project "Talent management in autonomous vehicle control technologies" (EFOP-3.6.3-VEKOP-16-2017-00001).

Institutional Review Board Statement: Not applicable.

Informed Consent Statement: Not applicable.

Data Availability Statement: Not applicable.

Conflicts of Interest: The authors declare no conflict of interest.

References

1. Sayer, J. *Adaptive Cruise Control (ACC) Operating Characteristics and User Interface: Standard J2399*; Society of Automotive Engineers: Warrendale, PA, USA, 2003.
2. Wróblewski, P.; Drożdż, W.; Lewicki, W.; Miązek, P. Methodology for assessing the impact of aperiodic phenomena on the energy balance of propulsion engines in vehicle electromobility systems for given areas. *Energies* **2021**, *14*, 2314. [CrossRef]
3. Wróblewski, P.; Kupiec, J.; Drożdż, W.; Lewicki, W.; Jaworski, J. The economic aspect of using different plug-in hybrid driving techniques in urban conditions. *Energies* **2021**, *14*, 3543. [CrossRef]
4. Norm. *SAE J3016:2016*; Taxonomy and Definitions for Terms Related to Driving Automation Systems for On-Road Motor Vehicles. SAE International: Warrendale, PA, USA. **2021** .
5. Catapult Connected Places. *Market Forecast for Connected and Autonomous Vehicles*; Technical Report; Department for Transport and Centre for Connected and Autonomous Vehicles, Smart Transport: Peterborough, UK, 2021.
6. Fang, X.; Tettamanti, T. Change in Microscopic Traffic Simulation Practice with Respect to the Emerging Automated Driving Technology. *Period. Polytech. Civ. Eng.* **2022**, *66*, 86–95. [CrossRef]
7. Goodall, N.J.; Lan, C.L. Car-following characteristics of adaptive cruise control from empirical data. *J. Transp. Eng. Part A Syst.* **2020**, *146*, 04020097. [CrossRef]
8. Leyn, U.; Vortisch, P. Calibrating VISSIM for the German highway capacity manual. *Transp. Res. Rec.* **2015**, *2483*, 74–79. [CrossRef]
9. Lu, Z.; Fu, T.; Fu, L.; Shiravi, S.; Jiang, C. A video-based approach to calibrating car-following parameters in VISSIM for urban traffic. *Int. J. Transp. Sci. Technol.* **2016**, *5*, 1–9. [CrossRef]
10. Li, Q.; Li, X.; Huang, Z.; Halkias, J.; McHale, G.; James, R. Simulation of Mixed Traffic with Cooperative Lane Changes. *Comput Aided Civ Inf.* **2021**. . [CrossRef]
11. Milanés, V.; Shladover, S.E. Modeling cooperative and autonomous adaptive cruise control dynamic responses using experimental data. *Transp. Res. Part C Emerg. Technol.* **2014**, *48*, 285–300. [CrossRef]
12. Papadoulis, A.; Quddus, M.; Imprialou, M. Evaluating the safety impact of connected and autonomous vehicles on motorways. *Accid. Anal. Prev.* **2019**, *124*, 12–22. [CrossRef] [PubMed]
13. The Highway-Chauffeur. Requirements and Conditions—Booth No. 04. Pegasus. In Proceedings of the PEGASUS Symposium 2019, Glasgow, UK, 10–12 April 2019; PEGASUS: Braunschweig, Germany, 2019.
14. Fellendorf, M. VISSIM: A microscopic simulation tool to evaluate actuated signal control including bus priority. In Proceedings of the 64th Institute of Transportation Engineers Annual Meeting, Dallas, TX, USA, 16–19 October 1994 ; Springer: Berlin/Heidelberg, Germany, 1994; Volume 32, pp. 1–9.
15. Behrisch, M.; Bieker, L.; Erdmann, J.; Krajzewicz, D. SUMO—Simulation of urban mobility: An overview. In Proceedings of the SIMUL 2011, the Third International Conference on Advances in System Simulation, ThinkMind, Barcelona, Spain, 23–29 October. 2011.
16. Halati, A.; Lieu, H.; Walker, S. CORSIM-corridor traffic simulation model. In Proceedings of the Traffic Congestion and Traffic Safety in the 21st Century: Challenges, Innovations, and Opportunities Urban Transportation Division, ASCE; Highway Division, ASCE; Federal Highway Administration, USDOT; and National Highway Traffic Safety Administration, USDOT, Chicago, IL, USA, 8–11 June 1997.
17. Cameron, G.D.; Duncan, G.I. PARAMICS—Parallel microscopic simulation of road traffic. *J. Supercomput.* **1996**, *10*, 25–53. [CrossRef]
18. Tettamanti, T.; Varga, I. Development of road traffic control by using integrated VISSIM-MATLAB simulation environment. *Period. Polytech. Civ. Eng.* **2012**, *56*, 43–49. [CrossRef]
19. Fellendorf, M.; Vortisch, P. Microscopic Traffic Flow Simulator VISSIM. In *Fundamentals of Traffic Simulation*; Springer: Berlin/Heidelberg, Germany, 2010; pp. 63–93.
20. Szoke, L.; Aradi, S.; Becsi, T.; Gaspar, P. Vehicle Control in Highway Traffic by Using Reinforcement Learning and Microscopic Traffic Simulation. In Proceedings of the 2020 IEEE 18th International Symposium on Intelligent Systems and Informatics (SISY), Subotica, Serbia, 17–19 September 2020; IEEE: Piscataway, NJ, USA, 2020; pp. 21–26.
21. Kawasaki, T.; Caveney, D.; Katoh, M.; Akaho, D.; Takashiro, Y.; Tomiita, K. *Teammate Advanced Drive System Using Automated Driving Technology*; Technical Report; SAE International: Warrendale, PA, USA. 2021.
22. Schneider, M. Automotive radar-status and trends. In Proceedings of the German Microwave Conference. Ulm, Germany, 5–7 April, 2005 , pp. 144–147.
23. *ISO 22179:2009*; Intelligent Transport Systems—Full Speed Range Adaptive Cruise Control (FSRA) Systems–Performance Requirements and Test Procedures. Norm, International Organization for Standardization: Geneva, Switzerland, September 2009.
24. Bernsteiner, S. *Integration of Advanced Driver Assistance Systems on Full-Vehicle Level-Parametrization of an Adaptive Cruise Control System Based on Test Drives*; Technical University Graz: Graz, Austria, 2016.
25. Li, Y.; Wang, H.; Wang, W.; Xing, L.; Liu, S.; Wei, X. Evaluation of the impacts of cooperative adaptive cruise control on reducing rear-end collision risks on freeways. *Accid. Anal. Prev.* **2017**, *98*, 87–95. [CrossRef] [PubMed]

26. *ISO 21202:2020*; Intelligent Transport Systems—Partially Automated Lane Change Systems (PALS)—Functional/Operational Requirements and Test Procedures. Norm, International Organization for Standardization: Geneva, Switzerland, April 2020.
27. Samiee, S.; Azadi, S.; Kazemi, R.; Eichberger, A. Towards a decision-making algorithm for automatic lane change manoeuvre considering traffic dynamics. *PROMET-Traffic Transp.* **2016**, *28*, 91–103. [CrossRef]
28. Nalic, D.; Li, H.; Eichberger, A.; Wellershaus, C.; Pandurevic, A.; Rogic, B. Stress Testing Method for Scenario-Based Testing of Automated Driving Systems. *IEEE Access* **2020**, *8*, 224974–224984. [CrossRef]
29. Ždánsky, P.; Balák, J.; Rástočný, K. Safety Integrity Evaluation of Safety Function. In Proceedings of the 2018 International Conference on Applied Electronics (AE), Pilsen, Czech Republic, 11–12 September 2018; IEEE: Piscataway, NJ, USA, 2018; pp. 1–6.
30. Shulman, M.; Deering, R. Vehicle safety communications in the United States. In Proceedings of the Conference on Experimental Safety Vehicles, Wahshington, DC, USA. 2007
31. Gutiérrez, J.; Condeço-Melhorado, A.; López, E.; Monzón, A. Evaluating the European added value of TEN-T projects: A methodological proposal based on spatial spillovers, accessibility and GIS. *J. Transp. Geogr.* **2011**, *19*, 840–850. [CrossRef]
32. Tihanyi, V.; Tettamanti, T.; Csonthó, M.; Eichberger, A.; Ficzere, D.; Gangel, K.; Hörmann, L.B.; Klaffenböck, M.A.; Knauder, C.; Luley, P.; et al. Motorway measurement campaign to support R&D activities in the field of automated driving technologies. *Sensors* **2021**, *21*, 2169. [CrossRef] [PubMed]
33. Li, H.; Makkapati, V.P.; Nalic, D.; Eichberger, A.; Fang, X. A Real-time Co-Simulation Framework for Virtual Test and Validation on a High Dynamic Vehicle Test Bed. In Proceedings of the 32nd IEEE Intelligent Vehicle Symposium, Nagoya, Japan, 11–15 July 2021; IEEE Press: Piscataway, NJ, USA, 2021.

 *energies*

MDPI

Article

Evaluation Methodology for Physical Radar Perception Sensor Models Based on On-Road Measurements for the Testing and Validation of Automated Driving

Zoltan Ferenc Magosi [1,*], Christoph Wellershaus [1], Viktor Roland Tihanyi [2], Patrick Luley [3] and Arno Eichberger [1]

[1] Institute of Automotive Engineering, Graz University of Technology, 8010 Graz, Austria; christoph.wellershaus@tugraz.at (C.W.); arno.eichberger@tugraz.at (A.E.)

[2] Department of Automotive Technologies, Faculty of Transportation Engineering and Vehicle Engineering, Budapest University of Technology and Economics (BME), 1111 Budapest, Hungary; tihanyi.viktor@kjk.bme.hu

[3] JOANNEUM RESEARCH Forschungsgesellschaft mbH, 8010 Graz, Austria; patrick.luley@joanneum.at

* Correspondence: zoltan.magosi@tugraz.at

Citation: Magosi, Z.F.; Wellershaus, C.; Tihany, V.R.; Luley, P.; Eichberger A. Evaluation Methodology for Physical Radar Perception Sensor Models Based on On-Road Measurements for the Testing and Validation of Automated Driving. *Energies* **2022**, *15*, 2545. https://doi.org/10.3390/en15072545

Academic Editors: Rui Esteves Araújo and Muhammad Aziz

Received: 31 January 2022
Accepted: 23 March 2022
Published: 31 March 2022

Abstract: In recent years, verification and validation processes of automated driving systems have been increasingly moved to virtual simulation, as this allows for rapid prototyping and the use of a multitude of testing scenarios compared to on-road testing. However, in order to support future approval procedures for automated driving functions with virtual simulations, the models used for this purpose must be sufficiently accurate to be able to test the driving functions implemented in the complete vehicle model. In recent years, the modelling of environment sensor technology has gained particular interest, since it can be used to validate the object detection and fusion algorithms in Model-in-the-Loop testing. In this paper, a practical process is developed to enable a systematic evaluation for perception–sensor models on a low-level data basis. The validation framework includes, first, the execution of test drive runs on a closed highway; secondly, the re-simulation of these test drives in a precise digital twin; and thirdly, the comparison of measured and simulated perception sensor output with statistical metrics. To demonstrate the practical feasibility, a commercial radar-sensor model (the ray-tracing based RSI radar model from IPG) was validated using a real radar sensor (ARS-308 radar sensor from Continental). The simulation was set up in the simulation environment IPG CarMaker® 8.1.1, and the evaluation was then performed using the software package Mathworks MATLAB®. Real and virtual sensor output data on a low-level data basis were used, which thus enables the benchmark. We developed metrics for the evaluation, and these were quantified using statistical analysis.

Keywords: automated driving; driver assistance system; virtual test and validation; radar sensor; physical perception model; virtual sensor model; digital twin

1. Introduction

In the field of driver assistance and active safety systems, an increasing level of automation has been introduced in public transport in recent years [1]. Automated vehicles will be able to detect and react to hazards and vulnerable road users faster and more appropriately than a human driver. This is expected to lead to a significant reduction in road accidents, as around 90 percent of accidents are primarily caused by human error, e.g., [2,3]. Taking the driver out of control will improve driving comfort by reducing the driver's workload, especially in monotonous situations, such as traffic jams, where the driver is mentally under-challenged, or in cases where the driver is overloaded and cannot fully manage the traffic situation. The human driver can also perform other productive or enjoyable activities during the journey, thus, reducing the opportunity cost of time spent in the car.

Automated vehicles can significantly increase access and mobility for populations currently unable or not allowed to use conventional cars [4]. In addition, automated driving functions can improve fuel economy by providing more subtle acceleration and deceleration than a human driver. However, a number of issues remain to be resolved before self-driving vehicles become a reality.

Motivation

Increasing the level of driving automation leads to an increase in system complexity. Therefore, the number of test cases necessary to proof the functional safety increases exponentially [5]. The literature reports that hundreds of millions of accident-free driving kilometres are needed to prove that the system is better than human vehicle control in terms of vehicle safety, e.g., [6]. Therefore, testing and validation efforts are being shifted towards virtual validation as well as X-in-the-Loop methods [7]. An essential requirement for testing ADAS/AD systems in virtual space is the realistic modelling of the virtual environment required for the system under test.

Virtual testing is particularly suitable for testing safety-critical scenarios that are difficult, costly, unsafe and impossible to reproduce on test tracks or roads [8]. The development and testing of driving assistance and automated vehicle control systems is performed step by step, from simple object detection to highly sophisticated functions, in the phases defined in the V-Model presented in ISO 26262-2:2018 [9]. Accordingly, the virtual environment, as well as the architecture and capabilities of the sensor models used, will vary according to the development phases [10].

To accelerate the test execution, a possible, and in recent years, very relevant solution is provided by the use of simulation tools on a virtual basis [11,12]. In the case of perception–sensor models in early development phases, virtual simulation based on ideal or phenomenological sensor models has become established in the industry. These models can be used to test and validate the fundamental operating principles of control architectures.

Using advanced perception–sensor models enables the testing of machine-perception and sensor-fusion algorithms in a later stage, enabling thus a first parameter tuning on a complete vehicle level before the first prototype is built.

Since sensor models have a limited ability to represent reality, careful consideration must be given to whether the model is a satisfactory replacement for the real sensor for validating the safety of ADAS/AD functions. However, there is no accepted methodology available that objectively quantifies the quality of perception–sensor models. In the present paper, a novel approach to assess the performance of virtual perception model is described.

The rest of the paper is structured as follows: Section 2 reviews the state-of-the-art technology, Section 3 describes the used method including the digital twin of the driving environment, the vehicle and the assessment approach. Section 4 presents the results of the new performance evaluation method using the IPG RSI model compared to an automotive radar sensor already proven on the market. Section 5 discusses the results of the sensor performance assessment and describes the limitations detected during the research and gives an outlook on future improvements.

2. State-of-the-Art

The simulation of perception sensors has been a part of worldwide research in recent years. Vehicles equipped with Automated Driving or Advanced Driver Assistance functions perform their tasks based on information provided by sensors that sense the environment, such as different types of radar sensors, front-, rear-, surround-view and night-vision cameras, LIDAR sensors, ultrasonic sensors [13]. The reliability of the warning and intervention hardware components and the algorithms that control them is strongly influenced by the quality of the information provided by the perception sensors. One of the goals of virtual testing is to model the behaviour of these sensors to match their behaviour under real environmental conditions.

In the simplest case, it is sufficient for the sensor model to give a perfect representation of the environment, using information artificially generated from the scenario, such as the relative distance, relative velocity, angular position, classification etc., taking into account the sensor's sensing properties, such as the maximum sensing range, horizontal and vertical coverage (FOV). This so called geometric or ideal sensor model is well suited to test the dynamic behaviour of complex systems at the whole vehicle level [14].

A more realistic modelling approach would be to implement physical sensor models simulating the behaviour of the complex interaction between sensors and the environment, such as the reflectivity, sight obstruction by other objects, multipath fluctuation, interference, damping etc.; however, the computational time and effort involved make the use of such models for vehicle testing impractical [15].

However, some sensor phenomena need to be modelled to test the performance of ADAS/AD systems with higher complexity. To test such systems under sub-optimal sensing conditions, sensor models are needed that reflect typical sensor phenomena based on the results of field tests, such as reduced range or increased noise levels in bad weather conditions, erroneous, missing or incomplete information on some sensed parameters, tracking errors, loss of objects etc. [16]. Models describing the phenomenological effects of different sensing technologies under similar conditions allow the performance and fault tolerance of the whole system chain to be investigated, taking into account an appropriate balance between the fidelity of the simulation, the complexity of the parameter settings and the computational power.

2.1. Classification of Virtual Sensor Models

Depending on the use case for virtual sensors, a general classification of sensor modelling approaches can be given. Using the basic simulation methodology, a model can be classified as a High-, Mid- or Low-Fidelity sensor model [10] or as black-, gray- or white-box [11]. As Peng in [17] and Schaermann in [18] denoted, there are three different modelling approaches for active perception sensors. These are the deterministic, the statistical and the field propagation approach; see Table 1. The deterministic modelling approach is based on mathematical formulations, which are represented by a multitude of parameters. If a sufficiently large amount of data is collected, the parameters can be trained and ideal sensor behaviour can be mapped.

The statistical approach is based on statistical distribution functions. This model is also called phenomenological sensor model and the realisations are drawn from a distribution function, which has to be determined before. This model architecture represents a good trade-off between computational effort and realism. This paper will focus on the validation of the third modelling approach, the field-propagation models. These models are simulating the propagation using Maxwell's equations. As these equations only can be solved for few geometries, a numerical approximation must be used.

In the case of a radar sensor, the propagation of the wave can be approximated with the Finite-Difference Time-Domain (FDTD). In order to achieve a real-time capable simulation, FDTD is too expensive in terms of computational power. However, other methods to approximate Maxwell's equations include the ray-optical approaches [17], simulating the propagation of electromagnetic waves by optical rays. Ray-tracing methods, like the published approach in [19], can model electromagnetic waves with various physical effects. For every radar transmitter–receiver pair, the wave propagation can be analysed to output the range and Doppler frequency for every detected target.

A major disadvantage of models based only on optics theory is that the effect of scattering is not included. Therefore, a scattering model must be implemented in the simulation. This can be either a stochastic scattering approach or a micro-facet-based scattering model as shown in [20]. By using these field-propagation models in the simulation environment, unprocessed environment sensor data with physical attributes are generated. This data is on a low-level in the signal processing chain, since it has not yet undergone any object detection or fusion algorithm.

Table 1. Classification of sensor modelling methods according to [17,18].

		Modeling Approach		
		Deterministic	Statistical	Field Propagation
data	object list	o	o	
level	low-level detection		(o)	o

2.2. Assessment Methods of Virtual Sensors

Virtual testing, for all its potential benefits, has limited fidelity because of its generalisation to the real environment. Sensor models represent reality only roughly or by focusing on a specific property of a sensor type. Therefore, it is necessary to assess whether a sensor model can sufficiently represent the real world to validate the safety of ADAS/AD systems. In the literature, several methods have been published on verification and validation methods for perception–sensor models [21–23]. All these approaches have their advantages and disadvantages and specific areas of application.

Depending on the general validation strategy and the accuracy of the available perceptual sensor models, different goals can be achieved with the virtual ADAS/AD validation. Therefore, it is difficult to quantify the performance of the different techniques since no accepted methodology to assess the performance of sensor models in the automotive spectrum exists today. One simple approach to assess the accuracy of a virtual sensor model can be to compare the performance of the ADAS/AD function within the virtual environment with its real world performance, running the same driving scenario [8]. This approach assumes that, for the assessment, models describing the ADAS/AD function under test are available in addition to sensor models of sufficient fidelity and that at least one prototype of the real hardware is available for real-world testing.

Since our hardware resources are limited to open-interface sensors freely available on the market, our sensor modelling efforts are designed accordingly. The present research introduces a novel approach that we call *Dynamic Ground Truth—Sensor Model Validation* (DGT-SMV) for performance assessment of perception–sensor models. The method is based on a statistical comparison of simulated and measured low-level radar data and aims to provide a quantifiable evaluation of the low-level radar-sensor model used. The method is presented in the next section.

3. Methodology

3.1. Dynamic Ground Truth Sensor Model Validation Approach

The DGT-SMV approach is depicted in Figure 1. The process starts with the definition of driving scenarios (*scenario definition*), which are related to phenomena of the evaluated sensor, such as multipath-propagation and separation capability [24,25]. In the next step, the tests are performed on a proving ground or in public traffic (*real test drive*), including an accurate measurement equipment.

The measurement data is used to label the recorded low-level sensor output (*measurement data labelling*). For providing a method for direct comparison, the test drives are then re-simulated in a detailed virtual representation using a digital twin of the environment and the investigated virtual sensor (*virtual sensor replay*). The sensor then produces the virtual low-level sensor output that is finally compared with the measured sensor output (*performance evaluation*) with statistical methods. The individual steps in the DGT-SMV process are described in detail in the next sections.

3.2. On-Road Measurements

Virtual testing can be conducted on a road section edited with a simple road editor in the initial development phase; however, commercially available simulation programs can be also used to re-simulate the real-world measurements of road sections. Ultra-High-Definition Maps [25] allow the simulation of existing real road geometries, thus, facilitating the realistic modelling of the virtual environment. This method allows the analysis of the complete functional chain from detection of environment perception sensors to intervention systems.

On the digital map, one can accurately measure the position, direction and movement of the test vehicle, as well as the reference distance to static objects, such as lane markings, guardrails and curbs using a high precision inertial measurement unit (IMU). As part of the measurement campaign in cooperation with the Department of Automotive Technologies of the Budapest University of Technology and Economics [25], several driving manoeuvres were performed to determine the validation criteria for radar-sensor models by collecting real radar-sensor data.

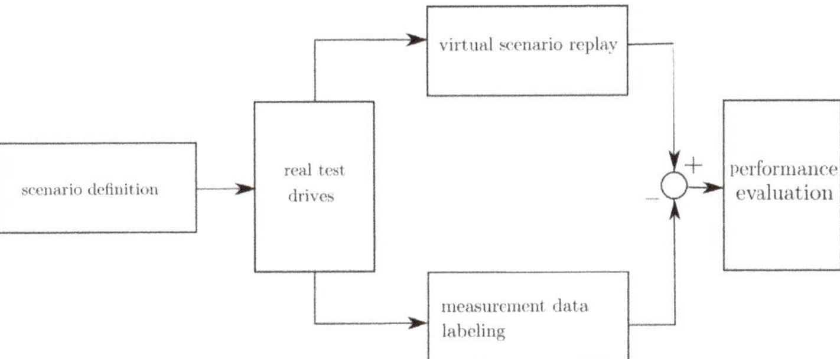

Figure 1. Dynamic ground truth sensor model validation process.

3.2.1. Driving Scenario

In our literature review, we did not find systematic validation and verification methods that would allow for the objective testing of available sensor models, in particular, with respect to the different phenomena that occur in different test scenarios during test drives with real sensors. The problem was first raised in the ENABLE-S3 [26] project. The ENABLE-S3 EU project aimed to develop an innovative V&V methodology that could combine the virtual and the real world in an optimal way. Several experiments were conducted to define and test validation criteria for sensor models. Three radar-specific phenomena were identified to be investigated in detail.

All of these phenomena derive from the physics of radar detection and are widely used to describe the performance of radar sensors. These are the ability to detect occluded objects (multipath propagation), the separation of close objects (separability) and the rapid fluctuation of the measured radar cross-sectional signal (RCS) over azimuth angles. As a result of this research, Holder et al. [24] concluded that validating and verifying sensor models and measurement data for repeatability and reliability is a difficult and complex task due to the highly stochastic nature of the radar output data. Furthermore, they found that radar-specific characteristics can be related either to the hardware architecture of the sensor under investigation (i.e., separability) or to signal propagation properties, such as multipath propagation, scattering and reflections (i.e., detection of occluded objects).

In this research work, the Continental ARS308 commercially available radar sensor was used for generating radar-sensor-measurement data to characterise the behaviour of the real radar sensor under different driving scenarios. Since we did not have detailed technical documentation for the radar sensor used in our experiments, which would allow us to infer its sensor performance and some expected hardware-related radar-specific phenomena, we decided to treat the sensor as a black box. Thus, we only used the information given in the sensor data-sheet to set the parameters in the virtual sensor model.

The measurement campaign on the M86 highway section in Hungary was conducted in cooperation with international industrial and academic partners. In this campaign, two important aspects of the assessment of ADAS/AD functions were considered. First, the mapping of the road geometry in order to produce a UHD map of the highway section.

Secondly, the creation of a ground truth database of all participating test cars using high accuracy Global Navigation Satellite System (GNSS). Detailed insights for the entire measurement campaign can be found in [25]. Taking into account the number of available potential target vehicles, a tailor-made manoeuvre catalogue was prepared, that includes a total of 17 manoeuvrers, with 12 manoeuvrers designed for measurement runs on a M86 highway section and five on the dynamic platform of the ZalaZone automotive proving ground [27].

As more target vehicles were available, the manoeuvres were performed with up to five vehicles, with different configurations in terms of distance, speed and acceleration. For completeness, it should be mentioned that there were also manoeuvres involving up to 11 vehicles and two trucks, for which high accuracy GPS data are also available. In this research work, we demonstrate the potential of our assessing method using one selected manoeuvre—the *Range test target leaving*, which is depicted in Figure 2.

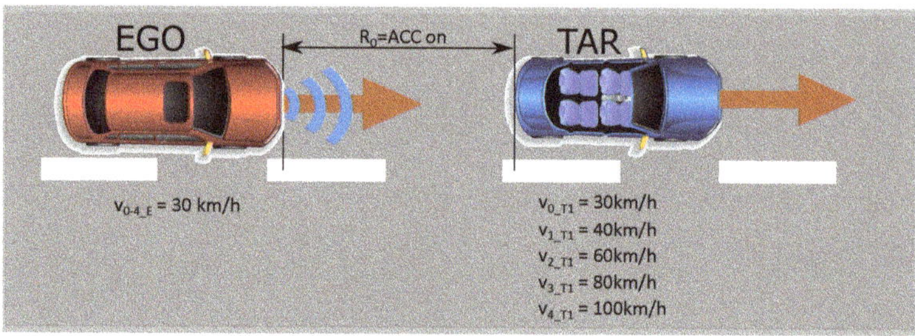

Figure 2. Target leaving with constant delta speed.

This driving scenario contains four driving manoeuvres with varying target vehicle speed parameters. The initial state is defined as follows. Both vehicles with activated FSRA (Stop and Go ACC) and follow time set to the minimum reached the initial speed, which was set to 30 km/h for this manoeuvre. The distance to the TARGET vehicle controlled by the FSRA system of the EGO vehicle is in steady state condition. After the initial conditions are reached, the driver of the target vehicle changes the set speed of the FSRA system from the initial 30 km/h to $v_{set\_TAR} = v_{1-4\_TAR}$ and leaves the EGO vehicle.

The measurement is considered complete when the distance between the vehicles has reached 250 m. To obtain the best measurement result, the angular orientation deviation (with respect to the direction of movement of the sensor) of the vehicles tested shall be kept below 1 degree. Unfortunately, the highway section used in this joint research project had a slightly curved characteristic, and therefore the angular orientation deviation continuously changes during the test runs and increases over 1 degree. This road geometry will lead to that the reflection points on a cumulative representation are shifted.

3.2.2. Vehicle Set-Up and Measurement System

The sensor performance evaluation process is based on the comparison of low-level detection points from real measurements using automotive radar sensors and the corresponding simulation. This requires high accurate ground truth reference data of the environment including static and dynamic objects. An appropriate approach to generate ground truth data in terms of high accuracy measurements is illustrated in Figure 3.

In order to detect as much information as possible of the surrounding of the car, the ego vehicle was equipped with the following sensors for environmental perception:

- Continental ARS 308 RADAR sensor configured to detect "targets", also referred as low-level data and providing a new data set for each scan period.
- Continental ARS 308 RADAR sensor, configured to detect "objects" also referred to as highly processed data, provides information on the output of the tracking algorithm over several measurement periods.

- Robosense RS-16 LIDAR sensor, provides the data point cloud of the 360° sensor field of view.
- MobilEye ME-630 Front Camera Module, provides information of traffic signs, traffic participants, lane markings etc.
- Video Camera, provides visual information of the driving scenarios, used during post processing.

Radar sensors mounted on ego-vehicles are shown in Figure 4. In addition to the environmental sensors, the DEWETRON-CAPS Measurement System [28] was also mounted on the target and the ego vehicle. The CAPS measurement system allows the implementation of a wirelessly connected topology of data acquisition units, consisting of one master and several slave measurement PCs. In addition to the time-stamped input and output hardware interfaces, the core of the CAPS system is the high accuracy inertial measurement unit (IMU).

In our experimental setup, we used the most advanced Automotive Dynamic Motion Analyzer (ADMA-G-Pro+) GPS/INS-IMU from GeneSys Ltd. The ADMA, combined with an RTK-DGPS receiver [29] and connected to the Hungarian Positioning Service, provided high accuracy dynamic state and position information of the test vehicles in real time. In addition, an accurate time measurement is derived from the GPS/PPS signal, allowing synchronous measurement of all connected data acquisition units.

As all measurement inputs are time-stamped, the measurement system ensures that all data streams received local or via WLAN connection are stored in time synchronisation. In addition, the robust WLAN connection allows for the real-time transmission of dynamic state and position information from the target vehicle to the ego vehicle. The transmitted data allows the driver to monitor the driving scenario online to rapidly ensure the quality of the measurement process.

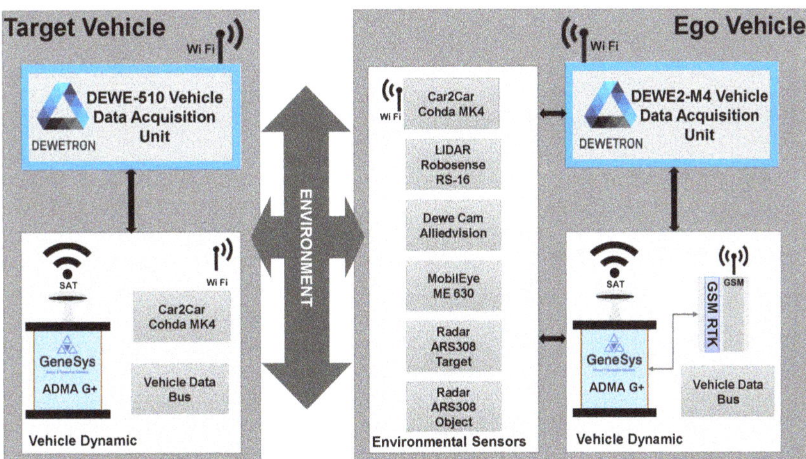

Figure 3. Schematic diagram of the measurement setup.

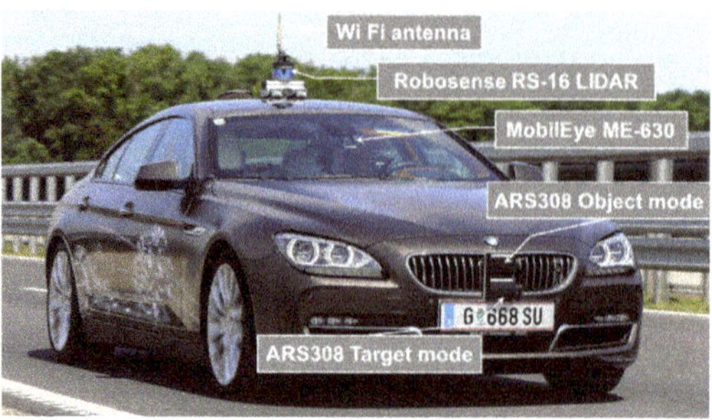

Figure 4. Test vehicle measurement setup.

3.3. Re-Simulation of Experiments

When on-road measurements are conducted, it is possible to generate exact the same scenarios in the virtual simulation using the recorded trajectory of the vehicles. IPG CarMaker® was selected for the simulation environment, as this software package provides a Virtual Vehicle Environment (VVE), which represents a Multi-Body-Simulation (MBS). This includes the equations of motions, kinematics and also for ADAS applications, sensor models. For building a Digital-Twin of a highway section, a detailed Ultra-High-Definition (UHD) map was provided by Joanneum Research Forschungsgesellschaft, which were also a partner in the consortium [25]. This map represents a highly accurate representation of the highway section in an OpenDrive file format. As the road geometry in IPG CarMaker® [30] is defined in a RD5 file format, the UHD map in in OpenDrive format has to be converted.

The virtual map includes the following road geometry items [25]:

- lane borders and markings,
- lane centre lines,
- curbs and barriers,
- traffic signs and light pole and
- road markings.

This detailed description of the environment makes it possible to reduce deviations between the real and virtual world to a minimum. In addition to the preparation of the virtual environment, the recorded trajectories must be transformed from the geodetic WGS-84 coordinate system to a metric coordinate system, which is used in the simulation. Since the operating radius of the experiments is smaller than 50 km, the curvature of the earth can be neglected, and the transformation can be performed on a plane metric coordinate system, which is spanned relative to a reference point [28].

IPG CarMaker distinguishes between two categories of vehicles, the ego vehicle and traffic vehicle. The first one represents the Vehicle Under Test (VUT), including a multi-body representation where all sub-systems can be changed by the user, e.g., mounting ADAS sensors to the vehicle, whereas the traffic vehicle only represents a motion model, which is, in our case, a single-track model. In order to make the traffic vehicle follow the previously recorded and afterwards transformed trajectory, the exact position in x- and y-direction was given to the vehicle at every time step.

The ego vehicle is controlled via the IPG Driver, a mathematical representation of the behavior of a human. This Driver performs any interaction to the car, e.g., steering or accelerating/braking. If the ego vehicle is now given a target trajectory, this would be approximated by the IPG Driver, just as a human driver would do. However, since in our case, an exact following of the recorded trajectories is absolutely necessary for the evaluation of the sensor model, a by-pass has to be performed on the driver model. This was done with a modification in the C-Code interface provided by the software vendor IPG. With this adaptation, the ego vehicle is now able to reproduce the same trajectory in the virtual environment as it was measured in the real world, ignoring any intervention by the IPG driver.

For the replay of the scenarios, a standard IPG-Car parameter set was used, including models of powertrain, tires, chassis, steering, aerodynamics and sensors. Since the ego vehicle exactly follows the recorded trajectory, no detailed parameter setting is required. The simulation software offers a number of different sensor models, which operate at different levels, ranging from ideal sensors to phenomenological sensors and to raw signal interfaces.

Since this paper focuses on the validation of low-level sensor models, the RSI radar-sensor model from IPG CarMaker V 8.1.1® [30] is considered and described in detail.

3.3.1. IPG RSI Radar Sensor Model

This sensor model provided by IPG CarMaker imitates the physical wave propagation by an optical ray-tracing approach. It includes the major effects of wave propagation, e.g.,

- Multipath/repeated path propagation.
- Relative Doppler shift.
- Road clutter.
- False positive/negative detections of targets.

Using this ray optical sensor model in a virtual environment requires the modelling of material properties of objects, such as the relative electric permittivity for electromagnetic waves and scattering effects. This parameters have a significant influence on the reflected direction and field strength of the reflected wave. The reflections are created by a detailed 3D surface in the visualisation. In the used set-up, the default values provided by the simulation tool for a 77 GHz radar were used.

3.3.2. Parameter Setting of the Sensor Model

Radar sensors are influenced by a multitude of parameters, which makes the parameter setting of such models complex. To ensure the comparability of the sensor model, the real hardware was treated like a black box so the sensor model was set with those parameters given in the data sheet provided by the manufacturer of the Radar sensor. To set the parameters for the atmospheric environment, temperatures and and in particular the data sheet based parameters: Field of View, Range, Cycle time, Max. Channels, Frequency, Separability Distance, Separability Azimuth, Separability Elevation, and Separability Speed was used.

With the additional offered two "design" parameters and scattering effect, CarMaker® gives users the possibility to fine tune the sensor model. However, one parameter given in the data sheet of the real hardware was not adjustable in the software package. The inaccuracy depending on the distance of the detected object was afterwards superimposed to the simulation results, as this leads to more robust results. To make this modified data visible in the results, this data is marked as *modified data* in Section 4 [30,31].

3.4. Labelling of Radar Measurement Data

In order to assign the individual reflection points of the radar sensor to the dynamic targets, a method that is already known in the field of object tracking is used, namely the gating technique. Using the ground truth information of the dynamic objects, target points only in a specific shaped area around an object of interest are considered. Figure 5 represents the gating area with the associated and not associated target points. The shape of the gating area can be variously designed, such as rectangular or elliptical [32].

In accordance with the shape of an average car, we used a rectangular shape. In this case, only dynamic objects are considered, as no static object information is available in the virtual map, e.g., bridge heads or overhead traffic signs. This means that the evaluation is limited to moving objects where the ground truth is measured with the RTK-GPS IMU measurement equipment but is also applicable to static objects, given that the ground truth is referenced.

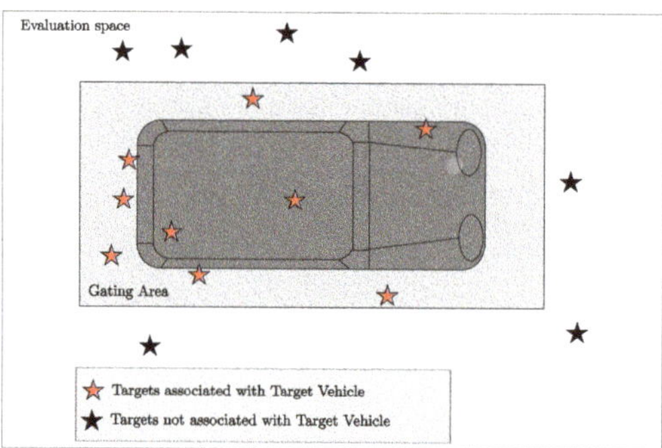

Figure 5. Gating area of the target vehicle with and with not associated reflection points according to [31].

3.5. Evaluation Procedure

Different evaluation metrics are given in literature, such as comparison of occupancy grid maps, statistical hypothesis testing, confidence intervals, correlation measurements and the generation of probability density functions [18,21,33,34]. In contrast to object list based deterministic sensor models, physical non-deterministic sensor models do not allow a direct comparison between experimental observations and simulation models. To make the highly stochastic process of a physical radar-sensor model comparable, including the physical attributes, e.g., the relative velocity or RCS value, statistical evaluation methods are best suited to describe the distribution of parameters in space and time.

Using previously labelled data, it is possible to evaluate them by statistical means in such a way that a quantitative statement can be made about the quality of the sensor model used in comparison to the real hardware. Introducing a reference point $\mathcal{P}_{ref}(x,y)$ on the target vehicle enables the calculation of the deviation on every radar detection point to the ground truth of the dynamic object, see Figure 6. Radar detection points are represented by the vector ζ_r for the measured sensor data and ζ_s for the simulated data. The deviation is calculated with

$$\zeta_{s,\Delta}(x,y) = \zeta_s(x,y) - \mathcal{P}_{ref}(x,y), \tag{1}$$

$$\zeta_{r,\Delta}(x,y) = \zeta_r(x,y) - \mathcal{P}_{ref}(x,y). \tag{2}$$

where $\zeta_{s,\Delta}$ representing the deviation of the simulation data and $\zeta_{r,\Delta}$ representing the deviation of the real sensor data to the reference point.

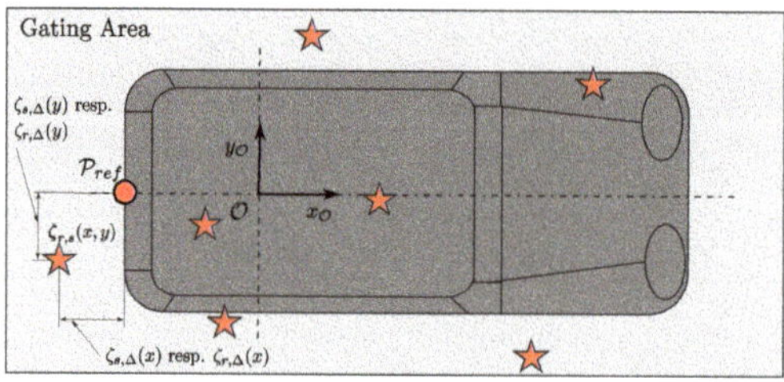

Figure 6. Reference point $\mathcal{P}_{ref}(x,y)$ on the dynamic object; Deviation target points real sensor to reference point $\zeta_{r,\Delta}(x,y)$ and deviation target points simulation to reference point $\zeta_{s,\Delta}(x,y)$ according to [31].

Assuming radar sensors are subject to a highly stochastic process, the detection points can be treated as realizations of a distribution function [35], p. 35. Using methods, including kernel density estimation (KDE), a probability density function (PDF) can be generated from the large number of realizations.

3.6. Validation Metrics for Comparing Probability Distributions

The validation of simulation models is based on the numerical comparison of data sets from experimental observations and the computational model output for a given use case. To quantify the comparison, validation metrics can be defined to measure the difference between the physical observation and the simulated output. Whether comparing measured physical quantities or virtual simulations, observed values contain uncertainties. In the presence of uncertainties, the observed values subject to validation are samples from a distribution of possible measured values, which are usually unknown.

To optimally quantify the difference or the similarity between distributions, we need the actual distributions. For the empirical data sets resulting from experimental observations (real radar sensor) and the output of the computational model (physical radar-sensor model), we do not know the actual distribution, or even its shape. Although one can always make assumptions (parametric) or estimate kernel density estimates (KDE), these are not quite ideal in practice, as their analysis is limited to specific types of distributions or kernels used. To stay as close to the data as possible, we therefore consider a non-parametric divergence measure. Non-parametric models are extremely useful when moving from discrete data to probability functions or distributions.

Non-parametric approaches are another way to estimate distributions. Such methods can be used to map discrete distributions of any shape. The simplest implementation of non-parametric distribution estimation is the histogram. Histograms benefit from knowledge of the data sets to be estimated and require fine-tuning to achieve optimal estimation results. In our application, this knowledge is available, since the bin width of the histogram can be determined according to the real sensor data sheet.

As stated above, a metric is a mathematical operator that gives a formal measure of the difference between experimental and model results. The metric plays a central role as it can be used to describe the fidelity of sensor models used to validate ADAS/AD functions. A low metric value means a good match and vice versa. According to [18] the metric can be defined by the following criteria: it must be intuitively understandable, applicable to both deterministic and non-deterministic data, a good metric defines a confidence interval as a function of the number of measured data and meet the mathematical properties of the metric.

The variables measured by perception sensors are usually non-parametric due to the highly stochastic nature of the output data [24]. Based on these properties, one possible description of the correspondence between synthetic and real perceptual data could be the comparison of their probability distribution functions. In the context of validating perception–sensor models, the most useful characterization appears to be the comparison of the distributions of random variables and the shapes of the corresponding observations. Random variables whose distribution functions are the same are called "distribution inequalities".

If the shapes of the distributions are not exactly the same, the difference can be measured using several possible measures. Maupin et al. [36] described a number of validation metrics for deterministic and probabilistic data that are used to validate computational models by quantifying the information provided by physical and simulated observations. In the context of this research, we proposed to use the Jensen–Shannon Divergence (JSD) [37], as it provides a quantified expression of the results of a comparison between two or more discrete probability distributions in a normalised manner.

The JSD, is a symmetrised version of the Kullback–Leibler Divergence described in detail in [18,36]. We consider a true discrete probability distribution $\mathcal{P}$ and its approximation $\mathcal{Q}$ over the values taken on by the random variable. The Jensen–Shannon Divergence calculated with

$$DJS(\mathcal{P}||\mathcal{Q}) = \frac{1}{2}DKL(\mathcal{P}||\mathcal{M}) + \frac{1}{2}DKL(\mathcal{Q}||\mathcal{M}) \tag{3}$$

where $\mathcal{M}$ is the mean distribution for $\mathcal{P}$ and $\mathcal{Q}$, as given by

$$\mathcal{M} = \frac{\mathcal{P} + \mathcal{Q}}{2} \tag{4}$$

The Jensen–Shannon Divergence uses the Kullback–Leibler Divergence to calculate a normalized measure. If $\mathcal{P}$ and $\mathcal{Q}$ describe the probability distribution of two discrete random variables, the KL divergence is calculated according to Equation (5).

$$DKL(\mathcal{P}||\mathcal{Q}) = \sum_{i=1} \mathcal{P}_i(x) log(\frac{\mathcal{P}_i x}{\mathcal{Q}_i x}) \tag{5}$$

Since the JS Divergence is a smoothed and normalised measure from the KL Divergence, it can be easily integrated into development processes. By definition, the square root of the Jensen–Shannon divergence describes the Jensen–Shannon distance.

$$DistJS(\mathcal{P}||\mathcal{Q}) = \sqrt{DJS(\mathcal{P}||\mathcal{Q})} \tag{6}$$

As both the divergence $DJS(\mathcal{P}||\mathcal{Q})$ and the distance $DistJS(\mathcal{P}||\mathcal{Q})$ are symmetric with respect to the arguments $\mathcal{P}$ and $\mathcal{Q}$ and the JS-Divergence is always non-negative, the value of $DJS(\mathcal{P}||\mathcal{Q})$ is always a real number in the closed interval of $[0; 1]$.

$$0 \leq DJS(\mathcal{P}||\mathcal{Q}) \leq 1 \tag{7}$$

If the value is 0, the two distributions $\mathcal{P}$ and $\mathcal{Q}$ are the same, if the value is 1, the two distributions are as different as possible. For better interpretation we present JSD in percentage values in the following. As $DistJS(\mathcal{P}||\mathcal{Q})$ fulfils the mathematical properties of a true metric [38], such as symmetry, triangle inequality and identity, the Jensen–Shannon Distance is a valid metric distance.

4. Results

In this chapter, the results of the statistical analysis are presented, showing the behaviour and the differences of the sensor model compared to the Ground Truth and the real sensor. In order to implement the DGT-SMV procedure and to illustrate the potential of the method, the driving scenario defined in Section 3.2.1 was used. As already described in Section 3.3.2, the modified data set with the super-positioned deviation was used for the evaluation of the RSI sensor model.

The data of both, the real and the simulated radar sensor, was accordingly to the radar properties split in a near range (0 to 60 m) and a far range (60 to 200 m) section. The used radar sensor can detect objects in the near range between 0 and 60 m in near and far range mode, which results in an overlap of the two sensor modes. For this reason, the far range was further subdivided into those data from the 0 to 60 m and 60 to 200 m for detailed analysis.

4.1. Comparison of Simulated and Measured Radar Signals

In Figures 7–9, visualizations of various statistical analysis of the near range radar and radar model are shown. Figure 7a,b presents the visualization of the radar detection points, which are associated to the corresponding dynamic target. The grey scale of each target point indicates its relative velocity, and the stroked line indicates the bounding box of the target vehicle. The contour lines in this plots are representing the multi-variant distribution of the reflection points. In Figure 8, the PDF's of the deviation to the reference point $\mathcal{P}_{ref}(x,y)$ in x- and y-direction of the realizations are shown.

The probability distribution can not only be used for the qualitative assessment of the distribution of the reflections but also serves as a basic prerequisite for the calculation of the Jensen–Shannon divergence. In Figure 8a, the deviation in the longitudinal direction, and in Figure 8b, the deviation in the lateral direction is shown. Figure 9 shows a PDF of the deviation of the relative velocity of each radar detection point in comparison to the Ground Truth relative velocity of the target vehicle.

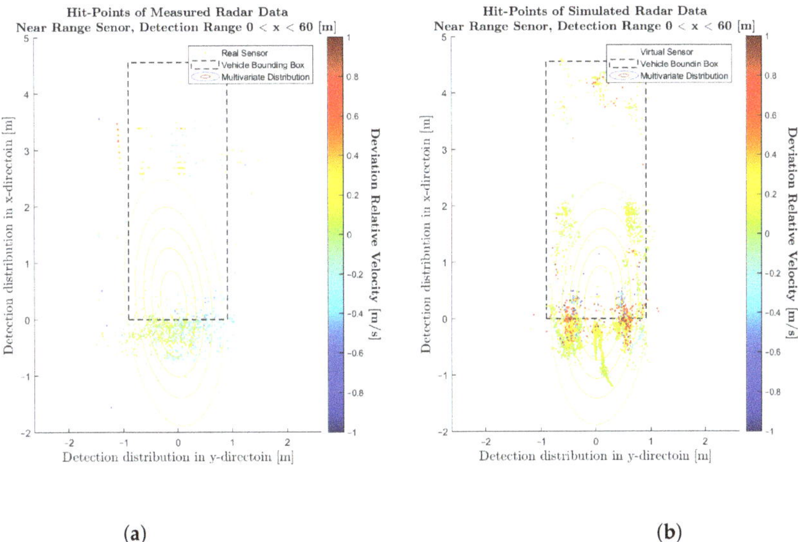

(a)   **(b)**

Figure 7. Evaluation of the detections in the range of 0 to 60 m of the near-range radar sensor and sensor model. (**a**) Scatterplot of detections for the real sensor. (**b**) Scatterplot of detections for the sensor model.

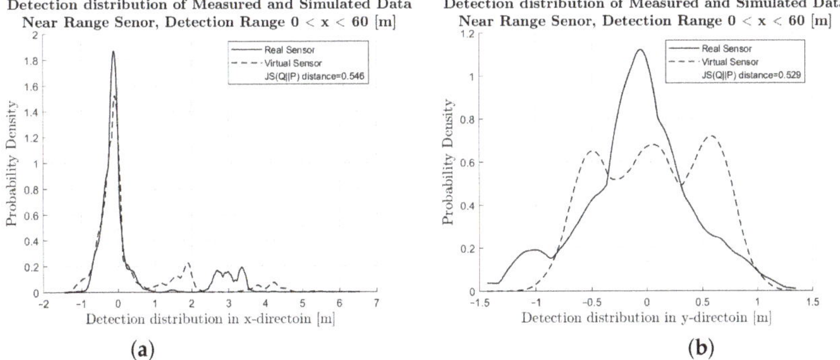

(a)   **(b)**

Figure 8. Evaluation of the detections in the range of 0 to 60 m of the near-range radar sensor and sensor model. (**a**) PDF of the deviation in the x-direction from $\mathcal{P}_{ref}(x,y)$ of the real sensor and sensor model. (**b**) PDF of the deviation in the y-direction from $\mathcal{P}_{ref}(x,y)$ of the real sensor and sensor model.

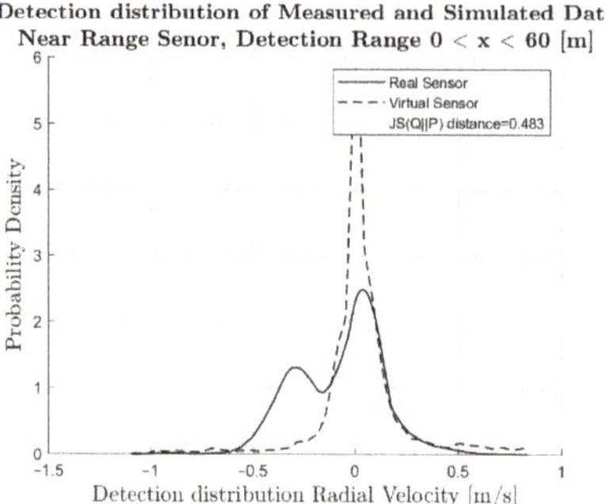

Figure 9. Evaluation of the detections in the range from 0 to 60 m of the near-range radar sensor and sensor model: PDF of the relative velocity in the *x*-direction from the reference velocity of the real sensor and sensor model.

In Figures 10–12, the visualization of the statistical analysis of the far range sensor in the near range section are shown. The results of the far range sensor for the far range section (60 to 200 m) can be found in Figures A1–A3.

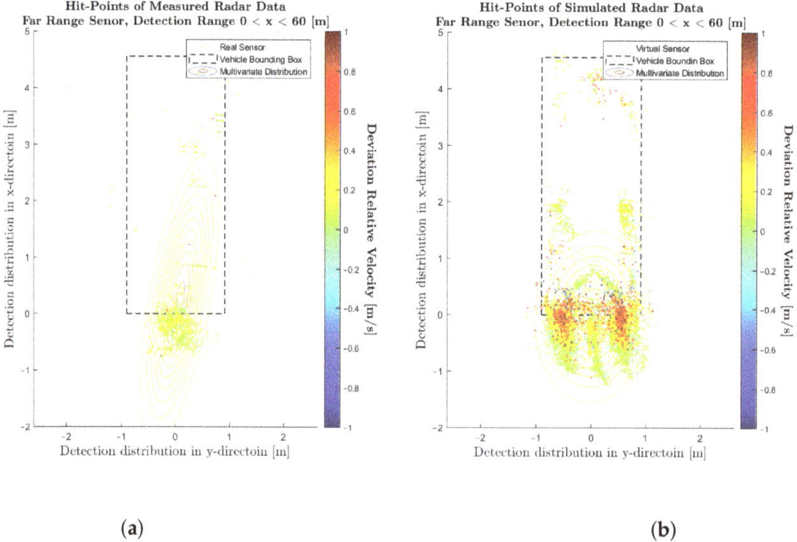

(a)   (b)

Figure 10. Evaluation of the detections in the range of 0 to 60 m of the far-range radar sensor and sensor model. (**a**) Scatterplot of detections of the real sensor. (**b**) Scatterplot of detections of the sensor model.

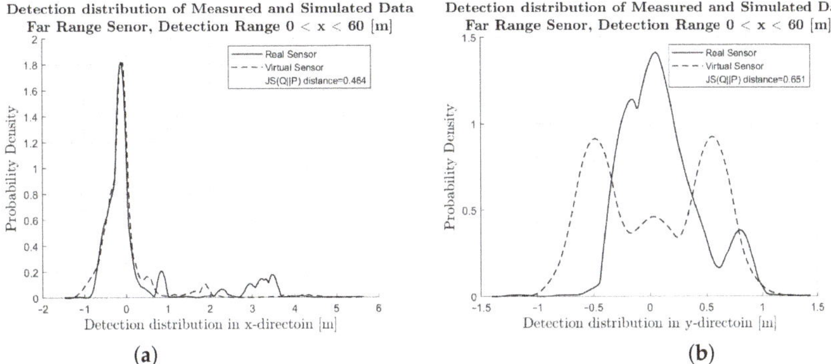

Figure 11. Evaluation of the detections in the range of 0 to 60 m of the far-range radar sensor and sensor model. (**a**) PDF of the deviation in the x-direction from $\mathcal{P}_{ref}(x, y)$ of the real sensor and sensor model. (**b**) PDF of the deviation in the y-direction from $\mathcal{P}_{ref}(x, y)$ of the real sensor and sensor model.

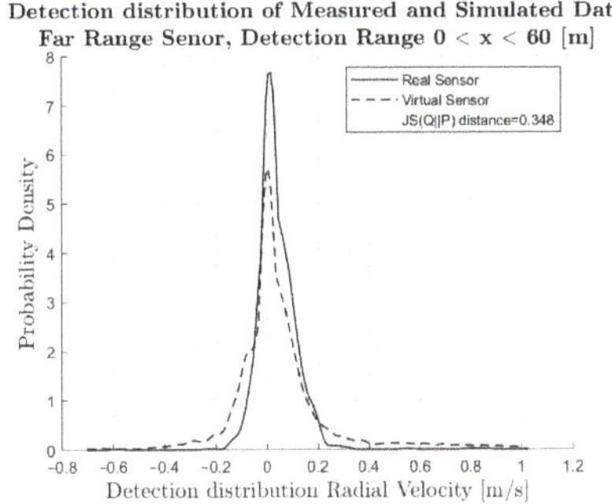

Figure 12. Evaluation of the detections in the range of 0 to 60 m of the far-range radar sensor and sensor model: PDF of the relative velocity in the x-direction from the reference velocity of the real sensor and sensor model.

4.2. Performance Metrics

When evaluating the performance of a virtual sensor for accuracy or fidelity, the correct performance metric should be selected to meet the requirements of the application. As described in [36], the data sets under comparison can be treated with or without uncertainty. Since, in our application, both experimental and predicted values are treated with uncertainty, comparison in the shape of a non-parametric discrete distribution is one promising solution.

The Jensen–Shannon Divergence (JSD) measures the distance between two discrete distributions by comparing the shape of two PDFs, one of which is the accuracy reference (real sensor data) and the other the output of a virtual model. JSD has two important features: first, JSD includes all the statistical information known about each distribution in the comparison. This means that the comparison is not limited to the average behaviour of the distributions. Second, it provides a real mathematical metric.

Since the Jensen–Shannon distance is a real mathematical metric, using the property that the value of $DistJS(\mathcal{P}||\mathcal{Q})$ is always a real number in the closed interval between 0 and 1, and if the value is 0, then the two distributions, $\mathcal{P}$ and $\mathcal{Q}$, are the same; otherwise they differ as much as possible, a quantitative comparison can be made between the sets of simulation.

In Tables 2–4, the JSD is expressed as a percentage for the near range as well as for the far range. The evaluated variables are $\zeta_{s,r\Delta}(x)$, defining the JSD metric for the relative distance in x, $\zeta_{s,r,\Delta}(y)$, for the relative distance in y and $\zeta_{s,r,\Delta}(v)$ for the relative velocity v.

Table 2. The results for the near-range radar sensor, detection range $0 < x < 60$ [m].

Evaluated Variable	JS-Distance in [%]
$\zeta_{s,r\Delta}(x)$	54.8
$\zeta_{s,r,\Delta}(y)$	53.1
$\zeta_{s,r,\Delta}(v)$	51

Table 3. The results for the far-range radar sensor, detection range $0 < x < 60$ [m].

Evaluated Variable	JS-Distance in [%]
$\zeta_{s,r\Delta}(x)$	46.5
$\zeta_{s,r,\Delta}(y)$	65.1
$\zeta_{s,r,\Delta}(v)$	34.2

Table 4. The results for the far-range radar sensor, detection range $60 < x < 200$ [m].

Evaluated Variable	JS-Distance in [%]
$\zeta_{s,r\Delta}(x)$	44.1
$\zeta_{s,r,\Delta}(y)$	52.8
$\zeta_{s,r,\Delta}(v)$	25.6

5. Discussion

Inspecting the results, a quick overview on the performance of the virtual sensor can immediately be achieved by the JSD metrics, where, for the deviations $\zeta_{s,r\Delta}$, 0% is perfect performance and 100% is the worst performance. In our example, it can be seen that, in the far range, the virtual sensor is more accurate in reproducing the relative velocity than in the relative distance. For the relative velocity, the JSD of $\zeta_{s,r\Delta}(v)$ is 34.2% up to 60 m and 25.6% up to 200 m. In the near range, the performance is worse at 51%.

For the relative distance, the better performance is seen in the x direction. The JSD of $\zeta_{s,r\Delta}(x)$ is 46.5% in the far range up to 60 m and 44.1% up to 200 m, for the near range 54.8% was observed. In the y direction, the related JSD values of $\zeta_{s,r\Delta}(y)$ are 65.1%, 52.8% and 53.1%, respectively.

This result is confirmed by visual inspection of the PDF illustrated in Figures 8, 9, 11 and 12 as well as the scatter plots in Figures 7 and 10. Comparing the shape of the PDFs, the strengths and shortcomings of the sensor model can be assessed, providing recommendations for parameter tuning and drawing conclusions on the validity of the results. The explanation of the results may be found in the specific modelling approach of the commercial radar-sensor model and is not part of this paper.

Limitations

The paper is subject to the following limitations:

- Limitations for dynamic objects: Since the UHD map in the simulation did not include any static objects, such as bridges, traffic signs, roadside barriers, vegetation and others, we only focused on the dynamic objects. The method can be enhanced for static objects in case the ground truth is annotated in the virtual sensor data.
- Limitations for the investigated radar phenomena: Here, we focused on a specific radar related phenomenon, the rapid fluctuation of the measured RCS over azimuth angles. Other phenomena as described in [26], such as multipath-propagation and separability were not covered here, since the real world driving tests included some limitations detected afterwards. The method can be extended to other phenomena, one has to define suitable driving scenarios and performance criteria.
- Limitations of specific benchmark results: Since no parameter tuning was performed in the IPG RSI model, the results obtained are not a direct indicator of the capabilities of the sensor model. However, the method can be used to improve the quality of the

modelling by fine-tuning the model parameters. Only after finding the best fit does the quality assessment become complete and can be directly compared with another model.

- Limitations for vehicle contours: According to the literature, the Jensen–Shannon divergence can be extended to a multivariate space with independent components, which allows for the comparison of multivariate random variables, making it possible to consider the contour of the vehicle. However, in this paper, we focused on the development of the methodology and data where the results are based on included one type of target vehicle. Hence, the difference of the rear wall of different vehicles can not be explicitly taken into account.

These limitations will be addressed in future research.

6. Summary

Despite the advantages of automated driving technologies with respect to safety, comfort, efficiency and new forms of mobility, only driver assistance of SAE levels 0 to 2, with the first applications in SAE L3, are on the market. One of the main reasons is the lack of proof in functional safety, which is due to the immense efforts required in real world testing. Virtual testing and validation is a promising option; however, the proof of realism of the simulation is not guaranteed at the moment.

One of the main obstacles is to reproduce the performance of machine perception in the simulation. Currently, there is a huge amount of development and research ongoing in providing virtual sensors. However, there is no accepted method for the proof of realism and prognosis quality for sensor models. In the present paper, we developed a method, the *Digital Ground Truth–Sensor Model Validation* (DGT-SMV), which is based on the re-simulation of actual test drives to thus allow for a direct comparison between the simulated and recorded sensor output. This approach requires defining suitable driving manoeuvres to reproduce the individual phenomena of the real sensor.

For the radar sensor, this is the multipath-propagation, separation ability and rapid fluctuation of the measured RCS over azimuth angles. The approach also requires accurate measurement equipment that records the ground truth of the driving scenario synchronously to the sensor data. After labelling the ground truth of the sensor output, a direct comparison between the simulated and recorded sensor output is possible.

For performance evaluation, we proposed a visual inspection of the simulated and recorded sensor output that we call *scatter plots* and, secondly, the transformation of these data with statistical methods based on Probability Distribution Functions to reveal the main performance of the virtual sensors. Finally, for a quick quantitative comparison, we proposed performance metrics based on the Jensen–Shannon distance. The method was applied on a commercially available sensor model (RSI radar-sensor model of IPG CarMaker) using real test drives on a closed highway in Hungary. For those tests, a high precision digital twin of the highway was available as well as the ground truth of the moving objects using RTK-GPS localization.

The results show that the DGT-SMV method is a promising solution for performance benchmarks of low-level radar-sensor models. In addition, the method can also be transferred to other active sensor principles, such as lidar and ultrasonic sensors.

Author Contributions: Conceptualization, Z.F.M.; methodology, Z.F.M.; software, C.W., P.L. and Z.F.M.; validation, Z.F.M. and C.W.; formal analysis, Z.F.M. and C.W.; investigation, Z.F.M. and C.W.; resources, V.R.T., A.E. and P.L.; data curation, Z.F.M. and C.W.; writing–original draft preparation, Z.F.M., C.W. and A.E.; writing–review and editing, Z.F.M., C.W. and A.E.; visualization, Z.F.M. and C.W.; supervision, A.E.; project administration, A.E. and V.R.T. All authors have read and agreed to the published version of the manuscript.

Funding: This research was not externally funded, however the preparatory work and the collection of measurement data was done in the funded Central System project, (2020-1.2.3-EUREKA-2021-

00001) and received funding from the NRDI Fund by the National Research, Development and Innovation Office Hungary. Open Access Funding by the Graz University of Technology.

Data Availability Statement: Data and software used here are proprietary and cannot be released.

Acknowledgments: The authors would like to express their thanks to the availability of measurement data as published in [25] and those who have supported this research and to the Graz University of Technology also.

Conflicts of Interest: The authors declare no conflict of interest.

Appendix A

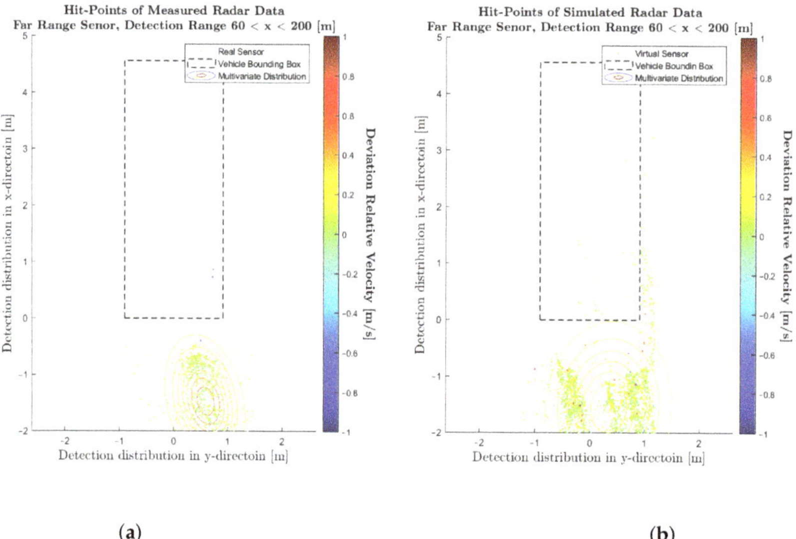

(a)   **(b)**

Figure A1. Evaluation of the detections in the range of 60 to 200 m of the far-range radar sensor and sensor model. (**a**) Scatterplot of the real sensor detections. (**b**) Scatterplot of the sensor model detections.

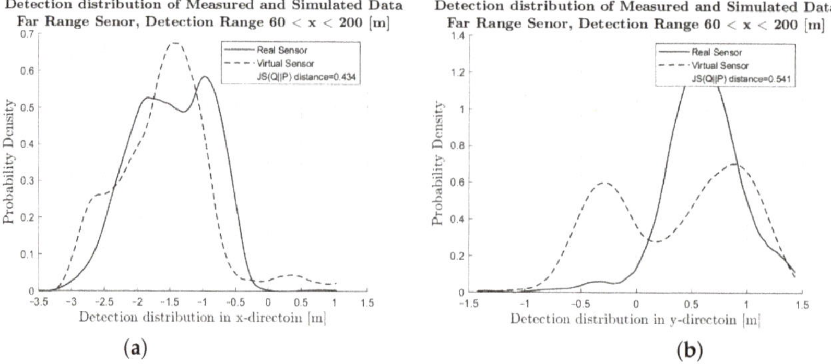

(a)   **(b)**

Figure A2. Evaluation of the detections in the range of 60 to 200 m of the far-range radar sensor and sensor model. (**a**) PDF of the deviation in the x-direction from $\mathcal{P}_{ref}(x,y)$ of the real sensor and sensor model. (**b**) PDF of the deviation in the y-direction from $\mathcal{P}_{ref}(x,y)$ of the real sensor and sensor model.

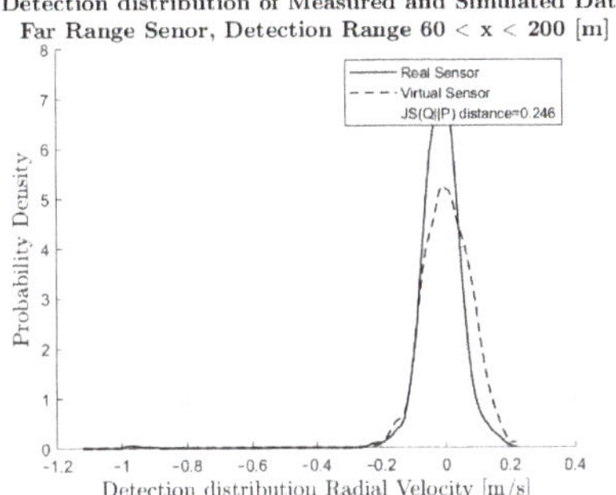

Figure A3. Evaluation of the detections in the range of 60 to 200 m of the far-range radar sensor and sensor model: PDF of the relative velocity in the *x*-direction from the reference velocity of the real sensor and sensor model.

References

1. Soteropoulos, A.; Pfaffenbichler, P.; Berger, M.; Emberger, G.; Stickler, A.; Dangschat, J.S. Scenarios of Automated Mobility in Austria: Implications for Future Transport Policy. *Future Transp.* **2021**, *1*, 747–764. [CrossRef]
2. DESTATIS. Verkehr. Verkehrsunfälle. In *Technical Report Fachserie 8 Reihe 7*; Statistisches Bundesamt: Wiesbaden, Germany, 2020; p. 47.
3. Dingus, T.A.; Guo, F.; Lee, S.; Antin, J.F.; Perez, M.; Buchanan-King, M.; Hankey, J. Driver crash risk factors and prevalence evaluation using naturalistic driving data. *Proc. Natl. Acad. Sci. USA* **2016**, *113*, 2636–2641. [CrossRef] [PubMed]
4. Tirachini, A.; Antoniou, C. The economics of automated public transport: Effects on operator cost, travel time, fare and subsidy. *Econ. Transp.* **2020**, *21*, 100151. [CrossRef]
5. Winner, H.; Hakuli, S.; Lotz, F.; Singer, C. (Eds.) *Handbuch Fahrerassistenzsysteme: Grundlagen, Komponenten und Systeme für Aktive Sicherheit und Komfort*; ATZ-MTZ-Fachbuch Book Series; Springer: Berlin/Heidelberg, Germany, 2015.
6. Kalra, N.; Paddock, S.M. Driving to safety: How many miles of driving would it take to demonstrate autonomous vehicle reliability? *Transp. Res. Part A Policy Pract.* **2016**, *94*, 182–193. [CrossRef]
7. Szalay, Z. Next Generation X-in-the-Loop Validation Methodology for Automated Vehicle Systems. *IEEE Access* **2021**, *9*, 35616–35632. [CrossRef]
8. Yonick, G. New Assessment/Test Method for Automated Driving (NATM): Master Document (Working Documents). 2021. Available online: https://unece.org/sites/default/files/2021-04/ECE-TRANS-WP29-2021-61e.pdf (accessed on 18 January 2022).
9. *ISO 26262-2:2018*; Road Vehicles—Functional Safety—Part 2: Management of Functional Safety. International Standardization Organization: Geneva, Switzerland, 2018.
10. Schlager, B.; Muckenhuber, S.; Schmidt, S.; Holzer, H.; Rott, R.; Maier, F.M.; Saad, K.; Kirchengast, M.; Stettinger, G.; Watzenig, D.; et al. State-of-the-Art Sensor Models for Virtual Testing of Advanced Driver Assistance Systems/Autonomous Driving Functions. *SAE Int. J. Connect. Autom. Veh.* **2020**, *3*, 233–261. [CrossRef]
11. Cao, P.; Wachenfeld, W.; Winner, H. Perception–sensor modeling for virtual validation of automated driving. *IT Inf. Technol.* **2015**, *57*, 243–251. [CrossRef]
12. Chen, S.; Chen, Y.; Zhang, S.; Zheng, N. A Novel Integrated Simulation and Testing Platform for Self-Driving Cars With Hardware in the Loop. *IEEE Trans. Intell. Veh.* **2019**, *4*, 425–436. [CrossRef]
13. Yeong, D.J.; Velasco-Hernandez, G.; Barry, J.; Walsh, J. Sensor and Sensor Fusion Technology in Autonomous Vehicles: A Review. *Sensors* **2021**, *21*, 2140. [CrossRef] [PubMed]
14. Safety First for Automated Driving. 2019. Available online: https://group.mercedes-benz.com/dokumente/innovation/sonstiges/safety-first-for-automated-driving.pdf (accessed on 3 March 2022).
15. Maier, M.; Makkapati, V.P.; Horn, M. Adapting Phong into a Simulation for Stimulation of Automotive Radar Sensors. In Proceedings of the 2018 IEEE MTT-S International Conference on Microwaves for Intelligent Mobility (ICMIM), Munich, Germany, 15–17 April 2018; pp. 1–4. [CrossRef]
16. Slavik, Z.; Mishra, K.V. Phenomenological Modeling of Millimeter-Wave Automotive Radar. In Proceedings of the 2019 URSI Asia-Pacific Radio Science Conference (AP-RASC), New Delhi, India, 9–15 March 2019; pp. 1–4. [CrossRef]
17. Cao, P. Modeling Active Perception Sensors for Real-Time Virtual Validation of Automated Driving Systems. Ph.D. Thesis, Technische Universität, Darmstadt, Germany, 2018.
18. Schaermann, A. Systematische Bedatung und Bewertung Umfelderfassender Sensormodelle. Ph.D. Thesis, Technische Universität München, München, Germany, 2019.
19. Gubelli, D.; Krasnov, O.A.; Yarovyi, O. Ray-tracing simulator for radar signals propagation in radar networks. In Proceedings of the 2013 European Radar Conference, Nuremberg, Germany, 9–11 October 2013; pp. 73–76.

20. Anderson, H. A second generation 3-D ray-tracing model using rough surface scattering. In Proceedings of the Vehicular Technology Conference, Atlanta, GA, USA, 28 April–1 May 1996; pp. 46–50. [CrossRef]
21. Sargent, R.G. Verification and validation of simulation models. In Proceedings of the 2010 Winter Simulation Conference, Baltimore, MD, USA, 5–8 December 2010; pp. 166–183. [CrossRef]
22. Oberkampf, W.L.; Trucano, T.G. Verification and validation benchmarks. *Nucl. Eng. Des.* **2008**, *238*, 716–743. [CrossRef]
23. Roth, E.; Dirndorfer, T.J.; von Neumann-Cosel, K.; Gnslmeier, T.; Kern, A.; Fischer, M.O. Analysis and Validation of Perception Sensor Models in an Integrated Vehicle and Environment Simulation. In Proceedings of the 22nd International Technical Conference on the Enhanced Safety of Vehicles, Washington, DC, USA, 13–16 June 2011.
24. Holder, M.; Rosenberger, P.; Winner, H.; D'hondt, T.; Makkapati, V.P.; Maier, M.; Schreiber, H.; Magosi, Z.; Slavik, Z.; Bringmann, O.; et al. Measurements revealing Challenges in Radar Sensor Modeling for Virtual Validation of Autonomous Driving. In Proceedings of the 2018 21st International Conference on Intelligent Transportation Systems (ITSC), Maui, HI, USA, 4–7 November 2018; pp. 2616–2622. [CrossRef]
25. Tihanyi, V.; Tettamanti, T.; Csonthó, M.; Eichberger, A.; Ficzere, D.; Gangel, K.; Hörmann, L.B.; Klaffenböck, M.A.; Knauder, C.; Luley, P.; et al. Motorway Measurement Campaign to Support R&D Activities in the Field of Automated Driving Technologies. *Sensors* **2021**, *21*, 2169. [CrossRef] [PubMed]
26. European Initiative to Enable Validation for Highly Automated Safe and Secure Systems. 2016–2019. Available online: https://www.enable-s3.eu (accessed on 20 December 2021).
27. Szalay, Z.; Hamar, Z.; Simon, P. A Multi-layer Autonomous Vehicle and Simulation Validation Ecosystem Axis: ZalaZONE. In *Intelligent Autonomous Systems 15*; Strand, M., Dillmann, R., Menegatti, E., Ghidoni, S., Eds.; Springer: Berlin/Heidelberg, Germany, 2019; pp. 954–963.
28. CAPS-ACC Technical Refernce Guide. 2013. Available online: https://ccc.dewetron.com/dl/52af1137-c874-4439-8ba0-7770d9c49862 (accessed on 12 January 2022).
29. An Introduction to GNSS, GPS GLONASS BeiDou Galileo and Other Global Navigation Satellite Systems. 2015. Available online: https://novatel.com/an-introduction-to-gnss/chapter-5-resolving-errors/real-time-kinematic-rtk (accessed on 18 January 2022).
30. IPG CarMaker. *Reference Manual(V 8.1.1)*; IPG Automotive GmbH: Karlsruhe, Germany, 2019.
31. Wellershaus, C. Performance Assessment of a Physical Sensor Model for Automated Driving. Master's Thesis, Graz University of Technology, Graz, Austria, 2021.
32. Wang, X.; Challa, S.; Evans, R. Gating techniques for maneuvering target tracking in clutter. *IEEE Trans. Aerosp. Electron. Syst.* **2002**, *38*, 1087–1097. [CrossRef]
33. Oberkampf, W.; Roy, C. *Verification and Validation in Scientific Computing*; Cambridge University Press: Cambridge, UK, 2010. [CrossRef]
34. Keimel, C. *Design of Video Quality Metrics with Multi-Way Data Analysis—A Data Driven Approach*; Springer: Berlin/Heidelberg, Germany, 2015.
35. Skolnik, M.I. *Introduction to Radar Systems*, 3rd ed.; McGraw-Hill: New York, NY, USA, 2001.
36. Maupin, K.; Swiler, L.; Porter, N. Validation Metrics for Deterministic and Probabilistic Data. *J. Verif. Valid. Uncertain. Quantif.* **2019**, *3*, 031002. [CrossRef]
37. Lin, J. Divergence measures based on the Shannon entropy. *IEEE Trans. Inf. Theory* **1991**, *37*, 145–151. [CrossRef]
38. Endres, D.; Schindelin, J. A new metric for probability distributions. *IEEE Trans. Inf. Theory* **2003**, *49*, 1858–1860. [CrossRef]

MDPI
St. Alban-Anlage 66
4052 Basel
Switzerland
Tel. +41 61 683 77 34
Fax +41 61 302 89 18
www.mdpi.com

Energies Editorial Office
E-mail: energies@mdpi.com
www.mdpi.com/journal/energies

* 9 7 8 3 0 3 6 5 4 5 0 3 5 *